BACTERIAL RESPONSES TO pH

The Novartis Foundation is an international scientific and educational charity (UK Registered Charity No. 313574). Known until September 1997 as the Ciba Foundation, it was established in 1947 by the CIBA company of Basle, which merged with Sandoz in 1996, to form Novartis. The Foundation operates independently in London under English trust law. It was formally opened on 22 June 1949.

The Foundation promotes the study and general knowledge of science and in particular encourages international co-operation in scientific research. To this end, it organizes internationally acclaimed meetings (typically eight symposia and allied open meetings, 15–20 discussion meetings, a public lecture and a public debate each year) and publishes eight books per year featuring the presented papers and discussions from the symposia. Although primarily an operational rather than a grant-making foundation, it awards bursaries to young scientists to attend the symposia and afterwards work for up to three months with one of the other participants.

The Foundation's headquarters at 41 Portland Place, London W1N 4BN, provide library facilities, open every weekday, to graduates in science and allied disciplines. The library is home to the Media Resource Service which offers journalists access to expertise on any scientific topic. Media relations are also strengthened by regular press conferences and book launches, and by articles prepared by the Foundation's Science Writer in Residence. The Foundation offers accommodation and meeting facilities to visiting scientists and their societies.

Information on all Foundation activities can be found at http://www.novartisfound.demon.co.uk

Novartis Foundation Symposium 221

BACTERIAL RESPONSES TO pH

1999

JOHN WILEY & SONS

Chichester · New York · Weinheim · Brisbane · Toronto · Singapore

Published in 1999 by John Wiley & Sons Ltd,
Baffins Lane, Chichester,
West Sussex PO19 1UD, England

National 01243 779777
International (+44) 1243 779777
e-mail (for orders and customer service enquiries): cs-books@wiley.co.uk
Visit our Home Page on http://www.wiley.co.uk
or http://www.wiley.com

Other Wiley Editorial Offices

John Wiley & Sons, Inc., 605 Third Avenue,
New York, NY 10158-0012, USA

WILEY-VCH Verlag GmbH, Pappelallee 3,
D-69469 Weinheim, Germany

Jacaranda Wiley Ltd, 33 Park Road, Milton,
Queensland 4064, Australia

John Wiley & Sons (Asia) Pte Ltd, 2 Clementi Loop #02-01,
Jin Xing Distripark, Singapore 129809

John Wiley & Sons (Canada) Ltd, 22 Worcester Road,
Rexdale, Ontario M9W 1L1, Canada

Novartis Foundation Symposium 221
ix+264 pages, 53 figures, 18 tables

Library of Congress Cataloging-in-Publication Data

Bacterial responses to pH / [editors, Derek J. Chadwick, Gail Gardew].
p. cm. – (Novartis Foundation symposium ; 221)
Includes bibliographical references and indexes.
ISBN 0-471-98599-6 (hbk : alk. paper)
1. Bacteria–Congresses. 2. Hydrogen-ion concentration–Congresses. 3. Extreme environments–Microbiology–Congresses. I. Chadwick, Derek. II. Cardew, Gail. III. Series.
[DNLM: 1. Bacteria congresses. 2. Hydrogen-Ion Concentration congresses. 3. Chemotaxis–physiology congresses. 4. Acids congresses. W1 N09285 v.221 1999 / QU 105 B131 1999]
QR97.A3B33 1999
579.3–dc21
DNLM/DLC
for Library of Congress 98-46836
CIP

British Library Cataloguing in Publication Data

A catalogue record for this book is available from the British Library

ISBN 0 471 98599 6

Typeset in 10½ on 12½ pt Garamond by Dobbie Typesetting Limited, Tavistock, Devon.
Printed and bound in Great Britain by Biddles Ltd, Guildford and King's Lynn.
This book is printed on acid-free paper responsibly manufactured from sustainable forestry, in which at least two trees are planted for each one used for paper production.

Contents

Symposium on Bacterial responses to pH, held at the Novartis Foundation, London 1–4 June 1998

This symposium is based on a proposal made by Mike Dilworth

Editors: Derek J. Chadwick (Organizer) and Gail Cardew

Participants

G. N. Bennett Department of Biochemistry and Cell Biology, Rice University, Houston, TX 77005-1892, USA

A. Bogachev (*Bursar*) Department of Bioenergetics, A. N. Belozersky Institute of Physico-Chemical Biology, Moscow State University, Moscow 119899, Russia

I. R. Booth Department of Molecular and Cell Biology, University of Aberdeen, Institute of Medical Sciences, Foresterhill, Aberdeen AB25 2ZD, UK

G. M. Cook Department of Microbiology, University of Otago, P O Box 56, Dunedin, New Zealand

M. J. Dilworth Centre for Rhizobium Studies, School of Biological Sciences and Biotechnology, Division of Science, Murdoch University, Murdoch, Perth, Western Australia, 6150, Australia

P. Dimroth Institute of Microbiology, ETH-Zurich, CH-8092 Zurich, Switzerland

W. Epstein Department of Molecular Genetics and Cell Biology, 920E 58th Street, University of Chicago, Chicago, IL 60639, USA

R. H. Fillingame Department of Biomolecular Chemistry, University of Wisconsin Medical School, Madison, WI 53706, USA

J.W. Foster Department of Microbiology and Immunology, University of South Alabama College of Medicine, Mobile, AL 36688, USA

A. R. Glenn Centre for Rhizobium Studies, School of Biological Sciences and Biotechnology, Division of Science, Murdoch University, Murdoch, Perth, Western Australia, 6150, Australia

C. Hill Department of Microbiology, University College Cork, Cork, Ireland

H. Kobayashi Faculty of Pharmaceutical Sciences, Chiba University, 1-33 Yayoi-cho, Inage-ku, Chiba 263-8522, Japan

M. Ito Department of Life Sciences, Toyo University, Oura-gun, Gunma, Japan

W. N. Konings Groningen Biomolecular Sciences and Biotechnology Institute, Department of Microbiology, University of Groningen, Kerklaan 30, 9751 NN Haren, The Netherlands

T. A. Krulwich Department of Biochemistry, Mount Sinai School of Medicine, New York, NY 10029, USA

A. Matin Department of Microbiology and Immunology, Stanford University School of Medicine, Sherman Fairchild Science Building, Stanford, CA 94305, USA

A. Moir Krebs Institute for Biomolecular Research, Department of Molecular Biology and Biotechnology, The University of Sheffield, Firth Court, Sheffield S10 2TN, UK

E. Padan The Hebrew University of Jerusalem, Department of Microbial and Molecular Ecology, Institute of Life Sciences, 91904 Jerusalem, Israel

S. F. Park Institute of Food Research, Reading Laboratory, Reading RG6 6BZ, UK

R. K. Poole (*Chair*) Krebs Institute for Biomolecular Research, Department of Molecular Biology and Biotechnology, The University of Sheffield, Firth Court, Sheffield S10 2TN, UK

R. G. Quivey Jr Department of Dental Research, Box 611, University of Rochester Medical Center, Rochester, NY 14642, USA

R. J. Rowbury Biology Department, University College London, Gower Street, London WC1E 6BT, UK

G. Schäfer Institute of Biochemistry, Medical University of Luebeck, 23538 Luebeck, Germany

V. Skulachev Department of Bioenergetics, A. N. Belozersky Institute of Physico-Chemical Biology, Moscow State University, Moscow 119899, Russia

J. L. Slonczewski Department of Biology, Kenyon College, Gambier, OH 43022, USA

J. B. Stock Department of Molecular Biology, Princeton University, Princeton, NJ 08544, USA

M. Verkovskaya Helsinki Bioenergetics Group, Institute of Biomedical Sciences, Department of Medical Chemistry, University of Helsinki, Helsinki, Finland

Introduction

Robert K. Poole

Krebs Institute for Biomolecular Research, Department of Molecular Biology and Biotechnology, The University of Sheffield, Firth Court, Sheffield S10 2TN, UK

Razors pain you;
rivers are damp;
acids stain you;
and drugs cause cramp.
(Dorothy Parker, *Not so deep as a well*, 1937)

One of the most remarkable properties of bacteria is their ability to survive extreme environmental stresses. As well as tolerating three of the above (hydrostatic pressure, extremes of pH and antibiotics), bacteria not only survive, but also flourish in, environments that are at extremes of temperature, pressure and salinity, or that may have vanishingly low or exceedingly high levels of potential nutrients (Postgate 1994).

Bacteria break the rules that macrobiologists take for granted and in so doing give microbiologists unique insights into the limits of life. These remarkable attributes of bacteria inevitably pose questions for the curious. Why do bacteria grow in such environments? How do they do so? What general principles of biology do these bizarre lifestyles teach us? How can we exploit our knowledge of these responses to pH to improve or indeed delay the growth of bacteria in such extremes? Is there any money in it? The purpose of this symposium is to answer all of these questions (with the possible exception of the last) and to generate new questions and hypotheses amenable to future experimental testing.

These questions are those of microbial physiology—the study of how bacteria and other micro-organisms *work*. Although rooted in the tradition of bacterial growth and intermediary metabolism, microbial physiology now confidently embraces molecular biology, genetics, biochemistry and indeed any discipline that can shed light on bacterial function. The speakers and discussants here represent all of these disciplines.

Textbooks half a century ago describe some of the physical conditions that affect bacterial growth. In some texts, a bell-shaped curve is shown as having two alternative abscissae: temperature and pH. 'An organism' is considered to have an optimum temperature for growth of 30–37 °C and an optimum pH of around

7. Gale & Epps (1942) were amongst the first to show that bacteria sense and respond to modest deviations from neutral pH. Many papers at this symposium will emphasize that, despite marked fluctuations in external pH, internal pH is remarkably constant. *Escherichia coli* manages to grow at pH values between about 5 and 8.5 (a 3000-fold change in H^+ concentration), but the internal pH is remarkably constant. This has led to much recent research into pH homeostasis — the mechanisms that bacteria adopt to sense external pH and then modulate gene expression to achieve the necessary ion fluxes or enzyme synthesis (Hall et al 1995). Not surprisingly, the pH extremophiles that have captured the imagination of many microbiologists are those that demonstrate the most dramatic abilities — thiobacilli that grow at pH 1 and certain alkalophilic bacteria that grow in soda lakes at pH values as high as 11 or more.

Undoubtedly, the most profound impact on our appreciation of the role of the proton in biology has come from Mitchell's chemiosmotic theory. According to this theory (Mitchell 1966), the energy-coupling cytoplasmic membranes of bacteria are topologically closed structures that are essentially impermeable to protons (as well as to ions generally). Energy-conserving metabolic highways are arranged so that during operation (e.g. respiration or photosynthesis) protons are translocated (either directly or indirectly) to generate an electrochemical gradient. It is this potential, and not a chemical intermediate, that is the driving force for energy-requiring processes such as ATP synthesis, solute transport and motility. The impact of these ideas on all areas of biology, and on bacterial energy transduction and metabolism in particular, is inestimable. The search for proton-driven mechanisms has inspired the work that has led to at least two Nobel Prizes in the past few years. The conceptual framework of many experiments and ideas presented in this symposium is that of chemiosmosis.

The bacteria that are the subject of this symposium are not only of academic interest. Manipulation of culture pH has long been practised for profit, most famously by Neuberg who showed that, after adding alkali to yeast cultures, glycerol appears as the major end-product of fermentation as an alternative to alcohol and CO_2 in the normal 'unsteered' fermentation. Many traditional fermentations rely on the ability of bacteria (and yeasts) to survive acidifying their own environments, as in sauerkraut production. Some bacteria, such as *Helicobacter*, which is able to endure the acid environment of the stomach, and *Salmonella typhimurium*, which is able to grow in macrophage phagolysosomes, are human pathogens. Dental plaque can provide a cosy environment for caries-forming bacteria, provided the acid that builds up can be tolerated. Many pH extremophiles have biotechnological or agricultural significance (Sharp & Munster 1986). *Thiobacillus* sp. and other acidophiles, which often have the additional ability to grow at high temperatures, are the basis of microbiological metal leaching — practised on a massive scale since Roman times, particularly for the mining of copper — and have also been considered for desulfurization of coal and gold mining. Certain enzymes, particularly proteases and α-amylases from alkalophilic bacteria are used in the manufacture of 'biological' washing powders and are active in alkaline solutions of soaps. Cyclodextrin glycosyltransferases from alkalophilic bacilli are increasingly used in foods, cosmetics and plastics. Low soil pH, exacerbated by mineral dissolution and aluminium release, restricts plant productivity in many areas of the world, and so the acid-tolerant growth and survival of rhizobia that form productive nodules on the roots of leguminous plants is a major concern.

I am confident that the formal presentations here will provide us all with an authoritative and up-to-date account of what has been achieved in this field. Most importantly, discussions will introduce supporting and conflicting views to challenge accepted concepts, prompt re-evaluation of data and point the way forward. No doubt we shall return to our laboratories to follow these pointers, and our noses, in pursuing this aspect of bacterial physiology.

References

Gale EF, Epps HMR 1942 The effect of the pH of the medium during growth on the enzymic activities of bacteria (*Escherichia coli* and *Micrococcus lysodeikticus*) and the logical significance of the changes produced. Biochem J 36:600–619

Hall HK, Karem KL, Foster JW 1995 Molecular responses of microbes to environmental pH stress. Adv Microb Physiol 37:229–272

Mitchell P 1966 Chemiosmotic coupling in oxidative and photosynthetic phosphorylation. Glynn Research Ltd, Bodmin

Postgate JR 1994 The outer reaches of life. Cambridge University Press, Cambridge

Sharp RJ, Munster MJ 1986 Biotechnological implications for microorganisms from extreme environments. In: Herbert RA, Codd GA (eds) Microbes in extreme environments. Academic Press, London, p 215–295

Problems of adverse pH and bacterial strategies to combat it

M. J. Dilworth and A. R. Glenn[1]

Centre for Rhizobium Studies, School of Biological Sciences and Biotechnology, Division of Science, Murdoch University, Murdoch, Perth, Western Australia, 6150, Australia

Abstract. This chapter aims to survey the problems faced by bacteria found in environments of adverse pH, to review strategies used to combat those problems and to ask how those strategies are implemented. At acid or alkaline pH, bacteria are challenged not just by excesses of H^+ or OH^- but also by excesses of metal ions (aluminium, heavy metals at acidic pH, Na^+ at alkaline pH), as well as shortages. Bacteria attempt to maintain their intracellular pH by minimizing membrane permeability to H^+ and other ions, buffering the cytoplasm, ameliorating the external pH through catabolism or selective substrate utilization, and developing ionic pumping systems. The amelioration of pH depends on the availability of substrate, and is unlikely in most naturally stressful environments. Ion pumping is expensive energetically, although the cost to growth is unknown. The response to adverse pH involves sensing systems and responsive regulatory systems. The adaptive acid tolerance response is now well known in enteric and other bacteria, but is there a widespread adaptive alkali tolerance response? What and where are the sensors? Whether they sense intracellular pH, extracellular pH or ΔpH is unclear, although an external sensory input seems essential. Is there one major sensory system responsive to pH or multiple systems with back-up mechanisms? What and where are the regulators? Is there one central regulator controlling all the responses or are there cascades of responses?

1999 Bacterial responses to pH. Wiley, Chichester (Novartis Foundation Symposium 221) p 4–18

When this symposium was initially proposed, we were asked why a symposium on the mechanisms by which bacterial cells respond to pH was relevant, why it was particularly relevant now, and why the Novartis Foundation rather than a learned society should run it. We responded that the mechanisms of resistance to adverse pH were highly relevant to medicine in relation to pathogenic invasion, to the food industry in terms of its potential to transmit pathogens and to agriculture in relation to managing soils of adverse pH to maintain productivity, particularly to

[1]Current address: Office of the Pro Vice Chancellor, University of Tasmania, Hobart, Tasmania, 7001, Australia.

counter human-induced acidification. We rather boldly claimed that the last few years had produced serious insights into how environmental sensor–regulator systems function in bacteria and that such systems also functioned in pH tolerance (at least for acidity). We noted that the disciplines contributing to such a broad topic did not come under the umbrella of any one learned society and that the Novartis Foundation was therefore a logical host. The symposium has gone ahead — we therefore presume that these answers satisfied the trustees and their referees, and hope that our discussions during this symposium will now justify those answers.

The tolerance of the extreme acidophiles and alkaliphiles to low or high pH represents fascinating adaptation to adverse environments. The mechanisms giving more modest acid or alkali tolerance to neutrophiles are no less interesting. These mechanisms, which allow them to run the gauntlet of stomach acids, cellular defence mechanisms or food acidity, are of considerable medical interest, while increasing soil acidity represents a major challenge to soil bacteria and through them to the productivity of our agricultural systems, particularly the crop and pasture legume components.

In this chapter we consider some of the problems facing bacteria experiencing non-optimal pH conditions, some of the ways in which organisms deal with them and some of the systems that are therefore needed. Many issues we raise will be dealt with more fully by later speakers, so the coverage will be general rather than detailed.

Extracellular pH effects

At the risk of stating the obvious, the use of pH to describe the H^+ concentration of a medium can for some obscure the logarithmic relationship: that each unit pH shift represents a 10-fold change in H^+ and OH^- concentrations. A transmembrane pH difference (ΔpH) of 1, 2 or 3 pH units therefore represents 10-, 100- or 1000-fold concentration gradients, with significant implications for the energetics of maintaining an approximately constant intracellular pH.

Macromolecules external to the cytoplasmic membrane on its outer surface (flagella, pili, chemoreceptors, other periplasmic proteins, cell walls, EPS, etc.) are clearly exposed to whatever the extracellular pH (pH_o) may be, and are logically the first point of adverse pH damage. Cells have limited options for protecting these structures, and must either produce modified versions able to resist the adverse pH or, when this is not feasible, dispense with the function. An example of a disposable property is the loss of flagellar motility (and therefore chemotactic response) under acidic pH conditions (Bowra & Dilworth 1981). Such cells cannot then relocate to more favourable conditions of substrate supply, pH or chemical environment. Even if flagellar activity remains, the function of the essential chemoreceptors may be damaged, with similar results.

Some serious problems arise where bacteria operate over wide pH ranges, say across 3–4 pH units. Over this sort of pH range, too, the ratio between ionized and non-ionized forms of substrates changes by orders of magnitude. How do cells regulate entry between permeant and non-permeant forms of substrates to control total entry? Processes such as solute transport systems are not disposable; can cells then make do with the limited non-optimal activity of their transporters at unfavourable pH, or do they have the capacity to induce synthesis of pH-adapted ones? Or do they simply up-regulate the synthesis of the now inefficient transport system to maintain some absolute rate of transport?

Interactions between pH and the inorganic environment

The problems faced by bacteria in natural environments like soils or waters at acid or alkaline pH are not simply due to higher concentrations of H^+ or OH^-, but to alterations in the availability of other ions as well. Bacterial persistence at a particular pH may therefore be as much dependent on mechanisms that counter these consequential changes as on mechanisms countering influx of the H^+ or OH^- ions themselves. Under acidic conditions, therefore, it is necessary to consider the impact of heightened availability of aluminium, copper, zinc, cadmium, manganese, etc., as well as reckon with potentially decreased availability of molybdenum. Although some of these metal ions, like aluminium (see Flis et al 1993), may not become serious threats at pH values above 5, others can and do interact with the stress imposed by H^+ *per se*. For example, mutants of root nodule bacteria that display increased acid sensitivity often show heightened sensitivity to metal ions like Cu^{2+}, Cd^{2+} and Zn^{2+} (Tiwari et al 1996a,b). Acid-tolerance mechanisms may therefore require resistance to such ions, perhaps involving induction of outward-directed metal ion pumps and repression of uptake systems, detoxification mechanisms, etc.

Under alkaline conditions, many of these same heavy metal ions, some required for growth, become increasingly less available; one characteristic of alkaliphiles is therefore likely to be possession of uptake systems for essential ions that can retrieve particular metals at low concentrations. Siderophore production and uptake would be expected as a means of acquiring Fe(III) when its concentration is low at high pH. Many naturally or artificially alkaline environments also have high Na^+ concentrations, with the requirement for adaptations to halotolerance or halophily as well as coping with the overall alkaline pH.

Other ions in the environment may modify the stress imposed by the actual pH. High external concentrations of Ca^{2+} markedly increase growth rate and lower the pH at which growth finally becomes impossible (Reeve et al 1993, Watkin et al 1997). How Ca^{2+} exerts these effects is still unknown. It is also possible that some other ionic effects on acid habituation—such as those of phosphate on acid habituation of *Escherichia coli* (Rowbury et al 1992)—could be associated with modulation of calcium concentrations. At low pH, replacement of Ca^{2+} by H^+ may interfere seriously with normal Ca^{2+} function or uptake. Whether Ca^{2+} can also hinder uptake of toxic ions like Cu^{2+} or Zn^{2+} remains to be seen.

This symposium examines tolerance to both acidic and alkaline conditions, with the aim of defining both possible similarities and differences in bacterial response to them. One might, because of the ancillary problems associated with acidic or alkaline environments, expect any detailed similarities to be extremely limited, even if the basic principles of stress detection and response may well be closely related. Our work with mutants of *Sinorhizobium meliloti* certainly gives little support to a close relationship between acid and alkali tolerance: all the acid-sensitive mutants grow at pH 10, the limit for the wild-type (Goss et al 1990).

Intracellular pH maintenance

It is now clear that the intracellular pH (pH_i) of bacterial cells tends to be maintained within limits which vary for acidophiles, neutrophiles and alkaliphiles (Booth 1985, Hall et al 1995). For a neutrophile such as *Escherichia coli*, pH_i is maintained at *c.* 7.6–7.8; for others it may vary more widely and fall markedly lower, particularly for fermentative bacteria (Booth 1985). Although pH_i values for acidophiles may be considerably lower than those for neutrophiles (Booth 1985) they are nonetheless considerably above those in the environment. With alkaliphiles, pH_i rises well above 7, but still not to the same extent as the pH_o.

The pH gradients typically experienced by acidophiles and alkaliphiles also have important implications for energy conservation mechanisms, since the latter clearly do not have scope for an acid-outside ΔpH-fuelled system, leaving them essentially dependent on the membrane potential element of proton motive force (Booth 1985).

Given that pH_i needs to be maintained constant, what mechanisms can bacteria use to minimize the stresses from non-optimal pH_o? Various solutions are given below.

(1) Decreased membrane permeability.
(2) Internal buffering.
(3) Amelioration of pH_o.
(4) Proton extrusion/uptake.
(5) Prevention of metal ion toxicity.

Defence against unfavourable pH

The first line of defence is obviously to minimize membrane permeability, to either prevent ingress of H^+ or its loss, depending on the pH. Although there is evidence that this solution is implemented (Booth 1985), and that membrane composition can vary with pH (Arneborg et al 1993), it clearly has limits; no membrane that has to allow for other ions to be taken up can be made impermeable to H^+ or OH^-; the greater the pH differential between pH_i and pH_o the greater the leakage.

Another often-mentioned solution involves cytoplasmic buffering to counteract potential upward or downward shifts in pH_i. Buffering due to macromolecular content must be limited, and continued ingress of H^+ (or OH^-) must presumably overcome it rapidly. Buffering due to synthesis of intracellular metabolites presumably involves continuous production of buffering materials and consequently a steadily increasing concentration of them. Although this might avert a temporary crisis, it is difficult to envisage it as a long-term strategy. Mutants with lesions in the isocitrate dehydrogenase (*icd*) locus in *Salmonella typhimurium*, which accumulate citrate and isocitrate, are more resistant to challenge at pH 3.3 (Hall et al 1995), although it is not clear that buffering is the answer.

Bacteria commonly modify pH_o through metabolic activities; the magnitude of the pH change is often not recognized. pH shifts around colonies on unbuffered agar media can be particularly large, and are potentially important when, for example, acid-sensitive mutants are being sought. The range of pH for the cut-off from growth to non-growth can be narrow (Tiwari et al 1992) and any local modification of pH may give misleading results.

Such catabolism-induced change to pH_o may be simply an inevitable consequence of using the substrate available, or a response to unfavourable pH. Catabolism of sugars or sugar alcohols generally results in acidification, whereas catabolism of organic acids or amino acids leads to rises in pH_o, and these are uncontrolled results of such metabolism.

By contrast, the decarboxylation of amino acids (Gale & Epps 1942), such as that of lysine to the diamine cadaverine via the *cad* system in *E. coli* (Meng & Bennett 1992, Watson et al 1992), or of arginine to agmatine via arginine decarboxylase (Stim & Bennett 1993), leads to an increase in pH_o, and is a response induced by low pH. In *Clostridium acetobutylicum*, where sugar fermentation initially results in acidification from production of acetic and butyric acids, further acidification can be prevented by reduction of butyric acid to butanol (Huang et al 1986). In *Lactobacillus plantarum*, a switch from lactic acid to acetoin production at low pH_o is a similar protective mechanism (Tsau et al 1992). It is not clear whether these help pH_i maintenance by decreasing internal acid production or decreasing H^+ ingress from a less acidic environment, or both.

In theory, bacteria in an environment containing substrates whose metabolism could increase or decrease pH_o could select the ones producing a favourable pH change. At least in some root nodule bacteria this option is not used. Cells given a mixture of glucose and fumarate did not use a greater proportion of fumarate at acidic pH (S. Castro, unpublished data 1996), nor a greater proportion of histidine from a glucose–histidine mixture. We do not know whether other organisms behave differently.

A major disadvantage to the amelioration strategy is the requirement for continuous access to suitable substrates in sufficient quantity. For enteric or other bacteria in nutrient-rich environments, such as the gut or elsewhere inside animals, amelioration may be practical. For many bacteria in natural pH-stressed environments, appropriate substrates are unavailable and amelioration is not achievable, though the better-supplied plant rhizosphere may have the substrate supply to allow effective amelioration.

Another and extremely important option involves energy-requiring systems to pump ions actively in such a way that pH_i can be maintained. Under acidic conditions, where H^+ removal is required, H^+/K^+ antiporters are known to function in, for example, *Enterococcus hirae* (Kobayashi et al 1984, 1986), whereas under alkaline conditions there are inducible systems for exporting Na^+ concomitant with H^+ import (Zilberstein et al 1982, Booth 1985); these systems will be more fully discussed by later speakers. The energetic cost of these pumping systems must presumably be reflected in yields of cell biomass, but few studies seem to have addressed the magnitude of this cost. In naturally pH-stressed environments, where carbon substrates for heterotrophic metabolism may be extremely limited, the energy costs involved in ion pumping may represent a major part of substrate consumption. At low pH, transport systems for removing unwanted ions like Cu^{2+} or Zn^{2+} may also be required and add to the energy cost. Systems for synthesizing, secreting and absorbing siderophores under iron-poor alkaline conditions must also add to the overall transport costs.

Response to adverse pH

Bacterial cells grown at a moderately acid pH or exposed to one for some time resist killing at low pH much better than cells grown at pH 7, a phenomenon called 'acid habituation' or the 'adaptive acid tolerance response' (ATR; see review by Hall et al 1995). This ATR is now known for a wide range of bacteria, including *E. coli* (Goodson & Rowbury 1989), *S. typhimurium* (Foster & Hall 1990, Foster 1991), *Aeromonas hydrophila* (Karem et al 1994) and *Listeria monocytogenes* (O'Driscoll et al 1996, 1997), as well as species of the root nodule bacteria *Rhizobium*, *Bradyrhizobium* and *Sinorhizobium* (O'Hara & Glenn 1994). The ATR requirements appear to vary considerably even between *E. coli* and *S. typhimurium*, particularly in the number of apparently different components (see Hall et al 1995 for review). Protein synthesis during the exposure to mildly acid conditions and greater acid resistance in cells in stationary phase appear to be fairly common denominators.

Although an adaptive ATR is now well researched (principally in Gram-negative bacteria), a similar response to alkali (Goodson & Rowbury 1989, Rowbury et al 1989) has attracted less attention.

Clearly a drop in growth pH for *E. coli* or *S. typhimurium* results in marked changes in gene expression associated with development of the ATR. In *S. meliloti* there is a similar requirement for protein synthesis for a successful ATR (O'Hara & Glenn 1994), but much of this protein synthesis appears to be 'precautionary', an increase in the mean generation time on transfer to lower pH, but no lag, implying that no new proteins are needed for immediate growth at the lower pH.

Within a bacterial species, there is a spectrum of genetically determined tolerance to acid or alkali. Are some strains acid or alkali tolerant because they possess some unique genes that other strains simply do not have, or are they tolerant because certain cellular proteins have evolved to confer tolerance where their sensitive counterparts have less well-adapted proteins? No broadly definitive answer seems available yet.

Because one can isolate acid-sensitive mutants from acid-tolerant parents the integrity of certain proteins must be essential for acid resistance; the same is likely to be true for alkali tolerance. One might expect some of these to be constitutively expressed, particularly those for detection and immediate response to pH change, whereas for others up-regulation or inducible expression may be adequate.

The wide range of genes whose protein products (of known and unknown function) are induced or repressed by medium pH (see list in Hall et al 1995), shows that major changes to gene expression must occur. Some responses may be only indirectly related to adverse pH; the greater availability of Fe(III) at more acidic pH results in repression of the synthesis of the siderophore vicibactin in *Rhizobium leguminosarum* bv. *viciae* WSM710 (Dilworth et al 1998). While most attention seems to have focused on genes induced by pH change, there should also be many whose repression will be required. A gene specifying a system for Cu^{2+}, Mn^{2+} or Zn^{2+} uptake could well be an embarrassment to the cell if the availability of these toxic metals increases at lower pH, and its expression would need to be curtailed. Antiport systems for H^+ efflux/uptake are obvious and there are known examples of inducible systems.

Sensory and regulatory systems for pH

In parallel with the responses of bacteria to other environmental stimuli, there is a requirement for pH stress to be sensed. The obvious questions are what is sensed, where and how?

The first two are essentially linked. The cell might sense pH_i, pH_o or ΔpH across the membrane. There is logical support for at least a sensing of pH_o, and some evidence for independent sensing of pH_i. Because many root nodule bacteria maintain a pH_i of *c.* 7.2 and fail to grow if it falls to 6.5 (O'Hara et al 1989), and full induction of many proteins in them may take 6–8 h (Dilworth et al 1983), a system giving early warning of impending pH danger would make biological sense. Waiting for a discernible change in pH_i may be too dangerous. However, this is not to say that monitoring pH_i closely may not be a satisfactory strategy, particularly for organisms with the capacity for rapid changes in gene expression. Monitoring ΔpH would also imply monitoring pH_i, since the cell would need to respond if the latter were to change for reasons other than pH_o. The finding of the gene *inaA* in *E. coli* (Slonczewski et al 1987) by virtue of its expression being induced by internal acidification created by addition of a weak permeant acid (benzoic acid) more strongly than by external acidification strongly suggests a sensing of pH_i. A decision on the questions of what is sensed and where is not yet possible. We strongly suggest, however, that sensing of pH_o should be one essential element. The 'how' of the sensing of pH is as yet unknown.

In *S. meliloti*, one typical sensor–regulator system has been identified, mutation in either component of which results in acid sensitivity (Tiwari et al 1996b). Are there others in the same organism? In our view, there probably are. When *E. coli* has multiple systems for the acquisition of just one essential element, iron, it would be surprising if its defences against the potentially lethal threat from pH change were not to have back-up. In *S. meliloti*, furthermore, mutants defective in *actS* (the gene encoding the putative sensor), which are normally acid sensitive, can be returned to acid tolerance by inserting multiple copies of *act*R, implying that there may be other sensors with which the ActR protein can productively interact.

The ActS/ActR system required for acid tolerance in *Rhizobium meliloti* shows remarkable similarity to the PrrA/PrrB and RegA/RegB systems in the photosynthetic *Rhodobacter* spp. (Tiwari et al 1996b). Yet the Prr and Reg systems in the photosynthetic bacteria appear to be involved in O_2 or redox sensing for controlling genes for photosynthesis (Eraso & Kaplan 1994, 1995), not with pH tolerance. A connection between sensor–regulators responding to redox/O_2 and acidic pH is at first sight unclear. In many heterotrophic bacteria, however, one result of limiting O_2 concentration or low redox status is a trend to fermentation and acid production, so the relationship may not be coincidental. Indeed, in *E. coli*, expression of β-galactosidase from *lacZ* fusions under the control of promoters of the *cad* genes is more responsive to pO_2 than to low pH (Watson et al 1992, Takayama et al 1994), despite the apparent function of the *cad* system in pH amelioration. While the expression of many low pH-dependent genes in facultative bacteria seems to require low redox status or anaerobiosis (see Hall et al 1995), our knowledge of pH-responsive regulatory systems is as yet too slight to draw firm conclusions.

Two final comments

In conclusion, we would like to emphasize the importance of the inorganic components of the medium in interacting with adverse pH to affect bacterial growth and survival, both positively and negatively. We also stress that for bacteria in naturally nutrient-poor environments with pH stress, amelioration of the external environment is unlikely to be a major factor in resistance.

Acknowledgements

We thank the Australian Research Council, the Australian Wool Corporation, and the Grains Research and Development Corporation for their financial support.

References

Arneborg N, Salskou-Iversen AS, Mathiasen TE 1993 The effects of growth rate and other growth conditions on the lipid composition of *Escherichia coli*. Appl Microbiol Biotechnol 39:353–357

Booth IR 1985 Regulation of cytoplasmic pH in bacteria. Microbiol Rev 49:359–378

Bowra BJ, Dilworth MJ 1981 Motility and chemotaxis towards sugars in *Rhizobium leguminosarum*. J Gen Microbiol 126:231–235

Dilworth MJ, McKay IA, Franklin M, Glenn AR 1983 Catabolite effects on enzyme induction and substrate utilization in *Rhizobium leguminosarum*. J Gen Microbiol 129:359–366

Dilworth MJ, Carson KC, Giles RGF, Byrne LT, Glenn AR 1998 *Rhizobium leguminosarum* bv. *viciae* WSM710 produces a novel cyclic hydroxamate siderophore, vicibactin. Microbiology 142:781–791

Eraso JM, Kaplan S 1994 PrrA, a putative response regulator involved in oxygen regulation of photosynthesis gene expression in *Rhodobacter sphaeroides*. J Bacteriol 176:32–43

Eraso JM, Kaplan S 1995 Oxygen-insensitive synthesis of the photosynthetic membranes of *Rhodobacter sphaeroides*—a mutant histidine kinase. J Bacteriol 177:2695–2706

Flis SE, Glenn AR, Dilworth MJ 1993 The interaction between aluminium and root nodule bacteria. Soil Biol Biochem 25:403–417

Foster JW 1991 *Salmonella* acid-shock proteins are required for the acid-tolerance response. J Bacteriol 173:6896–6902

Foster JW, Hall HK 1990 Adaptive acidification tolerance response of *Salmonella typhimurium*. J Bacteriol 172:771–778

Gale EF, Epps HMR 1942 The effect of the pH of the medium during growth on the enzymic activities of bacteria (*Escherichia coli* and *Micrococcus lysodeikticus*) and the logical significance of the changes produced. Biochem J 36:600–619

Goodson M, Rowbury RJ 1989 Habituation to normally lethal acidity by prior growth of *Escherichia coli* at a sub-lethal acid pH value. Lett Appl Microbiol 8:77–79

Goss TJ, O'Hara GW, Dilworth MJ, Glenn AR 1990 Cloning, characterization and complementation of lesions causing acid-sensitivity in Tn*5*-induced mutants of *Rhizobium meliloti* WSM419. J Bacteriol 172:5173–5179

Hall HK, Karem KL, Foster JW 1995 Molecular responses of microbes to environmental pH stress. Adv Microb Physiol 37:229–272

Huang L, Forsberg CW, Gibbins LN 1986 Influence of external pH and fermentation products on *Clostridium acetobutylicum* intracellular pH and cellular distribution of fermentation products. Appl Environ Microbiol 51:1230–1234

Karem KL, Foster JW, Bej AK 1994 Adaptive acid tolerance response (ATR) in *Aeromonas hydrophila*. Microbiology 140:1731–1736

Kobayashi H, Suzuki T, Kinoshita N, Unemoto T 1984 Amplification of the *Streptococcus faecalis* proton-translocating ATPase by a decrease in cytoplasmic pH. J Bacteriol 158:1157–1160

Kobayashi H, Suzuki T, Unemoto T 1986 Streptococcal cytoplasmic pH is regulated by changes in amount and activity of a proton-translocating ATPase. J Biol Chem 261: 627–630

Meng S, Bennett GN 1992 Regulation of the *Escherichia coli cad* operon: a system for neutralization of low extracellular pH. J Bacteriol 174:2659–2669

O'Driscoll B, Gahan CGM, Hill C 1996 Adaptive acid tolerance response in *Listeria monocytogenes*: isolation of an acid-tolerant mutant which demonstrates increased virulence. Appl Environ Microbiol 62:1693–1698

O'Driscoll B, Gahan CGM, Hill C 1997 Two dimensional polyacrylamide gel electrophoresis analysis of the acid tolerance response in *Listeria monocytogenes* LO28. Appl Environ Microbiol 63:2679–2685

O'Hara GW, Glenn AR 1994 The adaptive tolerance response in root nodule bacteria and *Escherichia coli*. Arch Microbiol 161:286–292

O'Hara GW, Goss TJ, Dilworth MJ, Glenn AR 1989 Maintenance of intracellular pH and acid tolerance in *Rhizobium meliloti*. Appl Environ Microbiol 55:1870–1876

Reeve WG, Tiwari RP, Dilworth MJ, Glenn AR 1993 Calcium affects the growth and survival of *Rhizobium meliloti*. Soil Biol Biochem 25:581–586

Rowbury RJ, Goodson M, Whiting GC 1989 Habituation of *Escherichia coli* to acid and alkaline pH and its relevance for bacterial survival in chemically polluted natural waters. Chem Ind 1989:685–686

Rowbury RJ, Goodson M, Wallace AD 1992 The PhoE porin and transmission of the chemical stimulus for induction of acid resistance. J Appl Bacteriol 72:233–243

Slonczewski JL, Gonzalez TN, Bartholomew M, Holt NJ 1987 Mud-directed fusions regulated by low pH in *Escherichia coli*. J Bacteriol 168:3000–3006

Stim KP, Bennett GN 1993 Nucleotide sequence of the *adi* gene, which encodes the biodegradative acid-induced arginine decarboxylase of *Escherichia coli*. J Bacteriol 175:1221–1234

Takayama M, Ohyama T, Igarashi K, Kobayashi H 1994 *Escherichia coli cad* operon functions as a supplier of carbon dioxide. Mol Microbiol 11:913–918

Tiwari RP, Reeve WG, Glenn AR 1992 Mutations conferring acid sensitivity in the acid-tolerant strains *Rhizobium meliloti* WSM419 and *Rhizobium leguminosarum* biovar *viciae* WSM710. FEMS Microbiol Lett 100:107–112

Tiwari RP, Reeve WG, Dilworth MJ, Glenn AR 1996a An essential role for *actA* in acid tolerance of *Rhizobium meliloti*. Microbiol 142:601–610

Tiwari RP, Reeve WG, Dilworth MJ, Glenn AR 1996b Acid tolerance in *Rhizobium meliloti* strain WSM419 involves a two-component sensor–regulator system. Microbiology 142:1693–1704

Tsau J-L, Guffanti AA, Montville TT 1992 Conversion of pyruvate to acetoin helps to maintain pH homeostasis in *Lactobacillus plantarum*. Appl Environ Microbiol 58: 891–894

Watkin ELJ, O'Hara GW, Glenn AR 1997 Calcium and acid stress interact to affect the growth of *Rhizobium leguminosarum* bv. *trifolii*. Soil Biol Biochem 29:1427–1432

Watson N, Dunyak DS, Rosey EL, Slonczewski JL, Olsen ER 1992 Identification of elements involved in transcriptional regulation of the *Escherichia coli cad* operon by external pH. J Bacteriol 174:530–540

Zilberstein D, Agmon V, Schuldiner S, Padan E 1982 The sodium/proton antiporter is part of the pH homeostasis mechanism in *Escherichia coli*. J Biol Chem 257:3687–3691

DISCUSSION

Konings: You mentioned that one of the responses to changes in pH is a reduction of proton permeability of the membrane, but what is the evidence for this?

Booth: I would like to respond to this question. Connor O'Byrne has studied the proton permeability of membranes from *Escherichia coli* O157 in three different conditions: during exponential growth; when they have been habituated by growth at low pH; and in stationary phase. He has found that there are clear differences in proton permeability in those three sets of cells, suggesting that adaptation does occur. What he would say, however, is that we don't know whether those changes involve the lipid or protein components of the membranes.

Konings: We have done similar analyses, and our results suggest that the membrane proteins are partly responsible for the changes in proton permeability.

Poole: If changes in membrane structure that alter proton permeability compensate for changes in extracellular pH, one would expect to see these long-term changes in cultures adapting relatively slowly to large changes in external pH. Has anyone studied the differences between chronic and acute responses to pH change?

Krulwich: We have done some preliminary experiments with alkaliphilic *Bacillus*. We determined the membrane lipid composition as a function of growth pH in mid-log phase. Some properties of the membrane lipids are expected to be involved in alkaline adaptation, but at pH 7.5 we didn't see significant changes in the quantities of various components. More recently, Ray Gilmore in my laboratory has done a series of 2D gels of proteins from steady-state cells at pH 7.5 that are suddenly shifted to pH 10.5. Some of the proteins expressed during this shift are associated with lipid, which are not necessarily membrane lipids; they could be antibiotics, for example. I don't yet know how to interpret this.

Rowbury: Tom McMeekin's group showed that when strains which are inherently acid sensitive are shifted to acidic pH their lipid compositions change markedly (Brown et al 1997). For example, the levels of cyclopropane fatty acids increase. I'm not sure whether this can explain the changes in proton permeability or not. Perhaps Ian Booth can comment on this. Additionally, strains that are inherently acid tolerant have membrane lipid compositions similar to cultures that have become inducibly acid tolerant.

Booth: This is an interesting paradox. Stationary phase cells have the highest concentration of cyclopropane fatty acids, but they also have the lowest capacity for pH homeostasis. The steady-state intracellular pH (pH_i) is often half a unit below that of exponential phase cells, even when those cells are energy replete. Therefore, the relationships between proton permeability, lipid structure and pH homeostasis are complex.

Stock: pH changes would cause global changes, affecting the rates of all reactions in the cell. Another factor that will induce such global changes is temperature. Has anyone looked at whether there is a relationship between pH homeostasis and the responses to changes in temperature?

Slonczewski: I'm not sure whether there are any correlations because an organism, such as *E. coli*, that maintains a high degree of pH homeostasis may have a completely different response to temperature. We have no evidence in our 2D gels that heat shock proteins are expressed during periods of long-term growth at extreme base or extreme acid.

Stock: But if you change the temperature, do you observe changes in homeostatic pH?

Slonczewski: The answer is no for *E. coli*. We measured intracellular pH (pH_i) at room temperature and at 37°C, and found that in both cases it was pH 7.6.

Poole: Is it coincidental that many bacteria and archaea grow at both high temperatures and at low pH values?

Konings: The proton permeability of the membrane increases with temperature, and this is what prevents growth at high temperatures. Unlike bacterial membranes, archaeal membranes are proton impermeable, even at high temperatures. *Sulfolobus acidocaldarius* at 80°C has the same proton permeability as *E. coli* has at 37°C. This explains why archaea can use protons as coupling ions at high temperature, whereas bacteria cannot. Instead they shift to using sodium-coupled transport systems.

Krulwich: Some years ago we asked David Ward to send us some samples from the hot alkaline lakes in Yellowstone National Park, USA, so that we could try to isolate some alkaliphiles. We were interested in oxidative phosphorylation and proton motive force-driven systems, so we selected under those circumstances, but we didn't find anything. I thought at the time that such an organism may not exist. However, since then some alkaline isolates have been reported, but I'm not sure that any of them grow on non-fermentable carbon sources at high temperatures under pH-regulated and static conditions, or whether they can only be grown under fermentative conditions, which are different physiological conditions.

Cook: Hugh Morgan's group in New Zealand have isolated a strain of *Bacillus* that grows optimally at pH 10.5 and 70°C, and it is dependent on sodium for both growth and transport of some amino acids (C.J. Peddie, G.M. Cook & H.W. Morgan, unpublished results 1998).

Krulwich: Does it grow on malate or succinate?

Cook: Yes, the organism grows on malate or succinate as the sole carbon energy source.

Konings: Bacillus stearothermophilus has a proton-translocating respiratory chain, a proton ATPase and an effective sodium/proton antiport system that translates the proton motive force to a sodium motive force. Therefore, this is an example of an organism that uses both sodium and protons.

Rowbury: I would like to address a point that Mike Dilworth made about pH sensing. He mentioned that there weren't any systems in which it has been studied in detail. However, I would like to ask Etana Padan about the NhaA system, because we do know a certain amount about how pH is sensed in this system.

Padan: I will be talking about this in detail tomorrow. However, I would like to point out now that there are differences between the systems employed in the periplasm, which is in direct contact with the changes in pH, and those in the cytoplasm. *E. coli* has a powerful pH homeostasis mechanism that is established immediately after a shift in pH, but in my opinion two mechanisms are involved: a homeostatic system in the cytoplasm and a different one in the periplasmic space.

Poole: In a Gram-negative bacterium what are the pH gradients across the outer membrane and across the inner membrane? If the extracellular pH (pH_o) is 8, for example, and pH_i is maintained at pH 7.0, can we assume that the pH of the periplasm matches the pH_o exactly?

Booth: The problem is that it is not an aqueous system. It is more like a gel, which means that you can't talk in terms of proton concentration, rather proton activity, and the reality is that we don't know what happens in the periplasm.

Slonczewski: Barry Rosen (unpublished results 1998) has looked at whether the periplasmic and membrane proteins are acid inducible by using alkaline phosphatase fusions. He found that the enzymes didn't function when the exterior was acidic, so he concluded that the periplasm was also acidic.

Stock: I have two points to make. The first is that I don't agree with this 'gel' concept. The periplasm is no different from the cytoplasm in terms of its being a gel. Secondly, in our experience it is possible to work out the periplasmic pH because it follows the Donnan potential across the outer membrane. This is dependent on the extracellular salt concentration.

Booth: When I talk about a 'gel' it's just to remind people that the protein concentrations in the periplasm are extremely high, i.e. of the order of 0.5 g/ml. The problem is that in a normal growth medium, with normal salt concentrations, the Donnan potential is taken up with potassium and sodium ions.

Stock: Protons will be affected by the Donnan potential in the same way as potassium and sodium are affected.

Booth: But the local pH will not. When you run ion-exchange columns, for example, in which you create a Donnan potential with negatively charged ions, which is just like the bacterial cell surface, then you can show that as the potassium or sodium concentration is raised the pH at that surface goes from being less acid to being more alkaline because the protons are displaced. In my opinion the periplasm is likely to be similar.

Stock: Under standard laboratory conditions, in which there are fairly high extracellular salt concentrations, there is a small Donnan potential across the outer membrane so that the pH in the periplasm is similar to that in the external medium.

Konings: I would like to make a point about solute transport. Mike Dilworth mentioned that the external pH can have a number of effects on transport activity, but what should also be taken into consideration is the effect of the availability of substrate because certain substrates, for example malate and oxalate, will be transported in different ionic states at different pH values.

Glenn: Does anyone here know whether different solute transport systems are expressed at different pH, especially as some of these organisms can grow over a pH range of four pH units?

Kobayashi: Enterococcus hirae has two sodium transport systems—the Na^+/H^+ antiporter and the sodium ATPase—that function at different pH values. The Na^+/H^+ antiporter functions at $pH < 8.0$, and at $pH > 8.0$ the sodium ATPase is induced.

Dilworth: But for a particular solute, e.g. malate or glucose, do the transporters change with pH?

Slonczewski: We are just beginning to look at this. We have found that the phosphotransferase system is induced under acid conditions, but we have not yet seen anything that is obviously pH related, in terms of uptake of acid.

Poole: I would like to introduce a topic that Mike Dilworth mentioned several times, which is the problem faced by alkaliphiles in obtaining metal ions, e.g. Fe^{3+}, that have low solubilities at pH 7.0 and must therefore be almost insoluble at pH 10. Can anyone comment on how alkaliphiles resolve this problem?

Slonczewski: I can't speak directly about Fe^{3+}, but it is clear that cells grown at alkaline pH tend to up-regulate a number of outer membrane components and periplasmic proteins that may be involved in the uptake and transport of various nutrients.

Krulwich: I'm more worried about the availability of magnesium at alkaline pH. I can't answer your question, but special siderophores may be involved.

Bennett: A related point is the problem of heavy metals and oxidation/reduction processes. Certain bacteria can reduce various heavy metals, which then form an insoluble mineral. This could be a mechanism by which bacteria protect themselves, but does anyone know whether it is modified by the acidity of the medium?

Dilworth: I am aware that bacteria may precipitate heavy metals as a means of protecting themselves, but I'm not sure whether pH-induced variation in this precipitation could explain low pH effects.

Poole: My understanding of the literature on this is that what happens is largely a function of the chemistry rather than a function of the way micro-organisms change the chemistry. Many of the interactions between metals and microbes can be explained adequately by inorganic chemistry.

I have one more point. Mike Dilworth mentioned that cells face large problems when trying to maintain neutral pH_i. Some of those problems relate to solute transport. Does anyone have any views on whether a neutral pH_i evolved before an acidic one or vice versa? There seem to be many advantages in having an acidic pH_i, in terms of obtaining essential metabolites such as organic acids and metal ions.

Matin: But there are finite limits to how low the pH can be because there are so many acid-labile constituents in the cytoplasm.

Poole: Yes, there are chemical restrictions on how low the pH can be, but it doesn't answer the question of whether a pH_i close to that lower limit is evolutionarily older than, say, a pH_i of 8.0.

Booth: If you take the view that fermentation preceded respiration then we know that in general fermentative organisms tolerate much wider variations in pH_i during growth. Therefore, one can argue teleologically that a tolerant pH_i came first, and that *E. coli* has become the least tolerant of the alkaliphiles.

Skulachev: But I'm not sure that fermentation evolved before other energy conservation mechanisms.

Slonczewski: One may also get an artificial view by just looking at *E. coli*, because *E. coli* has adapted to be a gut organism, and therefore its pH is closely related to that of the organisms it inhabits. Perhaps it would be better to consider free-living organisms.

Reference

Brown JL, Ross T, McMeekin TA, Nichols PD 1997 Acid habituation of *Escherichia coli* and the potential role of cyclopropane fatty acids in low pH tolerance. Int J Food Microbiol 37: 163–173

The regulation of intracellular pH in bacteria

Ian R. Booth

Department of Molecular and Cell Biology, University of Aberdeen, Institute of Medical Sciences, Foresterhill, Aberdeen AB25 2ZD, UK

Abstract. The regulation of intracellular pH (pH_i) in bacterial cells is achieved through control over cation (and anion) permeability. In addition to the active components of homeostasis there are contributions from essentially passive elements, such as the lipid composition of the membrane and the buffering capacity of the cytoplasm. Active homeostasis involves control over the movement of K^+, Na^+ and H^+. Alterations in the membrane permeability for any of these cations may cause perturbation of homeostasis. In *Escherichia coli* this is exemplified by the controlled activation of K^+ efflux systems by glutathione adducts leading to temporary acidification of the cytoplasm. This is achieved by sophisticated control over the KefB and KefC systems, and is tightly integrated with glutathione-dependent detoxification mechanisms. Such control over pH_i facilitates survival of the cell following exposure to toxic electrophiles. The components of pH homeostasis will be reviewed and the molecular mechanisms, and role of, the KefB and KefC systems will be discussed.

1999 Bacterial responses to pH. Wiley, Chichester (Novartis Foundation Symposium 221) p 19–37

Bacteria that grow in the acid-to-neutral pH range generally maintain their intracellular pH (pH_i) at a value more alkaline than that of the environment (Booth 1985). In contrast, bacteria that grow at alkaline pH exhibit pH_i values that are more acidic than that of the environment. Changes in pH_i may be important signals for cell physiology. Cells of many different species become habituated to extremely acidic pH when incubated for brief periods at slightly acidic pH (about pH 5.8–6.3; see Foster & Moreno 1999, this volume). Similarly, changes in pH_i alter cell behaviour (Kihara & Macnab 1981) and gene expression. As little as 0.2 pH units change can be sufficient to elicit significant changes in gene expression (Thomas & Booth 1992). In *Escherichia coli* one of the most important global survival regulons, the RpoS-regulated safety net (Hengge-Aronis 1993), is induced by mild acidification of the growth medium (Schellhorn & Stones 1992, Ferguson et al 1998). Therefore, cells may use perturbation of pH homeostasis to elicit changes in cell physiology.

The ability to maintain pH_i within a narrow range of values despite variations in the pH of the environment is termed 'pH homeostasis'. It has previously been suggested that the two major components of the homeostatic system can be termed 'passive' and 'active' (Booth 1985). The major components of the passive homeostatic system are the low permeability of the membrane to ions, particularly protons, and the buffering capacity of the cytoplasm. Active pH homeostasis is principally achieved through the controlled movement of cations across the membrane. The major cation fluxes are those of potassium, protons and sodium ions. The movement of sodium ions has a major role in regulation of the cytoplasmic pH of alkaliphiles (Krulwich et al 1997) and this will be discussed in greater depth in another contribution (see Krulwich et al 1999, this volume). Others (see Padan et al 1999, this volume) will consider, in depth, the role of Na^+ in other organisms. In the first part of this chapter an overview of the major systems involved in pH homeostasis will be given. Potassium transport plays a major role in pH homeostasis in *E. coli* and other bacteria living at neutral pH. One of the most significant advances in understanding the importance of potassium ions has come through the analysis of the KefB and KefC potassium efflux systems, and this forms the focus of second part of this chapter.

Passive homeostasis

In formulating the chemiosmotic hypothesis of energy transduction Mitchell (1961) laid specific emphasis on the proton impermeability of the membrane. The capacity for pH homeostasis is also diminished when the proton impermeability of the membrane is perturbed by specific ionophores (Harold & van Brunt 1978) or by less specific agents that disrupt the packing of the phospholipids (Booth & Stratford 1998, Stratford & Anslow 1998, M. Stratford, personal communication 1997). Classic studies on *Enterococcus faecalis* (Harold & van Brunt 1978) demonstrated the importance of the integrity of the permeability barrier of the membrane to protons, K^+ and Na^+ ions. The pH range for growth and survival was severely diminished by changes to membrane permeability (Harold & van Brunt 1978). In addition, studies in our own laboratory have shown that the survival of *E. coli* O157 at low pH can be severely compromised by ethanol and lactate. The failure of pH homeostasis in the presence of these agents is a major factor in the loss of viability (Jordan et al 1998). The permeability barrier to protons, and to other ions, therefore, is one of the major 'passive' components of the homeostatic system.

Changes in lipid composition have been noted in response to acid pH conditions and these may alter the proton permeability of the membrane. Growth at low pH altered the lipid composition of *Clostridium acetobutylicum* leading to a decreased ratio of unsaturated to saturated fatty acids and an increase in cyclopropane fatty acids (CPFAs; Lepage et al 1987). Phospholipids with amino head groups (e.g. alanyl- and lysyl-phosphatidyl glycerol) increase during incubation at low pH in *Staphylococcus aureus* (Koostra & Smith 1969, Gould & Lennarz 1970, Kocun 1970, Haest et al 1972) and this has been proposed as a barrier to proton entry (Haest et al 1972). Changes in the proton permeability of cell membranes after adaptation to mildly acidic pH and entry into stationary phase have been noted for *E. coli* (C. O'Byrne, personal communication 1998). CPFAs are known to increase in *E. coli* when cells enter the stationary phase and there is broad correlation between CPFA levels and acid tolerance (Brown et al 1997). Although no causative link has been established, such lipids could directly affect the proton permeability leading to a higher pH homeostatic capacity. Despite these changes in lipid composition the pH_i in stationary phase cells of *E. coli* is less well maintained than in exponential phase cells (Jordan et al 1998). Therefore, although we consider the membrane permeability to be essentially a passive mechanism for homeostasis, adaptive changes can take place that alter phospholipid composition, which may in turn affect the proton leak across the membrane.

We also define intracellular buffering capacity as a passive mechanism. The ability of cells to recover from perturbation of pH_i depends upon the magnitude of the change, which is determined by intracellular buffering capacity. In the neutral to slightly acidic pH range the buffering capacity of the cytoplasm has been estimated to be of the order of 50–100 nmol H^+/pH unit per mg cell protein (Booth 1985). For most cells that maintain pH_i in the neutral range the buffering capacity derives from the titration of the phosphate groups and carboxylates of physiologically abundant metabolites species. Glutamate and aspartate combined in cell proteins can have an effective concentration of approximately 300 mM. The contributions from protein-located aspartate and glutamate side-chains will depend on their exposure to solvent rather than being buried in the heart of proteins, so the full 300 mM may not be available as buffering capacity. The contribution of this component of buffering will increase as pH_i falls below pH 5.5 since the pK_a of these groups is in the range pH 3.9–4.1. However, the range of pH_i is generally between pH 6.0–7.6, which means that these groups can play only a limited role in buffering pH change in the cytoplasm.

Changing the anion composition of the cytoplasm can influence the survival of extreme perturbation of pH homeostasis. Mutants of *Salmonella typhimurium* selected for their ability to survive extreme acid shock were found to lack isocitrate dehydrogenase, and as a consequence accumulated citrate and isocitrate in the cytoplasm (Foster & Hall 1991). Accumulation of citrate in place of glutamate would raise the buffering capacity of the cytoplasm in the neutral range due to the higher p*K* value of citrate (p*K* values 2.1, 4.07 and 9.47 for glutamate, and 3.1, 4.76 and 6.4 for citrate). However, there remains no systematic evidence for changes to buffering capacity being a significant component of pH homeostatic mechanisms.

The limit of our understanding of the specific nature of the buffering capacity is further exemplified by the changes that take place after treatment of *E. coli* cells with acetate and other weak acids. The application of 8 mM acetate at pH 6 causes an immediate decline of pH_i from pH 7.6 to approximately pH 6.9 (Roe et al 1998). At the same time the glutamate pool is lowered and this helps to maintain the anion:cation balance in the cells. However, over the next 10 min both pH_i and the acetate pool rise slowly and the measured acetate pool corresponds closely with that predicted by the transmembrane pH gradient. Since there is no detectable change in the cation content of the cells the accumulation of acetate must be achieved by exchange for other anions. The glutamate pool reduction only accounts for approximately 25% of the acetate accumulated (60 mM drop in glutamate accompanying an increase in acetate of 240 mM). Therefore, there must be 180 mM anions that are not regularly identified but which may be components of the buffering capacity. These ions are probably the normal metabolic intermediates of pathways, but this awaits demonstration. However, depletion of these pools is clearly not a major determinant of growth rate since the inhibition caused by acetate can be overcome by the addition of a single amino acid, methionine (Han et al 1993, A. J. Roe, D. McLaggan, C. O'Byrne & I. R. Booth, unpublished data 1998). High anion pools are also required for the generation of cell turgor since the anions allow the cell to accumulate potassium. However, it seems possible that the evolution of such pools may have been driven, in part, by the need for cytoplasmic buffering capacity.

Active homeostasis

Active homeostasis implies the ability to sense, and to respond to, perturbations of pH_i. In order to set a specific value for pH_i the cell needs to be able to either acidify or alkalinize the cytoplasm beyond the desired set point depending upon the direction of the initial perturbation. Cation transport, specifically Na^+ and K^+, is essential to these processes.

Sodium

A role for Na^+ in pH homeostasis was originally proposed for *E. coli* and for the alkaliphiles, but is now generally held only to be true for the latter. Excellent molecular and physiological analyses of the Na^+/H^+ antiport systems of *E. coli* have been completed recently (see Padan et al 1999, this volume). These studies revealed that the Na^+/H^+ antiport systems are central to Na^+ homeostasis and maintenance of the Na^+ gradient for transport, but not for regulation of cytoplasmic pH (Padan & Schuldiner 1994). An important factor here is that mutants lacking Na^+/H^+ antiport activity are extremely sensitive to alkaline pH in the presence of Na^+. However, this is a result of Na^+ toxicity and not a failure of pH homeostasis. It has been observed that intracellular Na^+ levels rise as the external pH (pH_o) is raised, possibly under the influence of the membrane potential, which is also higher at alkaline pH and/or non-specific mechanisms that allow higher rates Na^+ entry only at alkaline pH. In contrast, the alkaliphiles have a more specific requirement for the Na^+/H^+ antiports for pH homeostasis and progress towards defining these systems has recently been attained (Krulwich et al 1997).

In the alkaliphile, *Exiguobacterium aurantiacum*, the addition of excess Na^+ to Na^+-depleted cells causes pH_i to become more acidic than the optimum set point, and this is followed by recovery to a more alkaline value (McLaggan et al 1984). Again, the steady-state involves a Na^+ gradient that is oriented inwards, i.e. normal Na^+ relations are re-established once pH adjustment is completed. In this study it was particularly obvious that there was no absolute correlation between Na^+ movement and pH_i; Na^+ ions could be expelled without net acidification (McLaggan et al 1991). Further net alkalinization after excessive acidification appeared to be both energy and K^+ dependent. Even in the alkaliphiles there are multiple cation circuits that function in the regulation of pH_i.

Potassium

Information on the precise control points that lead to homeostatic regulation of pH_i is limited. For *E. coli* we know that K^+ uptake is required for the maintenance of pH_i at an alkaline value. Potassium accumulation in *E. coli* is achieved by multiple systems and each of these can substitute for the others in generating an alkaline pH_i. There is a tendency, in the Frag series of *E. coli* strains, for pH_i to be more alkaline when the external K^+ concentration is raised so that non-specific routes of K^+ entry may become significant (Ferguson et al 1997). Similarly, when the high affinity Kdp system is active pH_i is raised (Ferguson et al 1996). Conversely, when K^+ efflux is accelerated by glutathione deficiency, which increases the activity of the KefB and KefC K^+ efflux systems (see below), the steady-state pH_i is reduced (Ferguson & Booth 1998). So the set point is probably not a fixed entity, but the result of fluxes that combine to define the steady-state pH_i, i.e. that a major determinant of pH_i is the rate of potassium cycling.

The need for external K^+ for pH homeostasis in *E. coli* cells is quite dramatic. Suspension of cells in K^+-free medium leads to a dramatic reduction of pH_i, which can be restored by the re-introduction of K^+ into the medium (Kroll & Booth 1981, Kroll & Booth 1983, A. J. Roe, D. McLaggan & I. R. Booth, unpublished data 1998). The pH_i of *E. coli* Frag1 fell from pH 7.6–7.7 to pH 6.9–7.0 when cells were re-suspended in K^+-free medium. Addition of K^+ led to a rapid rise in pH_i to pH 7.6. Cells expressing the Kdp system were able to sustain higher pH_i values, presumably due to their intrinsic scavenging ability for K^+. In both situations the recovery of alkaline pH_i requires glucose metabolism. The importance of potassium cycling for pH homeostasis is exemplified by recent experiments on cells subjected to an acid shift. An abrupt shift of *E. coli* cells from pH_o 7 to pH_o 6 leads to transient acidification of the cytoplasm (Kihara & Macnab 1981). Recovery of pH_i to a more alkaline value requires potassium transport systems to be active (A. J. Roe, D. McLaggan, C. O'Byrne & I. R. Booth, unpublished data 1998). Similarly, the acidification of the cytoplasm by the addition of acetate causes a transient acidification (Roe et al 1998) and the recovery is inhibited in mutants lacking the both Trk and Kdp K^+ uptake systems (A. J. Roe, D. McLaggan & I. R. Booth, unpublished data 1998). Net potassium gain by the cells cannot be demonstrated in either of these experimental regimes (A. J. Roe, D. McLaggan, C. O'Byrne & I. R. Booth, unpublished data 1998). This suggests that the return of pH_i to an alkaline value is occasioned by the normal alkalinizing trend of K^+ transport activity.

Failure of K^+ pool regulation may also perturb pH homeostasis. In *E. coli* cells there is a family of related membrane proteins, the KefA family, that plays a role in osmoregulation. There are five potential members of this family in *E. coli* and in many organisms there are at least two representatives. The precise role of these systems is far from clear and since they also vary considerably in structural complexity the members may also have evolved separate roles. So far, the creation of mutants in *E. coli* has added little to our understanding of the role of these systems in cell physiology. In contrast, a mutation in a similar system in *Erwinia chrysanthemi* exhibited an unusual failure of pH homeostasis. The mutant strain, selected for sensitivity to high osmolarity, was found to have a transposon insertion in a homologue of KefA, called Yjep. The *E. chrysanthemi* mutant fails to grow at alkaline pH when incubated in high K^+ medium with compatible solutes. Normally, cells incubated under these conditions would release K^+ during the accumulation of betaine. In wild-type *E. chrysanthemi* the exchange of intracellular potassium (and its counter ion glutamate) for betaine is associated with a rise in pH_i. In contrast, the mutant undergoes acidification, rather than alkalinization of the cytoplasm (Touze et al 1998). The absence of the Yjep protein, therefore, causes a specific defect in pH homeostasis. Thus, the KefA family may have an indirect role in pH homeostasis through the correct functioning of the osmoregulatory systems.

Protection of the cell by modulation of intracellular pH

E. coli cells demonstrate a marked capacity for pH homeostasis, which is specifically perturbed during the detoxification of electrophiles to protect the cell against damage. Electrophiles are toxic to the cell due to their ability to modify nucleic acids, proteins and lipids. The most common electrophile is probably methylglyoxal since this compound is readily generated as a by-product from glycolysis (Cooper & Anderson 1970). Cells that have lost control over carbon flux frequently produce methylglyoxal, and once this compound accumulates to high levels cell death ensues (Tötemeyer et al 1998). *E. coli* cells provoked to synthesize methylglyoxal can accumulate this compound to over 1 mM in the medium. In the absence of detoxification capacity cell death occurs at low concentrations (<0.1–0.2 mM). Detoxification occurs by several routes, but the most significant appears to be the formation of glutathione conjugates and their metabolism by glyoxalases I and II (MacLean et al 1998a). Mutants of *E. coli* that lack either glutathione or the glyoxalases are extremely sensitive to methylglyoxal (MacLean et al 1998a). The major effect of these compounds is to convert the electrophile to D-lactate. However, the intermediates in the detoxification pathway also have an important role in modulating pH_i (Ferguson et al 1995, 1996, 1997, 1998). The reaction of methylglyoxal with glutathione forms hemithiolacetal, which is the substrate for glyoxalase I. S-lactoyl-glutathione, the product of the reaction catalysed by glyoxalase I, is an activator of the K^+ efflux system KefB (MacLean et al 1998b). KefB and KefC are heteromeric K^+ efflux proteins, which may form cation channels through the cytoplasmic membrane. To date, sequences that are similar to that for the *E. coli* KefC protein, and which overall display similar domain organization, have been found in the enteric group, *Myxococcus xanthus*, *Haemophilus influenzae* and *Pseudomonas aeruginosa*, and a related protein is found in the archaebacterium, *Methanococcus jannaschii*.

The KefB and KefC proteins of *E. coli* are closely related in their primary sequence, in their organization and in their requirement for an ancillary protein, which is YabF for KefC and YheR for KefB. KefC and KefB differ in their activation specificity. Both are held in an inactive state by glutathione and are activated by glutathione adducts, but only KefB is strongly activated by adducts derived from methylglyoxal. The consequence of activation of KefB and KefC is that pH_i falls. The decline in pH_i is essential to survival of exposure to the electrophile. Cells lacking KefB and KefC lose viability more rapidly when exposed to methylglyoxal. However, the cells can be protected by artificially lowering pH_i by the addition of weak acids (Ferguson et al 1995, 1996, 1997). The acid must be added to cells within a few minutes of addition of methylglyoxal to prevent cell death. Two models have been advanced to explain the protection. Firstly, that the lowering of pH limits the damage of macromolecules; or secondly, that the change in pH_i activates existing repair functions. Protection by weak acids does not require new protein synthesis (Ferguson et al 1995). However, the observed protection by acidification of the cytoplasm has a parallel in the lowered mutagenesis of N-methyl-N′-nitro-N-nitrosoguanidine (Oktyabrsky et al 1993) observed in cells incubated with weak acids.

The actual value of pH_i must lie within a narrow range to allow protection. Cells that have been induced to express the Kdp system exhibit a higher pH value (about pH 8.0 compared with pH 7.7 for cells lacking Kdp) and although they exhibit a drop in pH_i on activation of KefB and KefC they are not protected (Ferguson et al 1996). Similarly, when cells are exposed to higher concentrations of K^+ pH_i is again raised and acidification by activation of KefB and KefC by methylglyoxal is not protective (Ferguson et al 1995). In each of these cases pH_i did not fall below pH 7.5 and the addition of acetate to lower the pH further could confer protection. Thus, there is a cut-off point close to the normal range within which pH_i is maintained that leaves cells sensitive to methylglyoxal. The activation of KefB and KefC is required to achieve a lower pH_i value for protection against methylglyoxal (and some other electrophiles). Thus, this is a specific instance of a cellular mechanism for perturbation of pH_i that mediates a distinct function in cell metabolism.

Conclusions

The essence of pH homeostasis is control over cation permeability at the cytoplasmic membrane. The key cations are H^+, K^+ and Na^+. Anions are also likely to be important, though principally as buffer components. Since 1985, progress on defining the components of the pH homeostatic machinery has enabled the elimination of the Na^+/H^+ antiport family as major elements of pH homeostasis in *E. coli*. We are at last beginning to identify components that may play significant roles in K^+ cycling, which underpins pH homeostasis in *E. coli*. The KefB and KefC systems have shown the importance of control over pH_i for cell survival.

Acknowledgements

The author's laboratory is supported by the Biotechnology and Biological Sciences Research Council, The Wellcome Trust and Unilever Research. The author wishes to thank all the members of his group, but with special thanks to Andy Roe for his recent work on pH homeostasis and potassium cycling.

References

Booth IR 1985 Regulation of cytoplasmic pH in bacteria. Microbiol Rev 49:359–378

Booth IR, Stratford M 1998 Acidulants and low pH. In: Gould GW, Russell NJ (eds) The molecular basis of food preservative action. Blackie, in press

Brown JL, Ross T, McMeekin TA, Nichols PD 1997 Acid habituation of *Escherichia coli* and the potential role of cyclopropane fatty acids in low pH tolerance. Int J Food Microbiol 37: 163–173

Cooper RA, Anderson A 1970 The formation and catabolism of methylglyoxal during glycolysis in *Escherichia coli*. FEBS Lett 11:273–276

Ferguson GP, Booth IR 1998 Importance of glutathionine for growth and survival of *Escherichia coli* cells: detoxification of methylgyloxal and maintenance of intracellular K^+. J Bacteriol 180:4314–4318

Ferguson GP, McLaggan D, Booth IR 1995 Potassium channel activation by glutathione-S-conjugates in *Escherichia coli*: protection against methylglyoxal is mediated by cytoplasmic acidification. Mol Microbiol 17:1025–1033

Ferguson GP, Chacko AD, Lee C, Booth IR 1996 The activity of the high affinity potassium uptake system Kdp sensitizes cells of *Escherichia coli* towards methylglyoxal. J Bacteriol 178:3957–3961

Ferguson GP, Nikolaev Y, McLaggan D, Maclean M, Booth IR 1997 Survival during exposure to the electrophilic reagent N-ethylmaleimide in *Escherichia coli*: the role of KefB and KefC potassium channels. J Bacteriol 179:1007–1012

Ferguson GP, Creighton RI, Nikolaev Y, Booth IR 1998 The importance of RpoS and Dps in the survival of exposure of both exponential and stationary phase cells to the electrophile, N-ethylmaleimide. J Bacteriol 180:1030–1036

Foster JW, Hall HK 1991 Inducible pH homeostasis and the acid tolerance response of *Salmonella typhimurium*. J Bacteriol 173:5129–5135

Foster JW, Moreno M 1999 Inducible acid tolerance mechanisms in enteric bacteria. In: Bacterial responses to pH. Wiley, Chichester (Novartis Found Symp 221) p 55–74

Gould RM, Lennarz WJ 1970 Metabolism of phosphatidylglycerol and lysyl phosphatidylglycerol in *Staphylococcus aureus*. J Bacteriol 104:1135–1144

Haest CWM, de Gier J, Op den Kamp JAF, Bartels P, van Deenan LLM 1972 Changes in the permeability of *Staphylococcus aureus* and derived liposomes with varying lipid composition. Biochim Biophys Acta 255:720–733

Han K, Hong J, Lim HC 1993 Relieving effects of glycine and methionine from acetic acid inhibition in *Escherichia coli* fermentation. Biotechnol Bioengin 41:316–324

Harold FM, van Brunt J 1978 Circulation of H^+ and K^+ across the plasma membrane is not obligatory for bacterial growth. Science 197:372–373

Hengge-Aronis R 1993 Survival of hunger and stress — the role of rpoS in early stationary phase gene regulation in E. coli. Cell 72:165–168

Jordan S, Glover J, Malcom L, Thomson-Carter FM, Booth IR, Park SJ 1998 Augmentation of killing of *Escherichia coli* O157 by combinations of low pH, lactic acid and ethanol. Appl Environ Microbiol, submitted

Kihara M, Macnab RM 1981 Cytoplasmic pH mediates pH taxis and weak acid repellent taxis of bacteria. J Bacteriol 145:1209–1221

Kocun FJ 1970 Amino acid containing phospholipids as major components of the phospholipids of *Streptococcus faecalis* 10CI. Biochim Biophys Acta 202:277–282

Koostra WL, Smith PF 1969 D- and L-alanylphosophatidyl glycerols from *Mycoplasma laidlawii* strain B. Biochem 8:4794–4806

Kroll RG, Booth IR 1981 The role of potassium transport in the generation of a pH gradient in *Escherichia coli*. Biochem J 198:691–698

Kroll RG, Booth IR 1983 The relationship between K^+ transport, pH and pH_i in *Escherichia coli*. Biochem J 216:709–716

Krulwich TA, Ito M, Gilmour R, Guffanti AA 1997 Mechanisms of cytoplasmic pH regulation in alkaliphilic strains of *Bacillus*. Extremophiles 1:163–169

Krulwich TA, Guffanti AA, Ito M 1999 pH tolerance in *Bacillus*: alkaliphiles versus non-alkaliphiles. In: Bacterial responses to pH. Wiley, Chichester (Novartis Found Symp 221) p 167–182

Lepage C, Fayolle F, Hermann M, Vandecasteele JP 1987 Changes in membrane lipid composition of *Clostridium acetobutylicum* during acetone–butanol fermentation: effects of solvents, growth temperature and pH. J Gen Microbiol 133:103–110

MacLean MJ, Tötemeyer S, Booth IR 1998a Analysis of *E. coli* mutants lacking glyoxalase II (*gloB*). J Bacteriol, in prep

MacLean MJ, Ness LS, Ferguson GP, Booth I R 1998b The role of glyoxalase I in the detoxification of methylglyoxal and in the activation of the KefB K^+ efflux system in *Escherichia coli*. Mol Microbiol 27:563–572

McLaggan D, Selwyn MJ, Dawson AP 1984 Dependence on Na^+ of control of cytoplasmic pH in a facultative alkalophile. FEBS Lett 165:231–237

McLaggan D, Selwyn MJ, Dawson AP, Booth IR 1991 Role of Na^+ in pH homeostasis by the alkalophilic bacterium *Exiguobacterium aurantiacum*. J Gen Microbiol 137:1709–1714

Mitchell P 1961 Coupling of phosphorylation to electron and hydrogen transfer by a chemiosmotic type of mechanism. Nature 191:144–148

Oktyabrsky ON, Golyasnaya NV, Smirnova GV, Demakov VA, Posokhina N Kh, Kholstova TA 1993 Acidification of *Escherichia coli* and *Salmonella typhimurium* cytoplasm reduces the mutagenic effect of N-methyl-N′-nitro-N-nitrosoguanidine. Mutat Res 293:197–204

Padan E, Schuldiner S 1994 Molecular physiology of Na^+/H^+ antiporters, key transporters in circulation of Na^+ and H^+ in cells. Biochim Biophys Acta 1185:129–151

Padan E, Gerchman Y, Rimon A, Rothman A, Dover N, Carmel-Harel O 1999 The molecular mechanism of the regulation of the NhaA Na^+/H^+ antiporter of *Escherichia coli*, a key transporter in the adaptation to Na^+ and H^+. In: Bacterial responses to pH. Wiley, Chichester (Novartis Found Symp 221) p 183–199

Roe AJ, McLaggan D, Davidson I, O'Byrne C, Booth IR 1998 Perturbation of anion balance during inhibition of growth of *Escherichia coli* by weak acids. J Bacteriol 180:767–772

Schellhorn HE, Stones VL 1992 Regulation of *katF* and *katE* in *Escherichia coli* K-12 by weak acids. J Bacteriol 174:4769–4776

Stratford M, Anslow PA 1998 Evidence that sorbic acid does not inhibit yeast as a 'classic weak-acid preservative'. Lett Appl Microbiol 26, in press

Thomas AD, Booth IR 1992 The regulation of expression of the porin gene *ompC* by acid pH. J Gen Microbiol 138:1829–1835

Tötemeyer S, Booth NA, Nichols WW, Dunbar B, Booth IR 1998 From famine to feast: the role of methylglyoxal production in *Escherichia coli*. Mol Microbiol 2:553–562

Touze T, Gousebet G, Roe AJ, Booth IR, Benassi A, Blanco C 1998 Analysis of a K^+ channel mutant in *Erwinia chrysanthemi*, in prep

DISCUSSION

Krulwich: The resemblance of the KefB and KefC systems to the NapA-type antiporter of *Enterococcus hirae* is striking. Is it possible, therefore, that the KefB and KefC systems arose as a result of gene duplication and that they respond to overlapping stresses?

Booth: KefB and KefC are part of a large family. Reizer et al (1992) originally pointed out that NapA, which is a sodium/proton antiporter in *E. hirae*, has structural resemblance to KefC. However, as the genome mapping projects continue we obtain more sequences, and we are beginning to discover that this statement is not true, rather they represent different branches of a large family. The question is in interpreting the branches of this family. If a core protein started to evolve in different directions a million years ago, then it's possible that the end-point function of that protein, despite the homology in certain sequence elements, is quite different. There seems to be three branches in this family: the KefB/KefC branch; a distinct branch that contains the iron regulator transporter protein in one of the magnetobacteria and other, related proteins; and the sodium/proton antiport branch. The sequence identity throughout the family resides in the first 70 residues of a 600 residue protein, and although I recognize the evolutionary significance of this, I am reluctant to speculate that because one is a sodium/proton antiporter and one is involved in stress tolerance they both have a role in stress tolerance or in sodium transport.

Padan: Are there any indications that KefB and KefC are channels? Secondly, we don't know the precise function of potassium in pH homeostasis, but does it affect the rate of electron transport?

Booth: In response to your first question, all our attempts to demonstrate channel activity have so far failed. Part of the reason for this is that we have tried to demonstrate this by patch clamp analysis, i.e. we ripped off small pieces of membrane from the surface of cells and looked at ion conductivity in response to various chemical signals such as glutathione. When we did these experiments we didn't realize that the ancillary subunit YabF is required for the activity of the system, and the problem with patch clamp analysis is that proteins which are only peripherally attached start to diffuse away. We are now trying to produce a chimera in which YabF is covalently attached, although it's proving difficult to make a linker that gives reconstituted activity when we attach YabF to the end of KefC.

In answer to your second question, acidification of the cytoplasm results in a twofold decrease in respiratory rate, and if potassium is then added the respiratory rate increases. I'm not sure whether this is because the pH is raised or because the membrane potential is reduced.

Stock: If KefB and KefC operated as channels wouldn't the response be dependent on the extracellular ion composition?

Booth: To address this question we incubated Kdp^- cells in 10 mM K^+ and in 0.2 mM K^+ (see Ferguson et al 1995). The intracellular pH (pH_i) is modulated in cells incubated in high concentrations of potassium. When we added methylglyoxal to activate KefB, we saw less activation, as we would have expected of a channel, but we did see some modulation of pH_i. In 0.2 mM K^+ we saw channel activation, potassium efflux and lowering of pH_i. These last cells are protected, whereas cells in 10 mM K^+ die rapidly. There is about a 10–100-fold difference in susceptibility. Therefore, the operation of this system can be blocked by increasing the potassium concentration in the medium. How does *Escherichia coli* find itself in the right potassium concentration in the presence of an electrophile when it wants to protect itself? Does it die the rest of the time?

This could be true, because we shouldn't think of homeostasis as being perfect, it's an approximate solution rather than a definitive solution.

Cook: Is there any effect on $\Delta\Psi$ when you add methylglyoxal?

Booth: Bakker & Mangerich (1982) showed that it increases.

Cook: Is methylglyoxal freely permeable in those experiments?

Booth: Yes. But we can do these experiments in two ways. Ackerman et al (1974) showed that it is possible to stimulate cells to make methylglyoxal. All these experiments can be done by adding methylglyoxal extracellularly or by generating it intracellularly, and it seems to be freely permeable.

Cook: Why doesn't it just diffuse out of the cells? Why do the cells have an elaborate mechanism for methylglyoxal efflux?

Booth: Because it is being generated internally. The levels increase gradually over a matter of hours. When the equilibrium concentration across the membrane reaches 0.3 mM cell growth stops, and at 0.6 mM the cells start to die. Above 0.6 mM the death rate increases. Therefore, cells experience a temporal change in methylglyoxal concentration. They excrete it but it builds up because there's less detoxification capacity compared with synthesis capacity.

Cook: Is the increased $\Delta\Psi$ due to a dissipation of the membrane potential?

Booth: If potassium flows out of the cell via a channel, one would expect the membrane to hyperpolarize anyway. This is one possible cause, but there are many others.

Skulachev: Can you exclude that there is a direct regulatory effect of $\Delta\bar{\mu}_{H^+}$ on the potassium channel?

Booth: No, and we don't want to. The reason I say that this is a distinct branch of the family is that there is a conserved voltage sensor sequence (three hydrophobic amino acids then Arg or Lys, followed by three hydrophobic amino acids then Arg or Lys) in members of the KefB/KefC family, but it's not present in the so-called regulatory family or the antiport family. This sequence results in five positive charges across the membrane, just like in a Shaker channel. We expect it to be voltage gated, but this is difficult to prove.

Bennett: Does this effect still occur in fermentative conditions?

Booth: We haven't done those experiments, but Russell (1993) suggests that it does.

Cook: Picking up on that point, Russell (1993) demonstrated that if you grow *Prevotella ruminicola* in continuous culture with an excess of glucose, there is an oscillating pattern of growth and cell death that has been correlated with methylglyoxal production.

Booth: And they excrete potassium.

Epstein: If these systems are non-specific, one would expect that when they open up they would take up sodium in exchange for potassium, and yet they acidify. Do the cells take up sodium under these conditions and, if they do, how much?

Booth: For each potassium ion that comes out, 0.7 is replaced by sodium and 0.3 by protons. If sodium is removed potassium efflux is limited; it is not possible to force the cells to take up protons in this way.

Epstein: If the pH is low and you remove sodium from the medium, will the internal pH decrease?

Booth: I've never tried to do that.

Slonczewski: Why has it been so difficult to pin down the role of potassium systems in pH regulation and pH detection?

Booth: There are a minimum of three uptake systems: Kup, Trk and Kdp. Two potassium efflux systems are associated with glutathione — KefB and KefC — and there are five osmoregulated potassium efflux systems. We have made a triplicate knockout, and we will be making a quadruplicate next week and a pentuplicate probably next month. Also, Verkhovskaya et al (1996) have detected another system in lipid vesicles, but we don't yet know the nature of this. Therefore, there are just so many systems for adjusting the amount of potassium in the cell that it becomes difficult to make a specific knockout which tells us something about pH homeostasis.

Slonczewski: That's true. But in the case of sodium there are several antiporters and it is possible to identify particular amino acids that are involved in pH regulation. Have there been any successes with looking at individual amino acids?

Booth: The only person who has attempted this systematically is Barry Rosen. He identified a *kha* mutant (Plack & Rosen 1980), but the problem with this was that it turned out to be suppressible by the addition of amino acids to the medium. One of the problems we have to bear in mind is that if acetate is added to these cells, the growth rate is inhibited by about 50%. If methionine is added the growth rate is practically restored. The change in pH_i has a negligible effect on the growth rate. The effect is caused by the depletion of methionine. Methionine biosynthesis is extremely susceptible to osmotic stress, pH stress and temperature stress, and this has immense ramifications for the cell. You run into problems when you do selections on complex media because you are supplying methionine. Another problem is that the cells begin to starve if you hit them so hard that their pH_i changes. Therefore, the impact on cell physiology is so great that it becomes difficult to isolate the relevant mutants.

Rowbury: What is the basis for the susceptibility of methionine biosynthesis to stress? Does it involve *O*-succinyl homoserine synthase?

Booth: That was what we expected. This enzyme has always been thought of as being a stress sensor, and it also seems to be affected by temperature. However, we did various complementation experiments, and it seems that the pH sensitivity is controlled by the tetrahydrofolate pathway. Last year, Crowley et al (1997) isolated a pH-sensitive mutant in *Streptococcus mutans* but when they analysed it they found it was formyl-tetrahydrofolate synthase.

Rowbury: Do you only work with *E. coli?* Because I was wondering whether S-adenosyl methionine is required to make stress sensors in *E. coli* and other Gram-negative bacteria.

Booth: We've only done the work in *E. coli.*

Poole: Some of your experiments were done with *E. coli* O157. Do you know of any differences between this and K12 with respect to pH homeostasis?

Booth: We were commissioned to find out whether O157 is more acid tolerant than other strains of *E. coli*. We found that the viability of laboratory strains at pH 3.0 decreases rapidly over 10 min. The viability of *E. coli* from the gut, however, decreases at a three- to fourfold slower rate and then it remains constant, so a fraction of the cells are acid tolerant. O157 has more or less the same death rate, but the acid-tolerant fraction is one log higher. By culturing *E. coli* in the laboratory, we seem to remove their variability. We culture to the point at which their protein composition lies within a narrow spectrum. If you look at the population of cells and you conceptualize the number of proteins in the cell, you might find that laboratory strains have narrow distribution profiles, whereas natural isolates have a broad profile. Organisms that are surviving are predicted to have broad protein profiles. We don't see any real differences between O157 and other strains, in terms of pH homeostasis, but paradoxically the most acid-resistant cells, i.e. the stationary phase cells, have a poorer capacity for pH homeostasis. Therefore, it's not just pH homeostasis, it's gene expression and population structure.

Matin: Do the survivors maintain a proton motive force?

Booth: The survivors represent 10^3 organisms within a population of 10^7, and we haven't yet developed a method for separating them, which we need to do before we can do such analyses.

Foster: Can't you just recover them by taking the survivors and growing them up again?

Booth: Yes, but they have altered their physiological state by then. These are not mutants — if we reculture that population we see the same profile again — they are physiological variants.

Cook: Do you culture under aerobic or anaerobic conditions?

Booth: We culture under aerobic conditions, although the same experiments have been done under anaerobic conditions.

Cook: If O157 has a smaller ΔpH than a laboratory strain of *E. coli*, such as K12, would this be a strategy for growth in the lower gut where there is a high concentration of fatty acids? Because rumen bacteria adopt this strategy to cope with fermentation acids (Russell 1991, Cook & Russell 1994).

Booth: It's possible, but no one has looked at the level of RpoS in bacteria taken from the gut, which would be an indicator of whether they are in exponential or stationary phase. If they are exponential, then their pH homeostasis capacity is fine. You would then argue that those cells would be at a disadvantage in terms of volatile fatty acids, whereas a true stationary phase cell would be at an advantage because it has a lower ΔpH and lower toxicity, but then it doesn't need as much acid to decrease its pH_i to the point where it dies.

Slonczewski: You mentioned that stationary phase cells have a lower capacity for pH regulation, but we don't see that in our experiments. We find that if we grow stationary cells to a couple of ODs, pellet them to 200 ODs and aerate them they immediately raise their pH_i to 7.6.

Poole: But does this have anything to do with going from stationary phase to non-stationary phase? What is the aeration doing, for example?

Slonczewski: These are non-growing cells.

Matin: Then why do they stop growing? One can easily engineer conditions in which one can predict why a cell would enter stationary phase, but I don't see why giving the cells some oxygen would restore pH homeostasis. Stationary phase cells are more resistant to acid stress, but the reason for this is not that they are better at maintaining a pH homeostasis, it's because of the expression of stress proteins.

Slonczewski: The question is, what do people mean when they say 'stationary phase cells'?

Matin: This is not a complicated question. It depends on how carefully you engineer the conditions. To a cell physiologist the stationary phase represents the exhaustion of a nutrient. But stationary phase can also be brought about by some other insult, such as an unfavourable temperature or pH.

Slonczewski: Some people define stationary phase as being when RpoS expression increases, which is around OD 0.4.

Matin: But the σS concentration can change as a result of many different stresses.

Booth: Rapidly growing cells have high levels of RpoS, even in minimal medium. If you knock out RpoS, you can show that their stress survival is strongly dependent on RpoS. Therefore, RpoS and stationary phase have to be separated.

Slonczewski: This is of interest to us because in the proteome we can identify proteins that have been defined as stationary phase proteins after the cells reach OD 0.4. When we looked at pH stress proteins, we had to scrap our early data because we were only seeing these so-called 'stationary phase proteins'.

Matin: If growth was stopped, say due to a toxic metabolite, one would see the expression of a few of the common stress proteins. But this doesn't mean that one was looking at stationary phase, say due to the exhaustion of a nutrient. It is important to keep in mind why growth stopped.

Booth: Did you do these studies in broth or in minimal medium?

Slonczewski: Mainly in broth.

Booth: But at about OD 0.4 one of the nutrients in broth runs out. One can then see a secondary growth phase. This is also the precise point at which RpoS transcription, not translation, changes. Therefore, broth is not a uniform medium to get from OD 0.05 to OD 1.0.

Konings: I would like to change the subject and pick up on Ian Booth's comments about the proton permeability of the membranes. The assumption of Mitchell (1981) is that membranes have to be impermeable to protons, so we measured the proton permeability of a variety of organisms including psychrophilic and thermophilic bacteria, archaea and the mesophilic bacteria *E. coli* and *Bacillus subtilis* (De Vrij et al 1988, Van de Vossenberg et al 1995). We found that in all cases the proton permeability increases with temperature (Fig. 1). I should mention that these studies were done with liposomes and also with membrane vesicles with the same results. We used liposomes of the same size to make these comparisons possible. In contrast to proton permeabilities, all the membranes have the same sodium permeabilities. This suggests that proton permeability involves a different mechanism than sodium permeability. One explanation is that protons can move via water molecules in the membranes and that sodium just moves via diffusion. The proton permeabilities of the psychrophiles, mesophiles and archaea, however, are about the same at their respective optimal growth temperatures. This suggests that the organisms have mechanisms to control the proton permeability of their membranes. There are a number of theories about the relationship between lipids and the physical properties of different membranes. The homeoviscous hypothesis suggests that the membrane has to have a certain viscosity, whereas the homeophasic adaptation hypothesis says that the membrane has to be in a fluid state. We have introduced the homeoproton permeability theory which suggests that the bacteria are able to adjust the lipid composition of their membranes in such a way that the proton permeability remains constant. In order to test this we grew *B. subtilis* at 13°C and at 50°C, which are the extremes of the growth temperatures of *B. subtilis*. We found that the proton permeabilities at both temperatures were the same. Therefore, *B. subtilis* appears to have a mechanism to control the proton permeability more tightly than the membrane fluidity because the membrane fluidity at 50°C is much greater than at 13°C.

Finally, I would like to talk about the thermophilic species *Bacillus stearothermophilus* grown at about 60°C. The proton permeability of *B. stearothermophilus* is about five- to 10-fold higher than that of *B. subtilis* at their respective optimum growth temperatures. *B. stearothermophilus* does not have a mechanism to control the proton permeability of its membrane, but it has a mechanism to adjust the proton motive force. The respiratory chain pumps protons, and the respiratory rate is 10–20-fold higher than in mesophilic *B. subtilis*. Therefore, pumping out protons as fast as possible compensates for the proton leak of the membrane. It is unlikely that proteins have a significant role in the regulation of proton

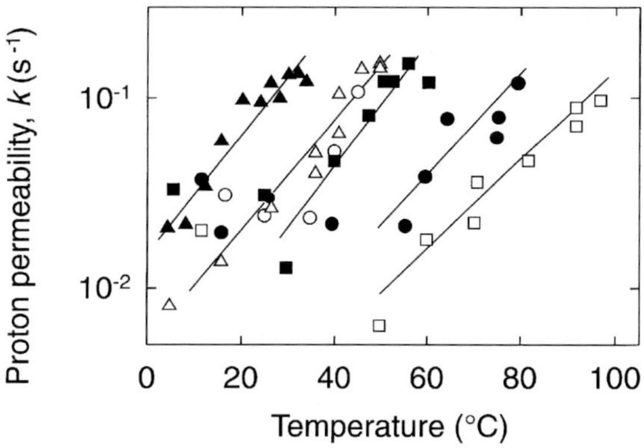

FIG. 1. Temperature dependency of the proton permeability of liposomes derived from various bacteria and archaea. The first order rate constant for the proton permeability (K_{H+}) was determined for liposomes composed of lipids derived from *Psychrobacter immobilis* (filled triangles), *Methanosarcina barkeri* (open triangles), *Escherichia coli* (open circles), *Bacillus stearothermophilus* (filled squares), *Thermotoga maritima* (filled circles) and *Sulfolobus acidocaldarius* (open squares). Taken from Van de Vossenberg et al (1995).

permeability, since the proton permeability of membrane vesicles is similar to that of liposomes made of their lipids.

Cook: But you're comparing an aerobe with an anaerobe.

Konings: Clostridium fervidus, which is an anaerobe and thermophile has the same high proton permeability of its membrane at the high growth temperature. This organism relies completely on sodium as a coupling ion in energy transduction and not at all on protons. Also, *Thermoalkalibacter borgoriae*, which grows optimally at pH 9.5 and 60°C, relies completely on sodium for growth.

Poole: Have similar measurements been made for potassium permeability?

Konings: Not as far as I know.

Schäfer: You quoted papers which showed that there is no significant difference in proton permeability of membrane vesicles and liposomes prepared from their lipids (De Vrij et al 1988, Van de Vossenberg et al 1995). It is possible that this contradiction can be explained by differences in experimental conditions. It is not easy to make lipid vesicles with reproducible permeabilities by just using extracts of lipids from membranes obtained during various growth states. Did you use extracts or purified lipid fractions?

Konings: We used lipids that were extracted from membrane vesicles, and the lipids were fractionated and mixed again.

Booth: The reason Martin Brand came back to look at this issue was because they developed a new method for extracting lipids that did not involve detergent. When they removed the detergent they found that this was the factor which caused the observation that lipid composition affects proton permeability. The new study described in my chapter in the absence of detergents showed that proton permeability in liposomes is independent of lipid composition. He now says that the detergent gets stuck and causes the problem.

Skulachev: In mitochondria there are special proteins called uncoupling proteins that are responsible for proton conductivity of the inner mitochondrial membrane. They transport anions of fatty acids from mitochondria, whereas protonated fatty acids return to the matrix via phospholipid regions of the membrane. The same function can be carried out by the ATP/ADP antiporter and some other anion carriers. Such effects are involved in thermoregulation, and they may also prevent superoxide formation by the respiratory chain.

Dimroth: It may be advantageous at higher temperatures to use sodium instead of protons as coupling ions for ATP synthesis, but we have to consider that a sodium ion concentration gradient (ΔpNa^+) may be more stable at high temperatures, even though proton permeability still exists. This may be a problem because proton permeability could decrease the $\Delta\Psi$ or even destroy it. We have shown that for the ATP synthase of *Propionigenium modestum* $\Delta\Psi$ is an obligatory driving force which cannot be compensated for by a large ΔpNa^+. If proton permeability increases with temperature, then how do these organisms maintain $\Delta\Psi$ at a sufficient level to make ATP?

Konings: The ATPase can be driven by $\Delta\Psi$ or by ΔpNa^+.

Dimroth: There are many claims in the literature that these are kinetically equivalent driving forces. But this has to be looked at again, because for the *P. modestum* ATP synthase at a ΔpNa^+ of 200 V ATP is not produced.

Konings: But you are using a mesophilic organism and it is not possible to do that experiment with a thermophile at its growth temperature. What is true for one may not be true for the other.

References

Ackerman RS, Cozzarelli NR, Epstein W 1974 Accumulation of toxic concentrations of methylglyoxal by wild-type *Escherichia coli* K-12. J Bacteriol 119:357–362

Bakker EP, Mangerich WE 1982 N-ethylmaleimide induces K^+-H^+ antiport activity in *Escherichia coli* K-12. FEBS Lett 140:177–180

Cook GM, Russell JB 1994 The effect of extracellular pH and lactic acid on pH homeostasis in *Lactococcus lactis* and *Streptococcus bovis*. Curr Microbiol 28:165–168

Crowley PJ, Gutierrez JA, Hillman JD, Bleiweis AS 1997 Genetic and physiological analysis of a formyl-tetrahydrofolate synthetase mutant of *Streptococcus mutans*. J Bacteriol 179:1563–1572

De Vrij W, Bulthuis RA, Konings WN 1988 Comparative study of the energy-transducing properties of cytoplasmic membranes from mesophilic and thermophilic *Bacillus* species. J Bacteriol 170:2359–2366

Ferguson GP, McLaggan D, Booth IR 1995 Potassium channel activation by glutathione-S-conjugates in *Escherichia coli*: protection against methylglyoxal is mediated by cytoplasmic acidification. Mol Microbiol 17:1025–1033

Mitchell P 1981 Bioenergetic aspects of unity in biochemistry: evolution of the concept of ligand conduction in chemical, osmotic, and chemiosmotic reaction mechanisms. In: Semenza G (ed) Of oxygen, fuels, and living matter. Part 1. Wiley, Chichester, p 1–160

Plack RH Jr, Rosen BP 1980 Cation/proton antiport systems in *Escherichia coli*. Absence of potassium/proton antiporter activity in a pH-sensitive mutant. J Biol Chem 255:3824–3825

Reizer J, Reizer A, Saier MH Jr 1992 The putative Na^+/H^+ antiporter (NapA) of *Enterococcus hirae* is homologous to the putative K^+/H^+ antiporter (KefC) of *Escherichia coli*. FEMS Micro Lett 94:161–163

Russell JB 1991 Intracellular pH of acid-tolerant ruminal bacteria. Appl Environ Microbiol 57:3383–3384

Russell JB 1993 Glucose toxicity in *Prevotella ruminicola*: methylglyoxal accumulation and its effect on membrane physiology. Appl Environ Microbiol 59:2844–2850

Van de Vossenberg JLCM, Ubbink-Kok T, Elferink MGL, Driessen AJM, Konings WN 1995 Ion permeability of the cytoplasmic membrane limits the maximum growth temperature of bacteria and archaea. Mol Microbiol 18:925–932

Verkovskaya ML, Verkovsky MI, Wikström M 1996 K^+-driven Na^+ transport driven by respiration in *Escherichia coli* cells and membrane vesicles. Biochim Biophys Acta 1273: 207–216

pH sensing in bacterial chemotaxis

Mikhail N. Levit and Jeffry B. Stock[1]

Department of Molecular Biology, Princeton University, Princeton, NJ 08544, USA

Abstract. Bacteria are able to sense a broad range of chemical and energetic stimuli and modulate their swimming behaviour to migrate to more favourable environments. Signal transduction in bacterial chemotaxis is mediated by a two-component system composed of a protein histidine kinase, CheA, and a response regulator, CheY. The phosphorylated response regulator, P ~ CheY, binds to a protein at the flagellar motor, FliM, to cause reversals in flagellar motor rotation. The level of P ~ CheY is controlled by the activity of the kinase CheA, which is in turn regulated by membrane receptors at the cell surface. Membrane receptors such as the aspartate receptor, Tar, are composed of two distinct regions: an extracellular sensing domain that binds stimulatory ligands, aspartate in the case of Tar; and an intracellular signalling domain that forms a complex with the protein kinase CheA. What is the mechanism of transmembrane signalling? How does aspartate binding to the sensing domain at the outside surface of the membrane translate into a change in kinase activity at the membrane cytosol interface? Recent results suggest that the mechanism depends on perturbations in lateral packing within an extensive array of receptors localized to patches at the cell poles. Receptor patching appears to depend on higher-order associations with the kinase CheA as well as an adaptor protein, CheW. It is difficult to assess the locus of pH effects within the context of even a simple signal transduction system like that involved in bacterial chemotaxis. Previous results with mutant strains have indicated that the serine receptor, Tsr, is critical for pH sensing, but *in vitro* results do not support such a straightforward interpretation of the genetic data.

1999 Bacterial responses to pH. Wiley, Chichester (Novartis Foundation Symposium 221) p 38–54

The signal transduction system that mediates chemotaxis in *Escherichia coli* provides a paradigm for understanding fundamental principles that underlie complex regulatory networks. Here we consider the sensing of pH. pH belongs to a category of general environmental parameters that are felt by virtually all proteins; other members of this class would include temperature, pressure and ionic strength. Changing these global stimuli will affect activities at all levels of a network, making it difficult to interpret or predict the effects of mutations. Mutations that lead to inverted responses are commonly observed (e.g. mutants

[1]This chapter was presented at the symposium by Jeffry B. Stock to whom correspondence should be addressed.

that are attracted to acid rather than repelled by it, which is the wild-type behaviour). Thus, any discussion of pH sensing in chemotaxis first requires a detailed knowledge of the entire chemotaxis sensory response apparatus.

Overview of the *Escherichia coli* chemotaxis system

Sensory information is relayed from the periplasm into the cell via a set of five closely related transmembrane receptors, each of which is specialized for sensing its own spectrum of stimuli (Fig. 1; for recent reviews see Stock & Surette 1996, Falke et al 1997, Manson et al 1998). To a first approximation, all five receptors are expressed constitutively: the aspartate (Tar) and serine (Tsr) receptors at approximately 2500 monomers per cell; and the receptors for ribose and galactose (Trg), peptides (Tap) and cellular redox potential (Aer) at about 10% of these values. Whereas the high abundance or major receptors, Tar and Tsr, sense aspartate and serine by direct binding, the low abundance or minor receptors, Trg, Tap and Aer, interact with environmental signals indirectly via auxiliary periplasmic binding proteins or respiratory chain intermediates. The receptors all share a common structural organization with an N-terminal sensing domain linked via a hydrophobic membrane spanning sequence to a highly conserved C-terminal cytoplasmic signalling region. This cytoplasmic portion is thought to be essentially an α-helical coiled-coil (Le Moual & Koshland 1996).

The receptors function by controlling the activity of a histidine protein kinase, CheA (Fig. 2). CheA phosphorylates itself on a histidine residue, and the phosphoryl group is then rapidly transferred to an aspartyl residue in the response regulator of the chemotaxis system, CheY. P$\sim$CheY binds to a protein, FliM, that is associated with the flagellar basal body where it acts as an allosteric regulator to offset the equilibrium between run (R) and tumble (T) motor states (Alon et al 1998). Thus, during chemotaxis the five sensory receptors continuously monitor the environment and use this information to control the activity of the kinase CheA. CheA activity controls the level of CheY phosphorylation to set the probability that a bacterium will tumble and change its course of motion. Attractants such as aspartate, serine, ribose, galactose and peptides cause the receptors to inhibit CheA activity so there is a decrease in P$\sim$CheY and cells tend to keep swimming towards attractants. Repellents such as indole, leucine and acid have the opposite effect, they cause an increase in P$\sim$CheY so that cells tend to tumble and swim away from these negative stimuli. Stimulus–response coupling in receptor–kinase complexes is mediated by CheW, an 18 kDa monomeric protein that binds to the signalling domain of the receptors as well as to CheA. Thus, the central signal transduction pathway from receptors to motor involves three proteins, W-A-Y.

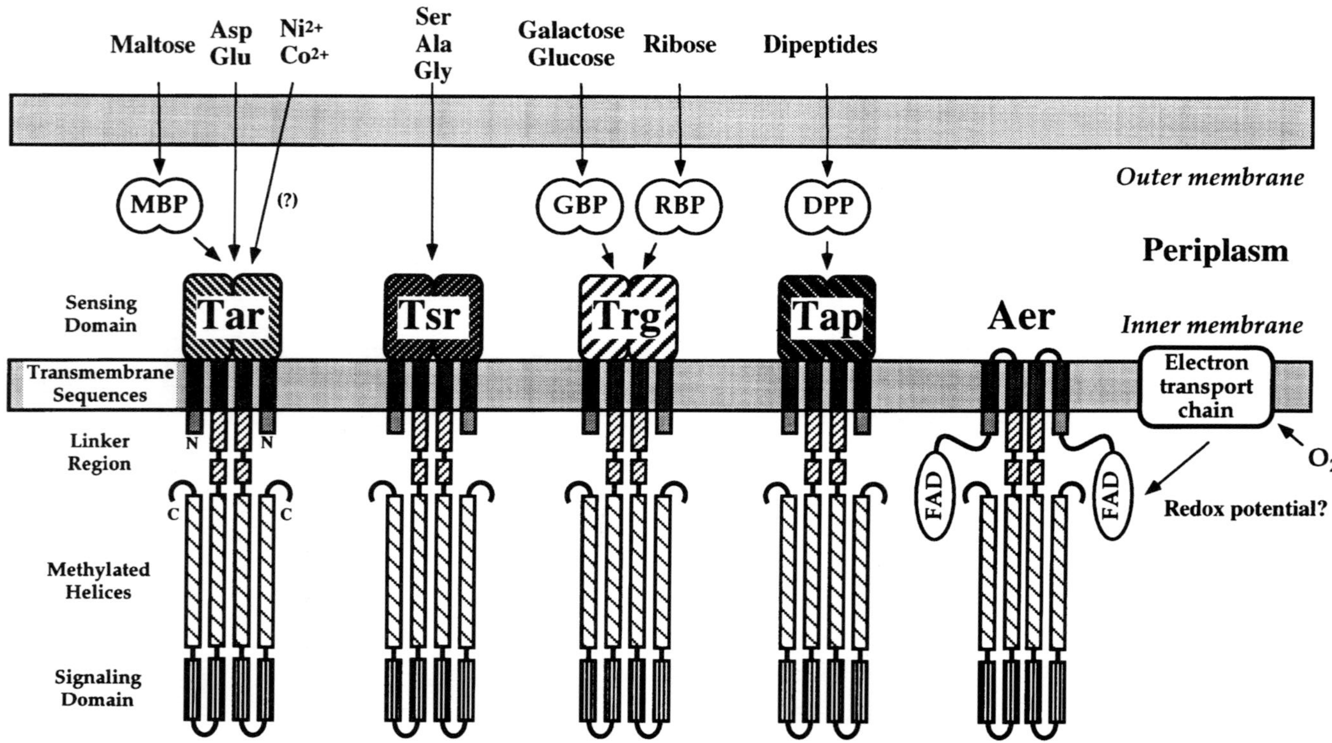

FIG. 1. *Escherichia coli* chemoreceptor family: Tar, for aspartate; Tsr, for serine; Trg, for ribose and glucose; Tap, for dipeptides; and Aer, for aerotaxis. MBP, RBP, GBP and DPP are the periplasmic binding proteins for maltose, ribose, galactose and dipeptides, respectively.

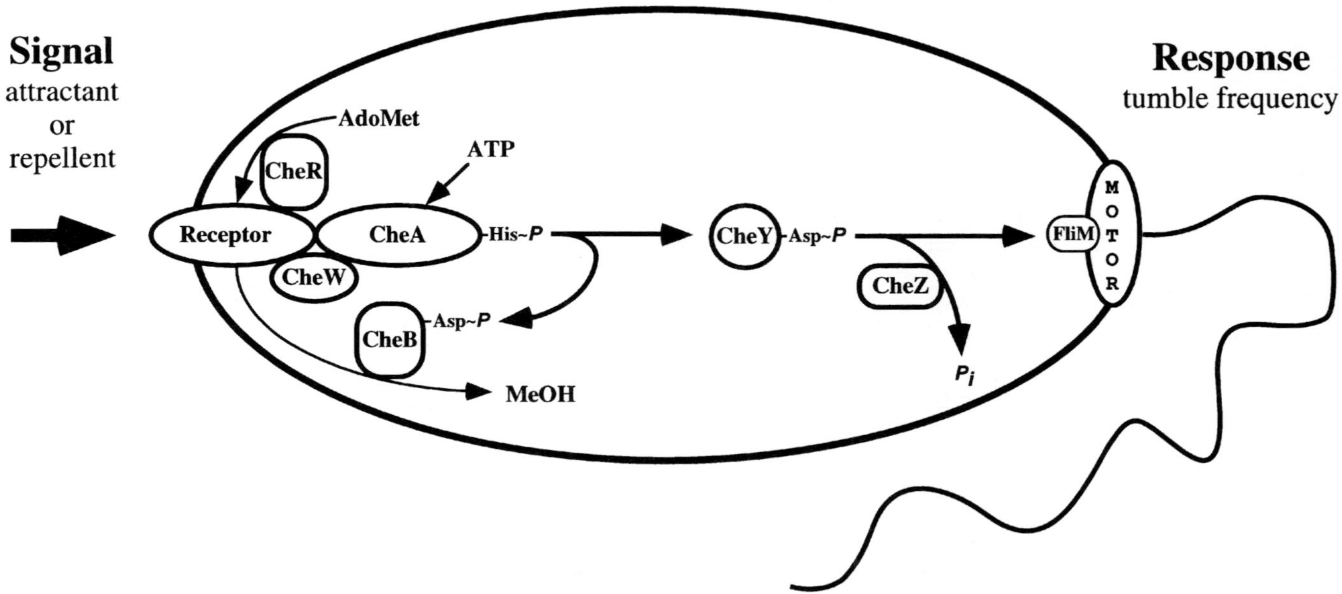

FIG. 2. Signalling pathways in the *Escherichia coli* chemotaxis system.

Other proteins of the chemotaxis system function to modulate the activity of the central W-A-Y pathway. CheR and CheB reversibly modify the receptors at specific glutamate residues localized within the methylated helices. CheR is an S-adenosylmethionine (AdoMet)-dependent methyltransferase that catalyses the methylesterification of these glutamates, and CheB is a methylesterase that catalyses the demethylation of the methylglutamates that are produced. Some glutamates that are eventually methylated are genetically encoded as glutamines; CheB can also act as an amidase to convert these to glutamates that are then subject to methylation by CheR. This glutamate modification biochemistry is unique to bacterial chemotaxis, and all chemotaxis receptors appear to be methylated. In fact, the chemotaxis receptors are termed MCPs for methylated chemotaxis proteins. Closely related homologues of all of these components — MCP, CheW, CheA, CheY, CheR and CheB — have been identified in all bacterial chemotaxis systems that have been investigated including Gram-positive bacteria such as *Bacillus subtilis* (Garrity & Ordal 1995) and even the archaebacterium *Halobacterium salinarum* (Rudolph et al 1995). In contrast, the CheZ protein, which facilitates the dephosphorylation of P ~ CheY, appears to be a special component of the chemotaxis systems of enterics such as *E. coli* and *Salmonella*.

Mechanism of regulation of the kinase CheA

CheA is a soluble protein composed of at least four structurally and functionally distinct domains (Fig. 3; Morrison & Parkinson 1994, Stock & Surette 1996). Starting at the N-terminus are the following: (i) the H or P1 domain, an approximately 130 amino acid, four helix bundle (Zhou et al 1995, Zhou & Dahlquist 1997) that contains the histidine phosphorylation site, His48; (ii) the YB or P2 domain, an approximately 70 amino acid open faced α/β sandwich that binds CheY or CheB with approximately equal affinity, $K_d \sim 0.2\,\mu$M (Li et al 1995, McEvoy et al 1998, Welch et al 1998); (iii) the C domain, an approximately 250 amino acid sequence that contains the highly conserved histidine kinase sequence motifs (N-, D-, F- and G-boxes; Stock et al 1995) and functions to bind ATP and catalyse the phosphorylation of His48; and (iv) the R domain, an approximately 150 amino acid sequence that is required to link CheA to CheW and the receptors. The H- and YB-domains are connected to one another and to the C-domain via flexible linkers like balls on a chain, with each functioning independently of the other (Morrison & Parkinson 1994, Zhou et al 1996). An isolated fragment of CheA corresponding to just the C-domain is able to phosphorylate a fragment corresponding to just the H-domain (Swanson et al 1993) with a V_{max} that is actually slightly greater than the rate obtained for autophosphorylation of H-domain within the context of the intact CheA protein (Levit et al 1996). The affinity of the C-domain for the H-domain appears to be relatively low at $K_m \sim 100\,\mu M$ (Levit et al 1996). Similarly, a fragment corresponding to the isolated YB-domain binds CheY as well as the intact CheA protein (McEvoy et al 1998, Welch et al 1998). A detailed analysis of the kinetics of CheA phosphorylation as well as biophysical analyses of the size of the protein under various conditions indicates an equilibrium between inactive monomers and active dimers, $K_d \sim 0.2\,\mu M$ (Surette et al 1996). The dimerization determinants appear to reside completely within the C-domain (Levit et al 1996).

From *in vitro* kinetic studies it is apparent that the rate of transfer of phosphoryl groups from P ~ CheA to CheY is rapid, $K_{obs} \sim 650\,s^{-1}$ (Stewart 1997), compared to the rate of CheA autophosphorylation, $K_{obs} \sim 0.2\,s^{-1}$ for purified CheA (Surette et al 1996) and up to $20\,s^{-1}$ for fully active CheA in ternary complexes with receptors and CheW (Liu et al 1997). The cytoplasmic concentration of CheY is sufficiently high, $\sim 50\,\mu M$ (Stock et al 1985, Alon et al 1998), so that every time CheA is phosphorylated it immediately transfers its phosphoryl group. The mechanism of sensory regulation in chemotaxis is therefore essentially the mechanism of regulation of CheA kinase activity. In active ternary complexes with CheW and receptors, CheA activity can be enhanced about 100-fold compared to CheA alone (Borkovich et al 1989, Liu et al 1997). Since, under physiological conditions, both CheA and the receptors are predominantly dimeric, it has been generally assumed that ternary complexes were composed of a 2:2:2 association of a receptor dimer, two CheWs, and a dimer of CheA (Fig. 4; Gegner et al 1992). It is not difficult to imagine how stimulatory ligands could bind to the periplasmic receptor sensory domain and cause a reorientation of the interface between subunits that would be transmitted to the signalling domain with its associated CheA. At the same time juxtaposition of negatively charged glutamates within the methylation helices could explain how changes in methylation provide an adaptive response mechanism (Stock et al 1991).

Transmembrane signalling through sensory receptor arrays

Using immuno-electron microscopy, Maddock & Shapiro (1993) showed that the MCPs in *E. coli* are clustered in large patches. Most cells appeared to have a

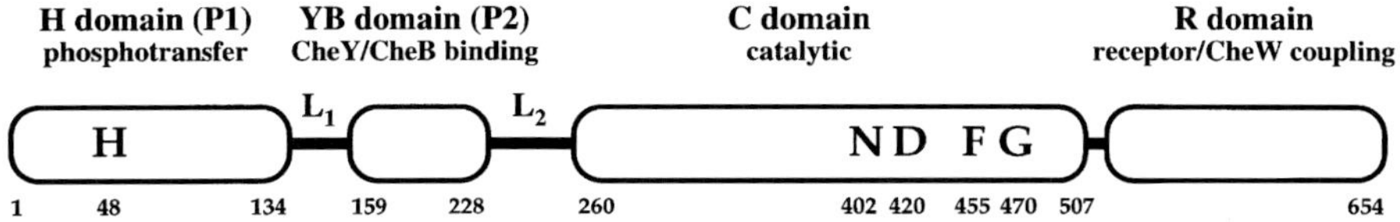

FIG. 3. Domain structure of CheA. Approximate beginnings and ends of domains of CheA from *Escherichia coli*, and positions of highly conserved amino acid residues (H, N, D, F and G) are indicated. L1 and L2 correspond to flexible linker sequences.

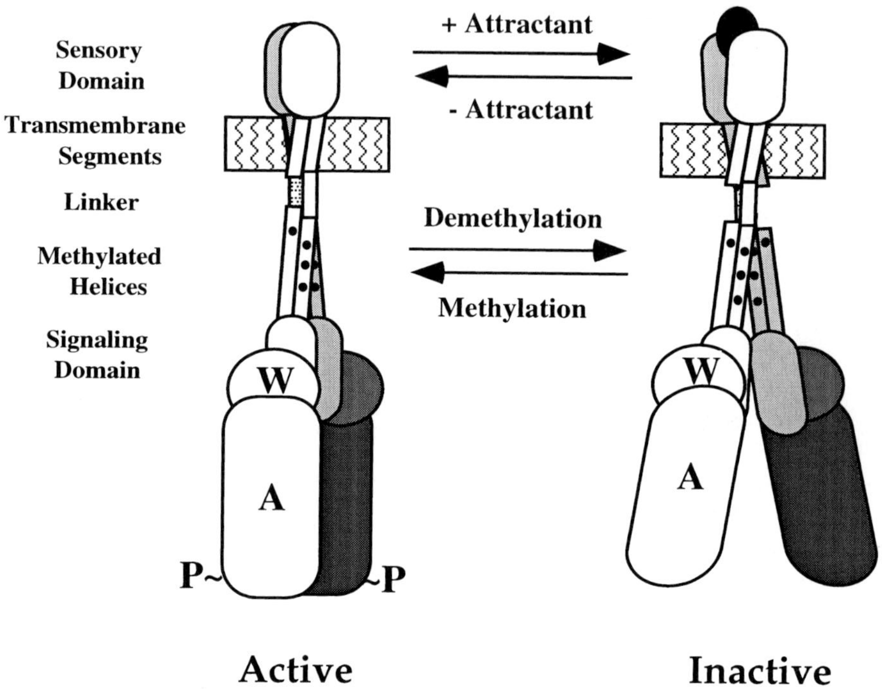

FIG. 4. A hypothetical transition of the receptor–CheW–CheA ternary complex between active and inactive conformations. This transition is modulated both by ligand binding (indicated by black oval at top) as well as by changes in electrostatic repulsion between helices resulting from the interconversion of anionic glutamyl side chains and neutral methyl glutamyl groups. Small black circles on methylated helices indicate sites of glutamate modification.

single patch, presumably containing thousands of receptor monomers, which is located at one of the poles. In addition to the MCPs, the patches contain CheW and CheA. Moreover, receptor patching is not seen in CheW or CheA mutants. Results with complexes formed between soluble fragments of the receptor signalling domain, CheW and CheA, indicate that the ternary complex is a large extended structure with a receptor dimer:CheW:CheA dimer stoichiometry of approximately 7:3:1 (Fig. 5; Liu et al 1997). There have been several observations with mutant receptors which suggest that signalling across the membrane involves higher-order lateral packing interactions (Stock 1996, Grebe & Stock 1998). Aspartate binding to the periplasmic sensing domain of the aspartate receptor introduces an asymmetry in the structure that could alter its ability to pack into an array that is compatible with the formation of active

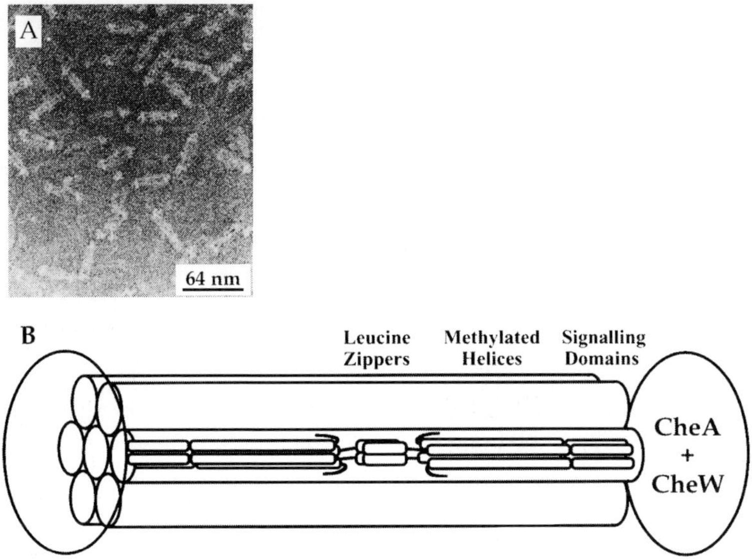

FIG. 5. (A) Electron micrograph of negatively stained complexes composed of CheA, CheW and the cytoplasmic signalling domain of Tar fused to a leucine zipper (Liu et al 1997). (B) Model of the single tetrameric unit of the Tar construct based on a previously proposed model for the cytoplasmic part of Tar (Le Moual & Koshland 1996). Approximately seven of these tetramers are aligned and held together through interactions with CheW and CheA.

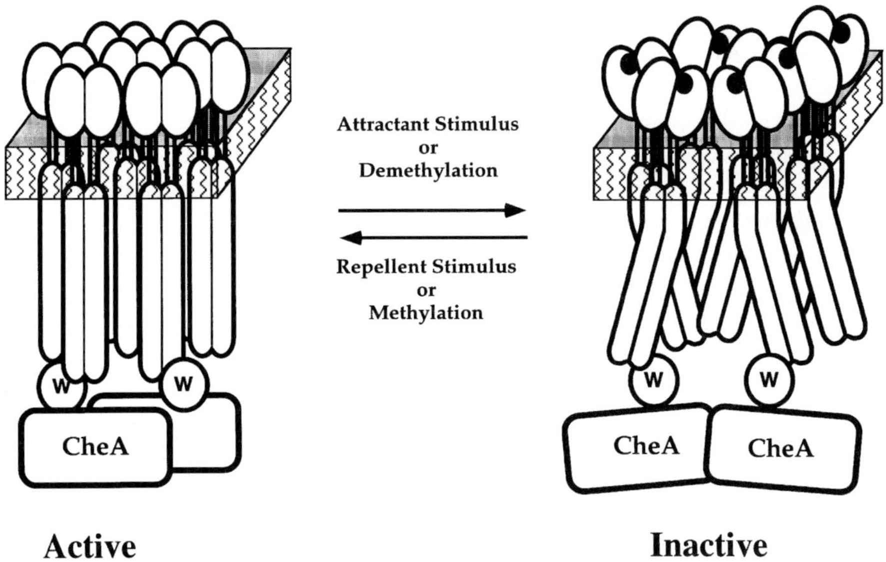

FIG. 6. Lateral packing interactions could be the basis for transmembrane signalling in receptor arrays.

signalling domain–CheW–CheA complexes at the opposing cytoplasmic surface of the membrane (Fig. 6).

Effects of pH

Weak acids such as acetate or benzoate cause bacteria to tumble incessantly (Tso & Adler 1974) as if CheY were highly phosphorylated under these conditions. These tumbly responses could be correlated in time and intensity with the acidification of the cytoplasm caused by the proton-conducting properties of these weak acid stimuli (Kihara & Macnab 1981, Repaske & Adler 1981, Slonczewski et al 1981, 1982, Conley et al 1994). An analysis of the responses in different mutants indicated a dependence on Tsr, and it has therefore been concluded that Tsr is an intracellular pH sensor (Tso & Adler 1974, Kihara & Macnab 1981, Repaske & Adler 1981, Slonczewski et al 1982). CheB mutants also failed to respond to acid stimuli (Kihara & Macnab 1981), implicating the methylation system in the responses. Actually, Tsr or CheB mutants are not completely blind to acid, but instead they give inverted responses so that these mutant cells perceive acid as attractant rather than repellent stimuli (Kihara & Macnab 1981). One possible explanation for the physiological effects of intracellular acidification would be a direct stimulation of the kinase activity of CheA in ternary complexes with Tsr. *In vitro* studies indicate that ternary complexes of CheA with Tsr or Tar have a pH dependence similar to that of purified CheA (Conley et al 1994) with an optimum of approximately 8.3 (Fig. 7). These findings could explain the attractant responses to acid in the absence of Tsr or CheB, but they suggest that the wild-type repellent response is not a simple effect of pH on Tsr signalling through CheA.

The inverted responses of CheB mutants may provide an important clue as to the underlying mechanism of acid sensing. There are two added pieces of information that relate to this question. First, in wild-type cells, acid responses are associated with a demethylation of the receptors (Slonczewski et al 1982); and second, the CheB methylesterase exhibits an acidic pH optimum *in vitro* (Borczuk et al 1986). Receptor demethylation has been shown to inhibit CheA in ternary complexes *in vitro* (Ninfa et al 1991), and to suppress tumbling *in vivo*, so it is not obvious why CheB activation by acid should lead to a tumbly response. A rapid demethylation of receptors is, in fact, generally associated with repellent responses—in this respect acid is like any other repellent stimuli. Thus, the addition of indole or leucine or the removal of aspartate or serine causes tumbly repellent responses that are also associated with rapid decreases in levels of receptor methylation. In fact, the responses to indole and leucine are inverted by mutation of Tsr, just like the responses to weak acids (Muskavitch et al 1978).

Unfortunately, most of the research on acid sensing in *E. coli* was performed before the discovery of the minor MCP that functions as a redox sensor, Aer

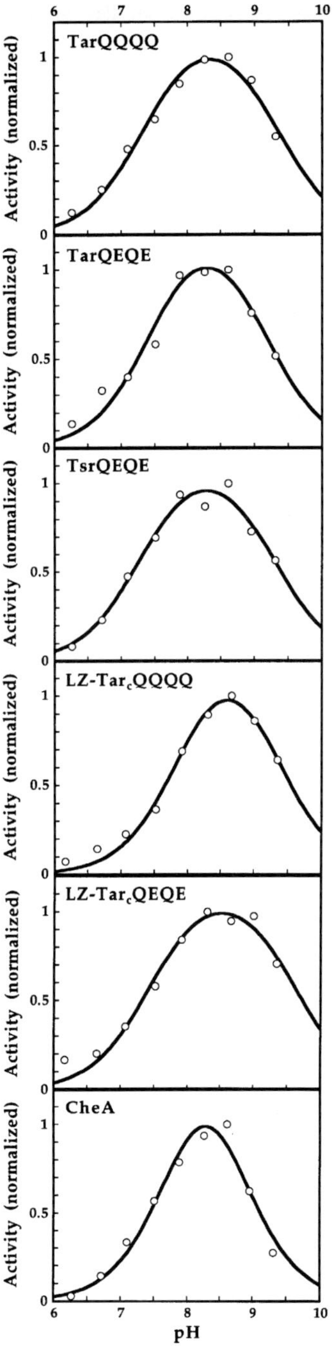

	pK_{a1}	pK_{a2}	pH_{opt}
TarQQQQ	7.3±0.1	9.3±0.1	8.3
TarQEQE	7.4±0.2	9.2±0.2	8.3
TsrQEQE	7.3±0.1	9.3±0.1	8.3
LZ-Tar$_c$QQQQ	7.9±0.1	9.3±0.1	8.6
LZ-Tar$_c$QEQE	7.4±0.1	9.6±0.2	8.5
CheA	7.8±0.2	8.8±0.2	8.3

FIG. 7. Effect of pH on activity of CheA alone and in complexes with CheW and various chemoreceptor preparations. Receptors in membranes: Tar, TarQQQQ and TarQEQE, or Tsr, TsrQEQE (Borkovich et al 1992); and soluble Tar constructs with the cytoplasmic domain fused to a leucine zipper, LZ-Tar$_c$QQQQ and LZ-Tar$_c$QEQE (Surette & Stock 1996, Liu et al 1997). Levels of receptor glutamate modification were varied by changing the level of amidation (equivalent to methylation) denoted as Q (glutamine) or E (glutamate). Activity of CheA alone was determined as the rate of autophosphorylation in the presence of (γ-^{32}P)ATP and activity of CheA in the complexes was determined as the rate of CheY phosphorylation as described earlier (Borkovich et al 1989). The solid lines represent the best fits of the data to the equation:

$$V = V_{max}/(1 + 10^{pKa1-pH} + 10^{pH-pKa2}),$$

describing the effect of pH on activity (V) of an enzyme when the first protonation of the enzyme (governed by pK_{a1}) results in an increase of its activity and the second protonation (governed by pK_{a2}) results in a decrease of its activity (Dixon & Webb 1979). The values of pK_{a1} and pK_{a2} as well as $pK_{optimum} = (pK_{a1} + pK_{a2})/2$ are summarized in the table.

(Bibikov et al 1997, Rebbapragada et al 1997). Aer, together with Tsr, is involved in aerotaxis in *E. coli*. Aer contains a flavin-binding domain, and it is presumably the redox state of the associated flavin that controls Aer activity. Like other minor receptors, Aer probably works in conjunction with the major receptors, Tsr and/or Tar, ostensibly in mixed ternary complexes. Of course, pH sensing and redox sensing are related through oxidative coupling between the respiratory chain and the electrochemical proton gradient. It is noteworthy that CheB mutants have inverted aerotaxis responses similar to the inverted responses they exhibit to acid. Thus, it is apparent that to understand pH sensing in bacterial chemotaxis, we may require a much better understanding of what has come to be termed 'energy sensing'—the integrated effects of pH gradients, membrane potential, redox potential and even perhaps the phosphorylation potential. The intriguing and inexplicable effects of CheB mutations suggest a hitherto unrecognized function for the methylation system in mediating this important class of chemotaxis responses. Perhaps this will provide an example where CheB is involved in excitation as well as adaptation.

Acknowledgements

We would like to thank Thorsten Grebe for providing preparations of membranes with the overexpressed chemoreceptors, and F.W. Dahlquist for the gift of pHSe5/tsr. This work was supported by a grant from the National Institute of Health (RO16M57773).

References

Alon U, Camarena L, Surette MG et al 1998 Response regulator output in bacterial chemotaxis. EMBO J 17:4238–4248

Bibikov SI, Biran R, Rudd KE, Parkinson JS 1997 A signal transducer for aerotaxis in *Escherichia coli*. J Bacteriol 179:4075–4079

Borczuk A, Staub A, Stock J 1986 Demethylation of bacterial chemoreceptors is inhibited by attractant stimuli in the complete absence of the regulatory domain of the demethylating enzyme. Biochem Biophys Res Commun 141:918–923

Borkovich KA, Kaplan N, Hess JF, Simon MI 1989 Transmembrane signal transduction in bacterial chemotaxis involves ligand-dependent activation of phosphate group transfer. Proc Natl Acad Sci USA 86:1208–1212

Borkovich KA, Alex LA, Simon MI 1992 Attenuation of sensory receptor signaling by covalent modification. Proc Natl Acad Sci USA 89:6756–6760

Conley MP, Berg HC, Tawa P, Stewart RC, Ellefson DD, Wolfe AJ 1994 pH dependence of CheA autophosphorylation in *Escherichia coli*. J Bacteriol 176:3870–3877

Dixon M, Webb EC 1979 Enzymes. Academic Press, New York

Falke JJ, Bass RB, Butler SL, Chervitz SA, Danielson MA 1997 The two-component signaling pathway of bacterial chemotaxis: a molecular view of signal transduction by receptors, kinases, and adaptation enzymes. Annu Rev Cell Dev Biol 13:457–512

Garrity LF, Ordal GW 1995 Chemotaxis in *Bacillus subtilis*: how bacteria monitor environmental signals. Pharmacol Ther 68:87–104

Gegner JA, Graham DR, Roth AF, Dahlquist FW 1992 Assembly of an MCP receptor, CheW, and kinase CheA complex in the bacterial chemotaxis signal transduction pathway. Cell 70:975–982

Grebe TW, Stock J 1998 Bacterial chemotaxis: the five sensors of a bacterium. Curr Biol 8:R154–R157

Kihara M, Macnab RM 1981 Cytoplasmic pH mediates pH taxis and weak-acid repellent taxis of bacteria. J Bacteriol 145:1209–1221

Le Moual H, Koshland DE Jr 1996 Molecular evolution of the C-terminal cytoplasmic domain of a superfamily of bacterial receptors involved in taxis. J Mol Biol 261:568–585

Levit M, Liu Y, Surette M, Stock J 1996 Active site interference and asymmetric activation in the chemotaxis protein histidine kinase CheA. J Biol Chem 271:32057–32063

Li J, Swanson RV, Simon MI, Weis RM 1995 The response regulators CheB and CheY exhibit competitive binding to the kinase CheA. Biochemistry 34:14626–14636

Liu Y, Levit M, Lurz R, Surette MG, Stock JB 1997 Receptor-mediated protein kinase activation and the mechanism of transmembrane signaling in bacterial chemotaxis. EMBO J 16:7231–7240

Maddock JR, Shapiro L 1993 Polar location of the chemoreceptor complex in the *Escherichia coli* cell. Science 259:1717–1723

Manson MD, Armitage JP, Hoch JA, Macnab RM 1998 Bacterial locomotion and signal transduction. J Bacteriol 180:1009–1022

McEvoy MM, Hausrath AC, Randolph GB, Remington SJ, Dahlquist FW 1998 Two binding modes reveal flexibility in kinase/response regulator interactions in the bacterial chemotaxis pathway. Proc Natl Acad Sci USA 95:7333–7338

Morrison TB, Parkinson JS 1994 Liberation of an interaction domain from the phosphotransfer region of CheA, a signaling kinase of *Escherichia coli*. Proc Natl Acad Sci USA 91:5485–5489

Muskavitch MA, Kort EN, Springer MS, Goy MF, Adler J 1978 Attraction by repellents: an error in sensory information processing by bacterial mutants. Science 201:63–65

Ninfa EG, Stock A, Mowbray S, Stock J 1991 Reconstitution of the bacterial chemotaxis signal transduction system from purified components. J Biol Chem 266:9764–9770

Rebbapragada A, Johnson MS, Harding GP et al 1997 The Aer protein and the serine chemoreceptor Tsr independently sense intracellular energy levels and transduce oxygen, redox, and energy signals for *Escherichia coli* behavior. Proc Natl Acad Sci USA 94:10541–10546

Repaske DR, Adler J 1981 Change in intracellular pH of *Escherichia coli* mediates the chemotactic response to certain attractants and repellents. J Bacteriol 145:1196–1208

Rudolph J, Tolliday N, Schmitt C, Schuster SC, Oesterhelt D 1995 Phosphorylation in halobacterial signal transduction. EMBO J 14:4249–4257

Slonczewski JL, Rosen BP, Alger JR, Macnab RM 1981 pH homeostasis in *Escherichia coli*: measurement by ^{31}P nuclear magnetic resonance of methylphosphonate and phosphate. Proc Natl Acad Sci USA 78:6271–6275

Slonczewski JL, Macnab RM, Alger JR, Castle AM 1982 Effects of pH and repellent tactic stimuli on protein methylation levels in *Escherichia coli*. J Bacteriol 152:384–399

Stewart RC 1997 Kinetic characterization of phosphotransfer between CheA and CheY in the bacterial chemotaxis signal transduction pathway. Biochemistry 36:2030–2040

Stock A, Koshland DE Jr, Stock J 1985 Homologies between the *Salmonella typhimurium* CheY protein and proteins involved in the regulation of chemotaxis, membrane protein synthesis, and sporulation. Proc Natl Acad Sci USA 82:7989–7993

Stock J 1996 Signaling across membranes: a one and a two and a . . . Science 274:370–371

Stock JB, Surette M 1996 Chemotaxis. In: Neidhardt FC, Curtiss R III, Ingraham JL et al (eds) *Escherichia coli* and *Salmonella*: cellular and molecular biology. ASM Press, Washington, DC, p 1103–1129

Stock JB, Lukat GS, Stock AM 1991 Bacterial chemotaxis and the molecular logic of intracellular signal transduction networks. Annu Rev Biophys Biophys Chem 20:109–136

Stock JB, Surette MG, Levit M, Park P 1995 Two-component signal transduction systems: structure-function relationships and mechanisms of catalysis. In: Hoch JA, Silhavy TJ (eds) Two-component signal transduction. ASM Press, Washington, DC, p 25–51

Surette MG, Stock JB 1996 Role of α-helical coiled-coil interactions in receptor dimerization, signaling, and adaptation during bacterial chemotaxis. J Biol Chem 271:17966–17973

Surette MG, Levit M, Liu Y et al 1996 Dimerization is required for the activity of the protein histidine kinase CheA that mediates signal transduction in bacterial chemotaxis. J Biol Chem 271:939–945

Swanson RV, Schuster SC, Simon MI 1993 Expression of CheA fragments which define domains encoding kinase, phosphotransfer, and CheY binding activities. Biochemistry 32:7623–7629

Tso WW, Adler J 1974 Negative chemotaxis in *Escherichia coli*. J Bacteriol 118:560–576

Welch M, Chinardet N, Mourey L, Birck C, Samama JP 1998 Structure of the CheY-binding domain of histidine kinase CheA in complex with CheY. Nat Struct Biol 5:25–29

Zhou H, Dahlquist FW 1997 Phosphotransfer site of the chemotaxis-specific protein kinase CheA as revealed by NMR. Biochemistry 36:699–710

Zhou H, Lowry DF, Swanson RV, Simon MI, Dahlquist FW 1995 NMR studies of the phosphotransfer domain of the histidine kinase CheA from *Escherichia coli*: assignments, secondary structure, general fold, and backbone dynamics. Biochemistry 34:13858–13870

Zhou H, McEvoy MM, Lowry DF, Swanson RV, Simon MI, Dahlquist FW 1996 Phosphotransfer and CheY-binding domains of the histidine autokinase CheA are joined by a flexible linker. Biochemistry 35:433–443

DISCUSSION

Foster: I have a question on chemotaxis. Is it possible to saturate the methyl-accepting sites by increasing the concentration of aspartate?

Stock: In most cases cells increase their levels of methylation so as to adapt exactly to the stimulus. For example, if the environment is approaching the aspartate saturation point, the cell continues to adapt to aspartate. However, at saturation, the aspartate receptor can no longer respond to added aspartate. At this point the cell behaves as though there is no aspartate present, in terms of its steady-state behaviour. Maltose is also detected by the aspartate receptors, through the maltose-binding protein. If you add maltose to receptors that are saturated with aspartate, you observe a second response followed by a further increase in methylation that leads to adaptation to maltose (Mowbray & Koshland 1987).

Foster: Even though it's saturated with aspartate?

Stock: Yes, because you're not saturating the methyl groups. There is a balance between methylation and stimulus, so as long as you keep adding more methyl groups the cell can still adapt. However, the serine receptor, which is the most abundant receptor, behaves differently. It has two binding sites, one specific for serine and the other for serine, alanine and other amino acids. Berg & Brown (1972) showed that if a large amount of serine is added the cells no longer adapt to serine. Apparently, in this case the adaptation capacity of the methylation system is swamped.

Matin: Through which receptor is glucose sensed?

Stock: Lengeler and his collaborators showed that enzyme I of the phosphotransferase system, in its dephosphorylated state, binds to CheA and inhibits it (Lux et al 1995). In evolutionary terms enzyme I is probably related to CheA; they are both dimers and they both autophosphorylate a histidine near the N-terminus. You might think from this that a mixed dimer complex is formed. However, the physiologically relevant CheA activity is in a complex with the receptors. Free CheA has too little activity to be relevant to chemotaxis, and it seems that almost all of the CheA is contained within complexes anyway. It is not unlikely that this signalling complex will turn out to contain enzyme I. This would then allow glucose responses to go through the methylated chemotaxis proteins (MCPs), because the MCPs are part of that complex. There are no known chemotaxis signalling systems in *Escherichia coli* that do not involve the MCPs. The oxygen response was originally thought to go through a separate mechanism, but a MCP has now been identified, Aer, that relates to oxygen sensing in some way (Rebbapragada et al 1997).

Matin: In two-component systems the response regulator generally acts on the DNA, but this one acts on a protein, CheY. Does CheY, therefore, have a different structure?

Stock: We've done detailed sequence analyses and CheY appears to have a typical response regulator domain. Response regulators that regulate transcription have a DNA-binding domain as well as a CheY-like domain (Stock et al 1990). This has been directly demonstrated for NarL (Baikalov et al 1996). The NarL regulatory domain lies over the DNA-binding domain, so it gets in the way of DNA binding. Phosphorylation or cleavage of the CheY-like response regulator domain leads to NarL activation. It's a protein–protein (or domain–domain) interaction that is affected by phosphorylation.

Fillingame: You mentioned that the receptors are organized into a large array. What is the functional advantage of this? Does it amplify the signal more rapidly?

Stock: A paper has just been published in *Nature* (Bray et al 1998) that discusses the signal amplification issue. From the high sensitivity of individual cells it is apparent that an entire receptor patch may be behaving as one unit. This would result in sensitivity amplification. However, no one has seen this *in vitro* yet, so it is still an issue that hasn't been resolved.

Another reason for the array mechanism may be its evolutionary versatility. Over 400 different kinases have been sequenced to date. They do not have conserved sensing domains, although they do have conserved kinase domains. It is even possible to make functional hybrids between the sensing domains of the *E. coli* aspartate receptor and the insulin tyrosine kinase domain of the human insulin receptor (Moe et al 1989). Thus, there appears to be little specificity in the effect of the sensing domain on kinase activity. This lack of conformational specificity would be evolutionarily advantageous because it wouldn't matter how the sensing domain was stimulated, as long as the packing interactions among kinase domains inside the cell were perturbed in some way a signal would be generated.

Fillingame: You said that within the patches there are five different types receptor, and it seems to me that these different types of receptor would interfere with the amplification of the conformational change from protein to protein.

Stock: Although their sensing domains differ, all of the receptor signalling domains are similar. There are two classes of receptors: those that are present at high levels, e.g. the aspartate, Tar, and serine, Tsr, receptors; and those that are present at low levels, e.g. the ribose receptor, Trg, which is present at less than 10% of the level of any of the aspartate and serine receptors. However, the response to ribose is equal to the response to aspartate and serine, and adaptation to ribose occurs just as well as it does to aspartate and serine. Adaptation to ribose, however, seems to be due to methylation of the aspartate and/or serine receptors. Thus, the major and minor receptors are almost certainly mixed together, but whether the aspartate and serine receptors are together in the same cluster or whether they form two distinct clusters remains to be determined.

Fillingame: Have you deleted all the receptors except the ribose receptors?

Stock: Hazelbauer and his collaborators have done that experiment (Feng et al 1997). The ribose receptor won't mediate a response without the aspartate and/or serine receptor being present.

Fillingame: Are there any indications that this cluster is arranged in a pericrystalline lattice?

Stock: We are looking at this. We know that the lattice we have isolated is regular, and we are now doing electron microscope analysis.

Konings: How do you visualize a chemotactic response if one set of receptors causes a positive response and another causes a negative response?

Stock: A negative stimulus and a positive stimulus cancel each other out, and two positive stimuli are additive. Therefore, the responses are integrated, but the question is, at what level are they integrated? They are integrated at the levels of CheY, but they may also be integrated within the receptor array.

Slonczewski: We showed a number of years ago that the detection of weak acids is mediated by Tsr, the so-called serine receptor protein. Has there been any progress on the molecular basis of that pH detection?

Stock: No, but the data are confusing. Responses to weak acids are mediated via serine receptors. However, the same responses are observed at pH 7.5. In addition, if the serine receptor is removed, then all of the weak acid as well as other repellent responses that were coming through the serine receptor now become attractant responses. No one has ever demonstrated binding of repellents to the serine receptor. The serine receptor is also involved in membrane potential sensing.

Slonczewski: My second question relates to the so-called two-component sensors, which have both components in one molecule, such as the CadC and ToxR pH detectors. In these cases, extracellular pH is detected by the extracellular component, and the internal part of the same protein has to find the DNA and turn on gene expression. Are there any models that explain how this works?

Stock: I would propose the membrane array model for any single transmembrane receptor protein. For instance, pH could affect the packing arrangement of a 2D array of ToxR. A complex structure is required to initiate transcription, so at least a ToxR dimer would be expected to be the active form. Response regulators that control transcription such as OmpR and NtrC polymerize and are attached to DNA. The take-home lesson from all the response regulator functions is that in nearly all cases they are involved in generating higher-order complexes.

Padan: Do you have genetic evidence for the receptor array? Because if you made mutants in different receptors you would expect to see transdominance.

Stock: There are numerous dominant mutants of the major receptors, but not the minor receptors. These results have been interpreted, not in terms of the array, but in terms of later processing events. The notion of a sensory array is relatively new, so older findings may have to be re-evaluated.

Booth: Does the array sequester specialized lipids? The emerging picture of eukaryotic protein kinase C is that it binds lipid domains around itself, which results in a specialized area of membrane. Protein kinase C then has the advantage of bringing substrate to it by this binding mechanism. If there was a specialized lipid structure in this case, then all these compounds that work by their ability to perturb phospholipid interactions would bring about changes in signalling.

Stock: It's possible. The reason why methylation is so crucial is that this system is sensitive to everything and methylation is required to keep it balanced.

References

Baikalov I, Schröder I, Kaczor-Grzeskowiak M, Grzeskowiak K, Gunsalus RP, Dickerson RE 1996 Structure of the *Escherichia coli* response regulator NarL. Biochemistry 35:11053–11061

Berg HC, Brown DA 1972 Chemotaxis in *Escherichia coli* analysed by three-dimensional tracking. Nature 239:500–504

Bray D, Levin MD, Morton-Firth CJ 1998 Receptor clustering as a cellular mechanism to control sensitivity. Nature 393:85–88

Feng X, Baumgartner JW, Hazelbauer GL 1997 High- and low-abundance chemoreceptors in *Escherichia coli*: differential activities associated with closely related cytoplasmic domains. J Bacteriol 179:6714–6720

Lux R, Jahreis K, Bettenbrock K, Parkinson JS, Lengeler JW 1995 Coupling the phosphotransferase system and the methyl-accepting chemotaxis protein-dependent chemotaxis signaling pathways of *Escherichia coli*. Proc Natl Acad Sci USA 92:11583–11587

Moe GR, Bollag GE, Koshland DE Jr 1989 Transmembrane signalling by a chimera of the *Escherichia coli* aspartate receptor and the human insulin receptor. Proc Natl Acad Sci USA 86:5683–5687

Mowbray SL, Koshland DE Jr 1987 Additive and independent responses in a single receptor: aspartate and maltose stimuli on the Tar protein. Cell 50:171–180

Rebbapragada A, Johnson MS, Harding GP et al 1997 The Aer protein and the serine chemoreceptor Tsr independently sense intracellular energy levels and transduce oxygen, redox, and energy signals for *Escherichia coli* behavior. Proc Natl Acad Sci USA 94:10541–10546

Stock JB, Stock AM, Mottonen JM 1990 Signal transduction in bacteria. Nature 344:395–400

Inducible acid tolerance mechanisms in enteric bacteria

John W. Foster and Matthew Moreno

Department of Microbiology and Immunology, University of South Alabama College of Medicine, Mobile, AL 36688, USA

Abstract. Enteric micro-organisms have developed several inducible mechanisms for surviving transient periods of extreme acid stress. *Salmonella typhimurium* possesses an acid tolerance response (ATR) induced in minimal medium by short exposures to mild acid stress. More than 50 acid shock proteins (ASPs) are induced during adaptation. Eight ASPs are regulated by the major iron regulatory protein, Fur, in an unusual iron-independent manner. The two-component regulator, PhoP, is an autoinduced ASP that controls the induction of three additional ASPs. The stress sigma factor σ^S is an ASP that regulates induction of eight ASPs. Acid induction of σ^S is due to its decreased proteolytic turnover via the ClpXP protease in conjunction with the two-component-type response regulator MviA (RssB in *Escherichia coli*). Mutations in any of these three regulators leads to a defective ATR. Repair of pH stress-induced DNA damage appears to require the Ada protein (O^6-methylguanine methyltransferase) since an *ada* mutant is both acid and alkaline sensitive. In contrast to *S. typhimurium*, *E. coli* and *Shigella* have acid resistance systems induced in complex media that include a glucose-repressed system protective at pH 2.5 without amino acid supplementation, a glutamate decarboxylase system that requires glutamate and an arginine decarboxylase system unique to *E. coli*.

1999 Bacterial responses to pH. Wiley, Chichester (Novartis Foundation Symposium 221) p 55–74

Acid is a common stress encountered by enteric bacteria such as *Salmonella typhimurium* and *Escherichia coli* during their pathogenic and commensal lifestyles. Obviously, the stomach and parts of the intestine are acidic environments that must be endured in order for these organisms to colonize and cause disease. As a facultative intracellular parasite, *S. typhimurium* will also encounter acidic conditions in macrophage phagolysosomes. Consequently, *S. typhimurium* and *E. coli* have evolved common and unique strategies for surviving transient encounters with potentially lethal acid conditions. Processes enabling acid tolerance appear to include inducible pH homeostasis systems and mechanisms of macromolecular protection and repair. This chapter will discuss the regulatory aspects of inducible acid tolerance and describe potential protection strategies.

The acid tolerance response of *Salmonella typhimurium*

S. typhimurium undergoes a programmed molecular response following exposures to moderately acidic conditions (pH 4.4). Figure 1 illustrates that adaptation to moderate acid conditions (pH 4.4) enables tolerance to more severe pH challenge (pH 3) and that acid adaptation requires protein synthesis. The acid tolerance response (ATR) includes the synthesis of over 50 acid shock proteins (ASPs) that provide a defence against even more acidic conditions (Foster & Hall 1990). In addition, we and others have noted that acid adaptation induces a cross protection against osmotic, heat and oxidative stresses (Lee et al 1995, Leyer & Johnson 1993). Reciprocal cross protection does not occur (i.e. osmotic shock does not induce acid tolerance). Thus, it seems that acid stress is a general stress indicator whereas heat, salt and H_2O_2 are more specific stress signals. Our laboratory has begun to dissect this response in *S. typhimurium*, and we have identified at least three regulatory subsets of ASPs. The subsets include eight proteins regulated by the alternative sigma factor σ^S encoded by *rpoS*, four proteins controlled by the iron regulatory protein, Fur, and four proteins

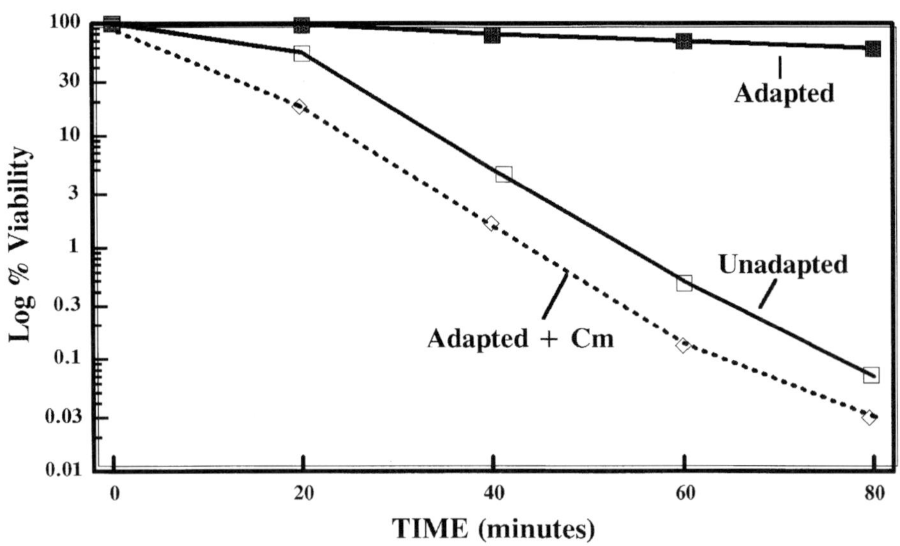

FIG. 1. The *Salmonella typhimurium* acid tolerance response. Cells (UK1) were grown in pH 7.7 minimal glucose media to a cell density of 2×10^8 cells/ml. Unadapted cells were immediately acid challenged by adjusting the medium pH to 3.1. Another aliquot of cells was adapted to pH 4.4 for 30 min and then acid challenged to pH 3.1. Viable counts were taken at the times indicated on the *x* axis. Open squares, unadapted. Filled squares, adapted. Open diamonds, cells to which chloramphenicol (Cm) was added during adaptation.

regulated by the two-component PhoPQ system. Mutations in any one of these regulators will confer an acid-sensitive phenotype. In addition to these regulatory subsets, we have recently discovered that O^6-methylguanine DNA methyltransferase, the product of the *ada* gene, is required for acid and alkaline tolerance but apparently not in its role as a transcriptional regulator.

Fur and acid tolerance

Fur is typically regarded as a repressor protein. At high intracellular concentrations, Fe^{2+} binds to apoFur converting it into an active repressor protein. The $FurFe^{2+}$ holorepressor binds to Fur-box sequences within the promoter regions of genes involved in acquiring iron from the environment. Mutants lacking Fur will constitutively express these genes. In apparent contrast to its role as a repressor, Fur may act as a positive regulator of a subset of ASPs (Foster & Hall 1992). Mutants lacking Fur will not synthesize these proteins and are acid sensitive. Experiments designed to separate the acid- and iron-sensing properties of this regulator have revealed three classes of *fur* mutations. One set of mutations cause the overexpression of Fe^{2+}-regulated genes (e.g. *iroA::lacZ* fusion) and cause an acid-sensitive phenotype; these are essentially *fur* null mutants (acid and iron blind). A second type of mutant exhibits normal acid tolerance but fails to repress the iron-regulated *iroA::lacZ* fusion (Fig. 2A & B). The mutation that caused this phenotype results in the substitution of an arginine for a histidine at residue 90 (H90R) generating an iron-blind but acid-sensing Fur protein (Hall & Foster 1996). The location of the change in the protein occurs in what is considered the most likely candidate site for iron binding. The third class of mutation resulted in the opposite phenotype, an acid-blind but iron-sensing Fur that did not function in acid tolerance but regulated *iroA::lacZ* normally (Fig. 2C & D). This mutation (H144L) occurs near the carboxyl end of Fur. The results support a model in which Fur can sense iron and acid separately and may in some instances act as a positive regulator.

PhoPQ is acid inducible and regulates an acid shock protein subset

PhoP, the regulator component of the two-component PhoPQ, was found to be ASP29 based upon N-terminal sequence analysis. A potential role for PhoP in acid tolerance was further suggested when it was discovered that four ASPs, including PhoP, were missing from an acid-adapted *phoP* mutant. Subsequent analysis of the acid tolerance phenotypes of *phoP* and *phoQ* mutants indicated that this system is not useful against acid stress involving organic acids but is useful against inorganic acid stress, especially when *rpoS*-dependent systems (see below) are not available.

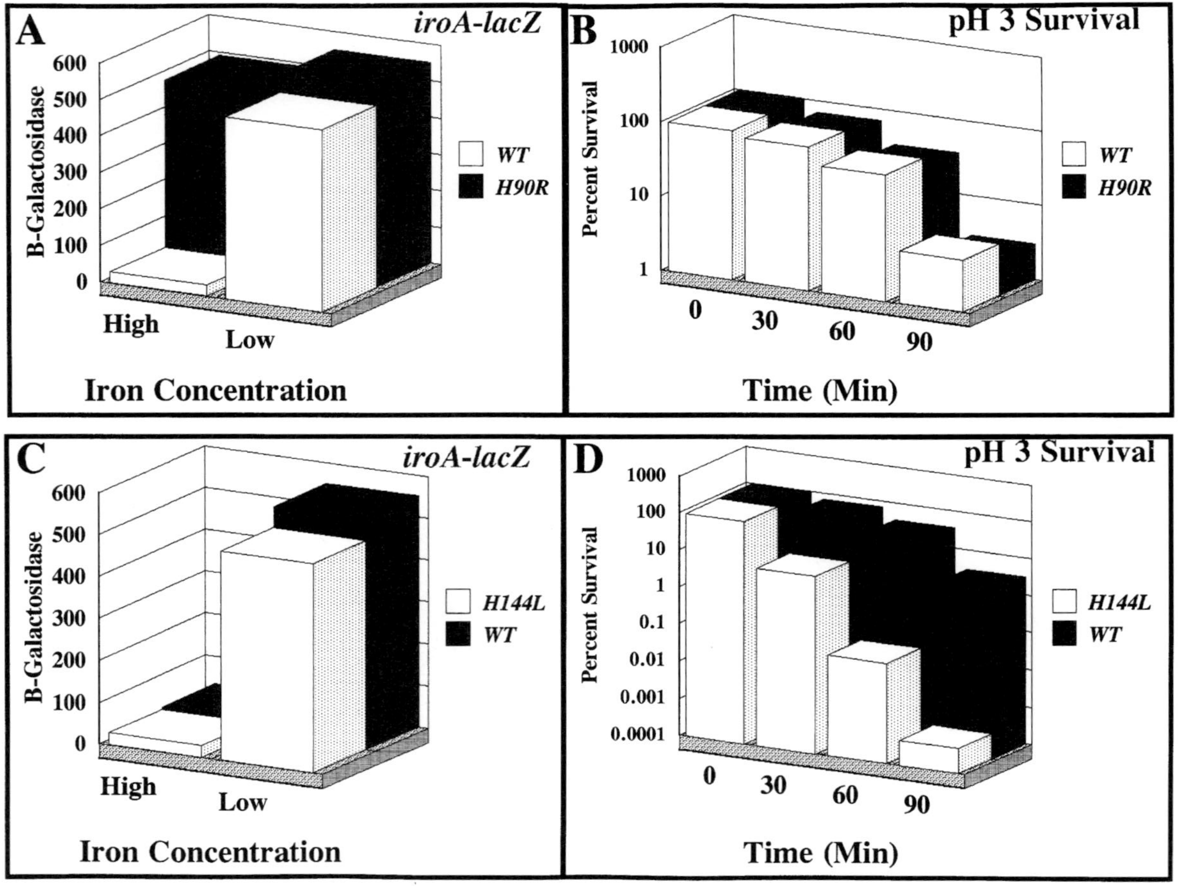
A
iroA-lacZ
B-Galactosidase
600
500
400
300
200
100
0
WT
H90R
High
Low
Iron Concentration
B
pH 3 Survival
Percent Survival
1000
100
10
1
WT
H90R
0
30
60
90
Time (Min)
C
iroA-lacZ
B-Galactosidase
600
500
400
300
200
100
0
H144L
WT
High
Low
Iron Concentration
D
pH 3 Survival
Percent Survival
1000
100
10
1
0.1
0.01
0.001
0.0001
H144L
WT
0
30
60
90
Time (Min)

Although PhoQ has been shown to sense extracellular Mg^{2+} (Soncini et al 1996), we have shown that it can also sense pH (Bearson et al 1998). As illustrated in Fig. 3, a *phoP-lacZ* fusion is induced by low Mg^{2+} when active PhoP and PhoQ proteins are present. However, the *phoP-lacZ* fusion can also be induced in the presence of high Mg^{2+} if cells are subjected to mild to moderate acid conditions. Since PhoPQ-dependent genes are required for the virulence of *Salmonella*, the acid inducible nature of this system takes on added significance. PhoPQ is required for survival of *Salmonella* within macrophage phagolysosomes. Consequently, one must wonder whether the relevant inducing condition for the PhoPQ regulon within macrophage phagolysosomes is Mg^{2+} and/or H^+ concentration. Evidence supporting a role for acid in this process can be found in a study indicating that phagolysosome acidification is actually required for *Salmonella* to survive within the macrophage environment (Rathman et al 1996).

Acid induction of sigma S and its role in acid tolerance

σ^S, encoded by the *rpoS* gene, has a well-established role in regulating stationary phase gene expression (Hengge-Aronis 1996). The dramatic increase in the level of σ^S during transition into stationary phase is associated with a heightened general stress resistance. However, recent work has shown that log phase cells subjected to acid stress will also induce σ^S, and that eight ASPs are σ^S dependent (Bearson et al 1996, Lee et al 1995). N-terminal sequence analysis has revealed the identity of four proteins: (1) OsmY (ASP74), a periplasmic protein of unknown function; (2) a Cu^{2+} Zn^{2+} superoxide dismutase (ASP71) 87% identical to *E. coli* periplasmic SodC; (3) YciE (ASP73), an 18.9 kDa *E. coli* protein of unknown function; and (4) ASP72, the N-terminal region of which is 90% identical to yajQ, a hypothetical 19 kDa *E. coli* protein. Figure 4 illustrates that *rpoS* mutants are extremely acid sensitive. In contrast to the PhoPQ-dependent system that primarily defends against inorganic acid stress, the σ^S-dependent system protects the cell against both organic and inorganic acid stresses (Baik et al 1996, Bearson et al 1998).

FIG. 2. Separation of iron-sensing and acid-sensing properties of Fur. (A, C) β-galactosidase activity (Miller units) of an *iroA-lacZ* fusion. (B, D) Acid tolerance of adapted cells as outlined in the legend to Fig. 1. White bars in A, B and dark bars in C, D represent wild-type (WT) *Salmonella typhimurium*. Dark bars in A, B represent H90R *fur* mutant. White bars in C, D represent H144L mutant. Iron concentrations were manipulated by adding 200 μM diethylamine triamine penta-acetic acid (DTPA) to chelate Fe^{2+} (low iron) or 60 μM $FeCl_2$ (high iron). Cells tested for acid tolerance were adapted for one cell doubling at pH 5.8.

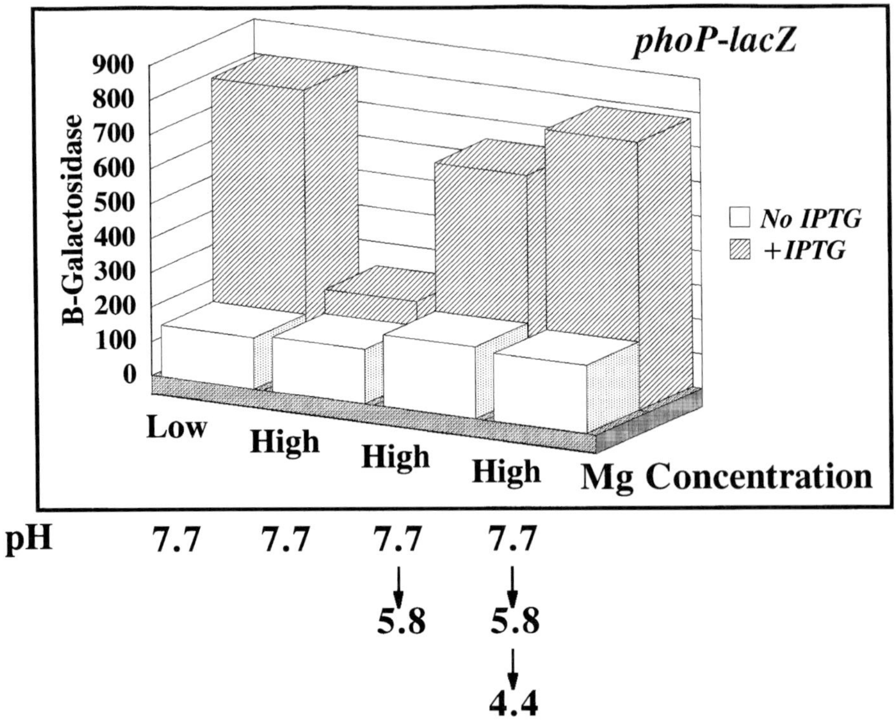

FIG. 3. Acid induction of a *phoP-lacZ* fusion. Cells containing a *phoP-lacZ* fusion and a plasmid containing the *phoPQ* operon under the control of the *lac* promoter were examined for β-galactosidase activity (Miller units) under different Mg^{2+} concentrations and pH conditions. Regulation required production of PhoP and PhoQ accomplished by adding the gratuitous *lac* operon inducer isopropyl thiogalactoside (IPTG; dark bars). Shifts in pH were accomplished by growing cells at pH 7.7 to 2×10^8 cells/ml then HCl added to adjust pH to pH 5.8. Following growth of the pH 5.8 culture to 4×10^8 cells/ml, aliquots were either assayed or further adjusted to pH 4.4 for 30 min. $MgSO_4$ was added at 10 μM (low) or 10 mM (high).

Regulation of sigma S

An important focus of this research involves the mechanism by which acid stress induces the synthesis of σ^S. Efforts to answer this question have revealed a mutation that causes elevated resistance to acid stress as well as increased production of σ^S and the σ^S-dependent ASPs (Bearson et al 1996). The gene responsible, *mviA* (*m*ouse *vi*rulence), controls virulence and possesses homology to the OmpR family of two-component regulatory systems (Benjamin & Hall 1993, Benjamin et al 1996, Benjamin et al 1991). The *mviA* gene is also known as *rssB* (Muffler et al 1996) and *sprE* (Pratt & Silhavy 1996). The current model for MviA control over σ^S production is that MviA modulates proteolytic turnover of

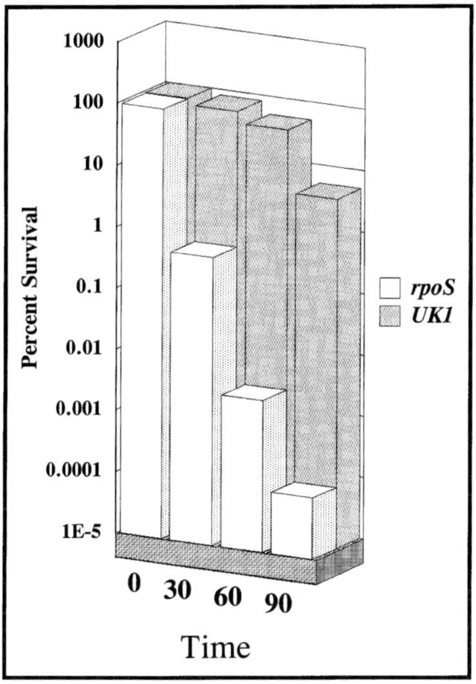

FIG. 4. Effect of *rpoS* on inducible acid tolerance. Acid tolerance response of an *rpoS* mutant (white bars) and wild-type *Salmonella typhimurium* (UK1, dark bars). Cells were adapted at pH 4.5 and challenged at pH 3.1 as described in Fig. 1.

σ^S by the ClpXP protease. Since the N-terminal portion of MviA possesses homology to the receiver end of response regulators but the C-terminal end does not resemble transcriptional regulators, it is believed that MviA functions by protein–protein interactions either with σ^S or ClpXP (Fig. 5). The result of this proposed interaction is rapid turnover of σ^S in the absence of stress. Introducing stress is believed to inactivate MviA which leaves ClpXP unable to degrade σ^S.

The relevance of this system to virulence is evident in studies that indicate over or under production of σ^S will attenuate virulence of *S. typhimurium* (Fang et al 1992). To be a successful pathogen, *Salmonella* must exert precise control over σ^S-dependent gene expression during infection. As *Salmonella* progresses through its host, some body sites will present a considerable level of stress to the bacterium while other sites will present little to no stress. Part of how this system contributes to virulence may be through regulating *in vivo* growth of the bacterium. Survival of the organism as it passes through high stress body sites will require higher levels of σ^S-dependent stress proteins and probably slower

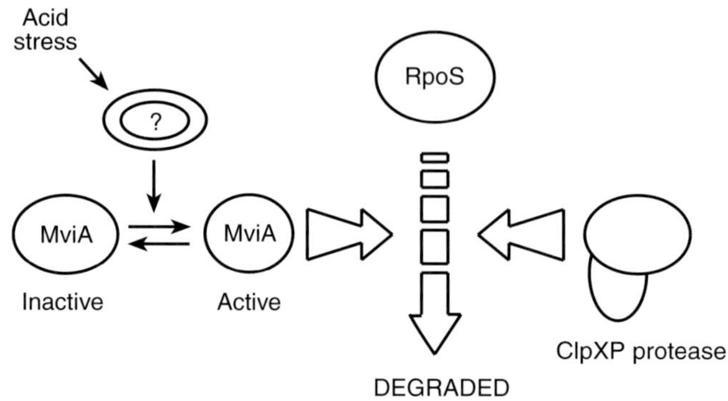

FIG. 5. Under normal non-stress growth conditions, an unknown sensor kinase is thought to phosphorylate and activate MviA, which enables ClpXP protease to degrade σ^S. Under stress conditions, such as low pH, MviA(RssB) is in an inactive form that prevents σ^S degradation.

growth than it would want or need in low stress sites. Clearly, a lack of σ^S in high stress situations will be detrimental to *Salmonella* since the organism will be more susceptible to the stresses. In contrast, excess σ^S, as occurs in *mviA* or *clpP* mutants, would be detrimental to the cell in body sites where it should grow quickly.

The effect of σ^S on growth rate is evident even in log phase cells grown in minimal glucose media (M. Moreno & J.W. Foster, unpublished observations 1998). The generation time of cells lacking σ^S (*rpoS::Ap*) is 5 min faster than wild-type cells (50 vs. 55 min, respectively). Figure 6 illustrates that overproducing σ^S through mutations in *mviA* or *clpP* slows growth and causes a small colony phenotype. In an effort to reveal how σ^S might modulate growth, suppressors of small colony growth have been identified in several genes including *pstB* (via overexpression of PhoB), *rcsB* (a two-component regulator of capsule production and the cell division regulator *ftsZ*) and *dsbB* (disulfide oxidoreductase). However, how these mutations suppress the σ^S small colony phenotype is not yet known, other than they apparently do not affect σ^S production.

ada mutants are acid and alkaline sensitive

We have accumulated evidence that acid stress can cause DNA damage. For example, *polA* mutants proved to be acid sensitive when tested in an *rpoS* background at pH 3.3 (Foster & Bearson 1994, Wilmes-Riesenberg et al 1996). In addition, DNA extracted from unadapted acid stressed cells appeared more fragmented than did DNA extracted from similarly stressed adapted cells (J. Lin & J. W. Foster, unpublished observations 1996). Acid stress could conceivably

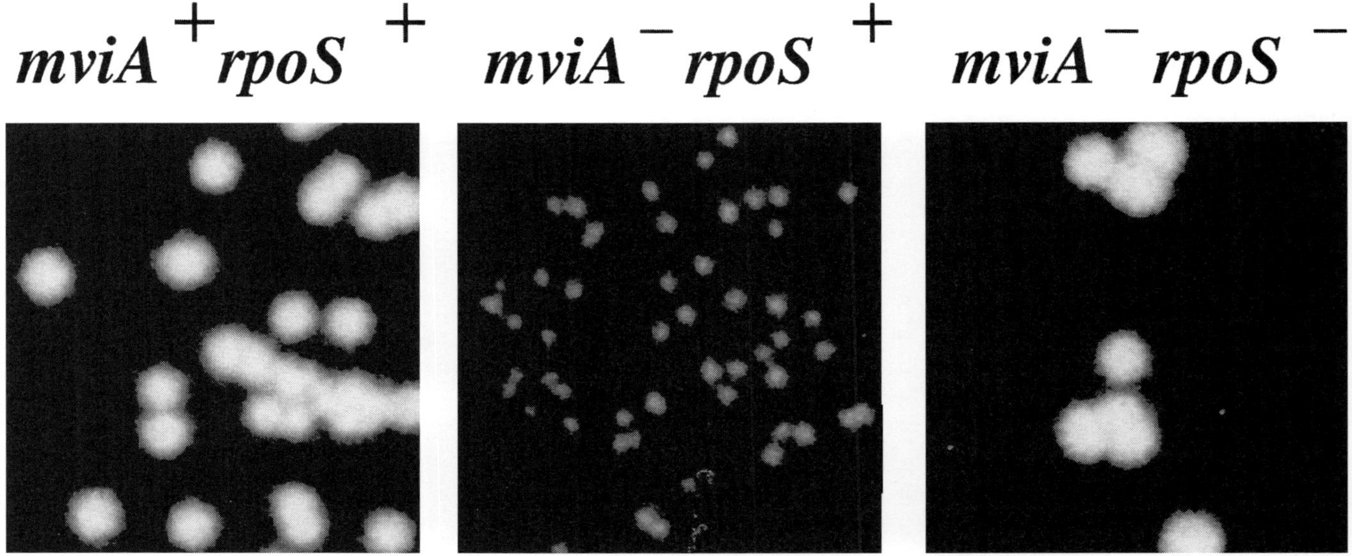

FIG. 6. Overproduction of σ^S causes a small colony phenotype. Colonies of cell types indicated above each panel were grown for similar lengths of time on Luria–Bertani (LB) complex media.

generate DNA damage via inappropriate methylation of adenine by S-adenosyl methionine (Barrows & Magee 1982) or via endogenously formed methylating agents created by nitrosidation at low pH (Taverna & Sedgwick 1996). Enzymes associated with repairing methylation damage include the Ogt and Ada DNA methyltransferases which can demethylate O^6-methylguanine and O^4-methylthymine, and the enzyme 3-methyladenine glycosylase, the product of *alkA*, which can demethylate 3-methyladenine and 3-methylguanine. Ada is also involved as a positive regulator of *alkA* (reviewed in Rupp 1996). When tested for effects on acid tolerance, *ada* but not *ogt* mutants of *S. typhimurium* (Hakura et al 1991, Yamada et al 1995) were acid sensitive (Fig. 7A; Park et al 1998). We discovered that the *ada* mutant was not only acid sensitive but also alkaline sensitive, exhibiting poor growth at pH 7.7 in minimal media (Fig. 7B). The little growth that was observed at pH 7.7 was shown to depend upon an active σ^S since an *ada rpoS* double mutant failed to grow at all at that pH. Two-dimensional analysis of polypeptides produced by the *ada* mutant revealed that σ^S-dependent proteins were induced at pH 7.7 but not pH 7.0, confirming that σ^S-dependent systems were activated under slightly alkaline conditions (J. W. Foster, unpublished observation 1998). The reason for the acid- and alkaline-sensitive nature of the *ada* mutant is not known, but is presumably due to defects in the repair of acid- and alkali-damaged DNA.

Inducible pH homeostasis mechanisms

Constitutively expressed housekeeping pH homeostasis systems involving potassium/proton antiporters are thought to maintain cytoplasmic pH near neutrality over a wide range of external pH values, with the maximum ΔpH of around 2 pH units. However, below external pH 4.5, these systems begin to fail. Adaptation at pH 5.8 before submitting the culture to pH 3.5 challenge will induce an ill-defined system that helps maintain a ΔpH of around 2 pH units. Whether that system involves inducible proton pumps and/or enhanced buffer capacity is under investigation. However, certain amino acid decarboxylase systems, such as lysine decarboxylase, appear to help maintain internal pH in the presence of their substrate (Park et al 1996). These systems essentially produce both a buffer in the form of a decarboxylation product and a pump that exchanges that product for fresh substrate in the medium.

Stationary phase acid tolerance

In addition to the log phase mechanisms of acid tolerance described above, *Salmonella* also possesses stationary phase mechanisms (Lee et al 1994). There is a σ^S-dependent stationary phase system of general stress resistance not induced by

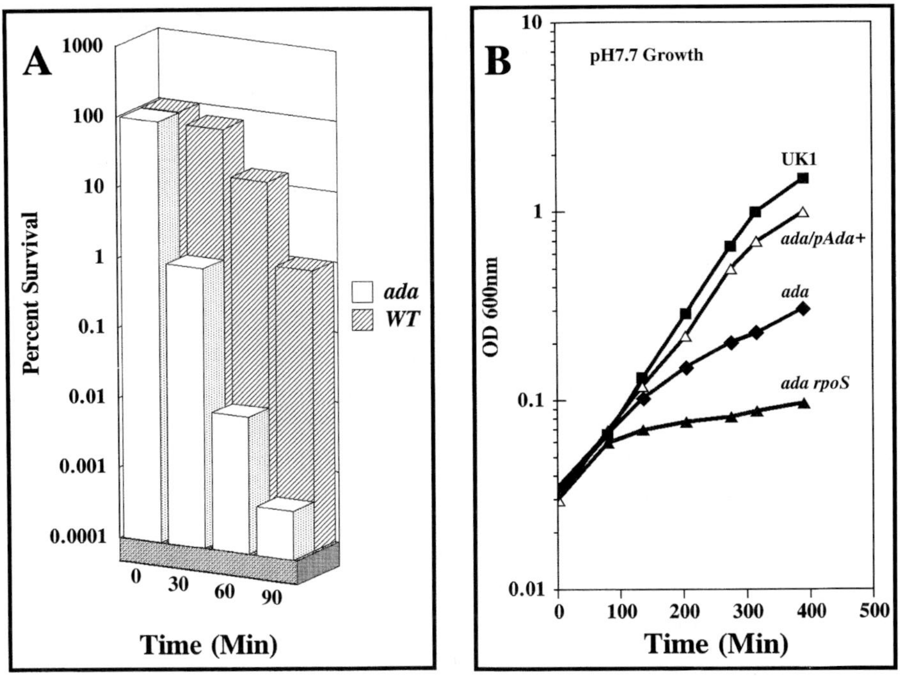

FIG. 7. Effect of *ada* (O^6-guanine methyltransferase) on acid and alkaline tolerance. (A) Acid tolerance response of an *ada* mutant and wild-type (WT) *Salmonella typhimurium* cells. Cells were adapted at pH 4.5 as described in Fig. 1. (B) Growth of wild-type UK1 *S. typhimurium* (filled squares), *ada* (filled diamonds), *ada rpoS* (filled triangles), and an *ada* mutant containing an ada^+ plasmid (open triangles).

low pH because σ^S levels are already elevated due to entry into stationary phase. However, there is a second system, independent of σ^S, that requires induction by acid pH. Little is known regarding this system other than it is also independent of Fur protein.

Acid resistance of *Escherichia coli* O157:H7 and *Shigella flexneri*

Conflicting reports in the literature regarding the relative degrees of acid resistance of *E. coli, Shigella flexneri* and *S. typhimurium* led us to compare these organisms directly. When tested using stationary phase cells in pH 2.5 minimal media, *Shigella* exhibited extremely poor acid resistance, whereas *E. coli* and *S. typhimurium* were similarly tolerant. The situation was dramatically different when complex media was used. In this situation, *Shigella* and *E. coli* were far superior to *S. typhimurium* (Lin et al 1995). The reason for this difference was attributed to the presence, in *Shigella,* of a glutamate-dependent acid resistance system and, in *E. coli,* the presence of glutamate-dependent and arginine-dependent acid resistance systems. In addition, growth of either organism to stationary phase in complex medium produced a glucose-repressible acid resistance system that would protect cells in minimal media even without glutamate or arginine. Figure 8 summarizes the acid resistance systems. The glutamate-dependent system has since proven to be glutamate decarboxylase including a potential glutamate: γ-aminobutyric acid antiporter (γ-aminobutyric acid is the decarboxylation product of glutamate) encoded by *gadBC* (Hersh et al 1996, Waterman & Small 1996). The arginine-dependent system involves arginine decarboxylase encoded by *adiA* (Lin et al 1995). *S. typhimurium* does not possess

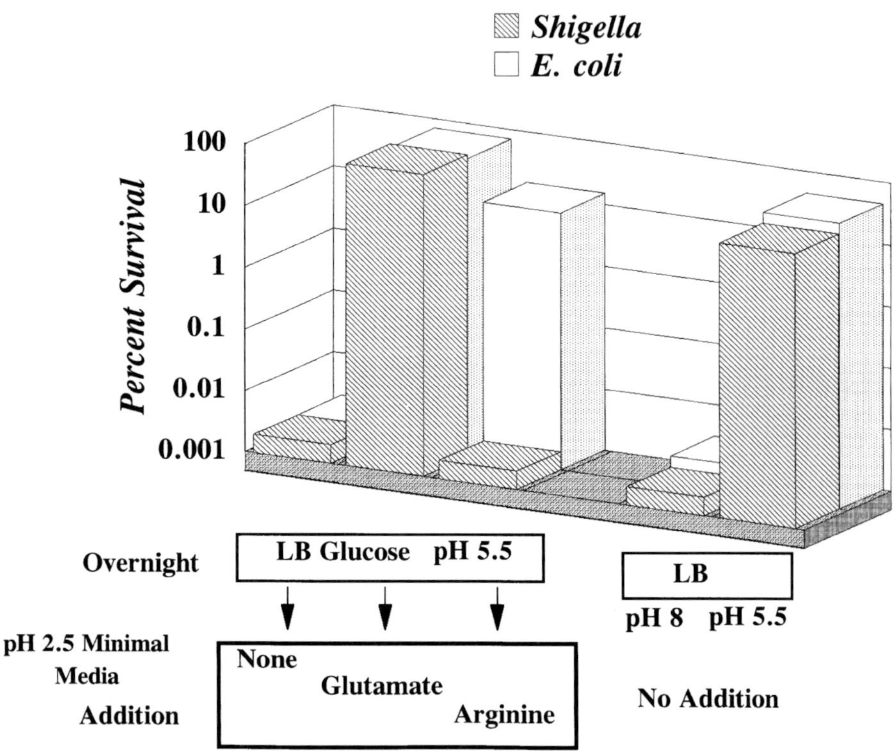

FIG. 8. Acid resistance systems of *Escherichia coli* and *Shigella flexneri*. Cells were grown overnight in Luria–Bertani (LB) glucose, LB pH 8 buffered with MOPS or LB pH 5.5 MES-buffered media as indicated. Cells were then diluted to approximately 2×10^6 cells/ml in pH 2.5 minimal E glucose media containing either no addition or supplemented with 1.4 mM glutamate or 1.2 mM arginine. Viable counts were determined one hour after acid challenge.

glutamate decarboxylase but does produce arginine decarboxylase yet, curiously, does not exhibit an arginine-dependent acid tolerance. The reason for this is not apparent. Table 1 illustrates the three forms of acid resistance in *E. coli* and the effects of *adiA*, *gadC*, *rpoS* and *crp* mutations on the arginine, glutamate and oxidative acid resistance systems, respectively.

E. coli O157:H7 is one of the enterohaemorrhagic *E. coli* that has recently emerged as an important food pathogen. Although infection has been associated with many different foods, it has primarily been associated with contaminated and undercooked hamburgers. Since it is believed to infect at a low dose due to an exceptional level of acid resistance, we performed a comparative analysis of the three acid resistance systems. We have found that both commensal *E. coli* and O157:H7 exhibit outstanding acid resistance but that, in general, O157 strains are not better than commensal strains in this regard. Once induced, the acid resistance systems will persist at refrigerator temperatures for at least one month (Lin et al 1996). This is important when one considers that processed hamburgers are usually refrigerated for some period of time before consumption. Once the acid resistance systems are induced, organisms that contaminate food products can for long periods of time remain poised to survive stomach acidity even when refrigerated.

Summary and future directions

S. typhimurium, *S. flexneri* and *E. coli* have developed several strategies for surviving acidic environments. *S. flexneri* appears to rely heavily on the glutamate and glucose repressed stationary phase systems of acid resistance. It appears virtually defenceless in log phase. *S. typhimurium* possesses log and stationary phase acid tolerance systems. *E. coli,* on the other hand, appears to possess the log and stationary phase systems of *S. typhimurium* and the glutamate system of *Shigella*. It also exhibits an arginine acid resistance system, a system that is unique to *E. coli*.

TABLE 1 Effects of various mutations on the acid resistance phenotype of *Escherichia coli* K12

	Per cent survival at pH 2.5 at 2 h		
Mutation	*Glutamate*	*Arginine*	*Oxidative*
Wild-type	75	25	19
gadC	<0.01	17	15
adi	63	<0.01	12
rpoS	5	0.9	<0.01
crp	55	9	0.01

Cells were grown and tested for acid resistance as described in the legend to Fig. 8.

Many questions remain regarding the operation and regulation of these acid tolerance systems. In *S. typhimurium*, we would like to know which specific ASPs are required for acid tolerance. How does Fur protein sense acid pH and does it truly act as an positive regulator of specific genes? With respect to σ^S, how does MviA sense acid stress and then modulate the ability of ClpXP to degrade σ^S? Questions about the acid resistance systems of *E. coli* include how they are induced by acid, can they be inhibited by analogues and why don't they function in *Salmonella*? Finally, unravelling the intricacies of acid stress responses will reveal associations between stress responses and the virulence of these organisms.

Acknowledgements

The body of work described in this article was supported by grants from the National Science Foundation (DCB-89-04819) and the National Institutes of Health (GM48017). The authors wish to thank Patricia Couling for her expert secretarial assistance in helping prepare this manuscript. J.W.F. is especially grateful for the diligence and skill of the many outstanding graduate and postdoctoral students who have ventured into the acid realm.

References

Baik HS, Bearson S, Dunbar S, Foster JW 1996 The acid tolerance response of *Salmonella typhimurium* provides protection against organic acids. Microbiology 142:3195–3200

Barrows LR, Magee PN 1982 Nonenzymatic methylation of DNA by S-adenosylmethionine *in vitro*. Carcinogenesis 3:349–351

Bearson BL, Wilson L, Foster JW 1998 A low pH-inducible, PhoPQ-dependent acid tolerance response protects *Salmonella typhimurium* against inorganic acid stress. J Bacteriol 180: 2409–2417

Bearson SMD, Benjamin WH Jr, Swords WE, Foster JW 1996 Acid shock induction of *RpoS* is mediated by the mouse virulence gene *mviA* of *Salmonella typhimurium*. J Bacteriol 178:2572–2579

Benjamin WH Jr, Hall PD 1993 The *Salmonella typhimurium* virulence gene *mviA* is a response regulator gene related to the *ompR* gene family. Am Soc Microbiol 93rd Gen Mtg B-67:38

Benjamin WH Jr, Yother J, Hall P, Briles DE 1991 The *Salmonella typhimurium* locus *mviA* regulates virulence in *Ity*^{s} but not *Ity*^{r} mice: functional *mviA* results in avirulence; mutant (nonfunctional) *mviA* results in virulence. J Exp Med 174:1073–1083

Benjamin WH Jr, Wu X, Swords WE 1996 The predicted amino acid sequence of the *Salmonella typhimurium* virulence gene *mviA*$^+$ strongly indicates that MviA is a regulator protein of a previously unknown *S. typhimurium* response regulator family. Infect Immun 64:2365–2367

Fang FC, Libby SJ, Buchmeier NA et al 1992 The alternative α factor (RpoS) regulates *Salmonella* virulence. Proc Natl Acad Sci USA 89:11978–11982

Foster JW, Bearson B 1994 Acid sensitive mutants of *Salmonella typhimurium* identified through a dinitrophenol selection strategy. J Bacteriol 176:2596–2602

Foster JW, Hall HK 1990 Adaptive acidification tolerance response of *Salmonella typhimurium*. J Bacteriol 172:771–778

Foster JW, Hall HK 1992 Effect of *Salmonella typhimurium* ferric-uptake regulator (*fur*) mutations on iron and pH-regulated protein synthesis. J Bacteriol 174:4317–4323

Hakura A, Morimoto K, Sofuni T, Nohmi T 1991 Cloning characterization of the *Salmonella typhimurium ada* gene, which encodes O^6-methylguanine-DNA methyltransferase. J Bacteriol 173:3663–3672

Hall HK, Foster JW 1996 The role of Fur in the acid tolerance response of *Salmonella typhimurium* is physiologically and genetically separable from its role in iron acquisition. J Bacteriol 178:5683–5691

Hengge-Aronis R 1996 Back to log phase: σ^S as a global regulator in the osmotic control of gene expression in *Escherichia coli*. Mol Microbiol 21:887–893

Hersh BM, Farooq FT, Barstad DN, Blankenhorn DL, Slonczewski JL 1996 A glutamate-dependent acid resistance gene in *Escherichia coli*. J Bacteriol 178:3978–3981

Lee IS, Slonczewski JL, Foster JW 1994 A low-pH-inducible stationary-phase acid tolerance response in *Salmonella typhimurium*. J Bacteriol 176:1422–1426

Lee IS, Lin J, Hall HK, Bearson B, Foster JW 1995 The stationary-phase sigma factor σ^S(RpoS) is required for a sustained acid tolerance response in virulent *Salmonella typhimurium*. Mol Microbiol 17:155–167

Leyer GJ, Johnson EA 1993 Acid adaptation induces cross protection against environmental stresses in *Salmonella typhimurium*. Appl Environ Microbiol 59:1842

Lin J, Lee IS, Frey J, Slonczewski JL, Foster JW 1995 Comparative analysis of extreme acid survival in *Salmonella typhimurium*, *Shigella flexneri* and *Escherichia coli*. J Bacteriol 177:4097–4104

Lin J, Smith MP, Chapin KC, Baik HS, Bennett GN, Foster JW 1996 Mechanisms of acid resistance in enterohemorrhagic *Escherichia coli*. Appl Environ Microbiol 62:3094–3100

Muffler A, Fischer D, Altuvia S, Storz G, Hengge-Aronis R 1996 The response regulator RssB controls stability of the sigma^S subunit of RNA polymerase in *Escherichia coli*. EMBO J 15:1333–1339

Park KR, Giard J-C, Eom JH, Bearson SD, Foster JW 1998 The *Salmonella typhimurium* TyrR protein and tyrosine receptor are required for the expression of *hyaB* and *aniC* under low pH and anaerobic conditions, submitted

Park Y-K, Bearson B, Bang SH, Bang IS, Foster JW 1996 Internal pH crisis, lysine decarboxylase and the acid tolerance response of *Salmonella typhimurium*. Mol Microbiol 20:605–611

Pratt L, Silhavy T 1996 The response regulator, SprE, controls the stability of RpoS. Proc Natl Acad Sci USA 93:2488–2492

Rathman M, Sjaastad MD, Falkow S 1996 Acidification of phagosomes containing *Salmonella typhimurium* in murine macrophages. Infect Immun 64:2765–2773

Rupp DW 1996 DNA repair mechanisms. In: Neidhardt FC, Curtiss III R, Ingraham JL et al (eds) *Escherichia coli* and *Salmonella:* cellular and molecular biology, 2nd edn. ASM Press, Washington, DC, p 2277–2294

Soncini FC, Vescovi EG, Solomon F, Groisman EA 1996 Molecular basis of the magnesium deprivation response in *Salmonella typhimurium*: identification of PhoP-regulated genes. J Bacteriol 178:5092–5099

Taverna P, Sedgwick B 1996 Generation of an endogenous DNA-methylating agent by nitrosation of *Escherichia coli*. J Bacteriol 178:5105–5111

Waterman SR, Small PLC 1996 Characterization of the acid resistance phenotype and *rpoS* alleles of shiga-like toxin-producing *Escherichia coli*. Infect Immun 64:2808–2811

Wilmes-Riesenberg MR, Bearson B, Foster JW, Curtiss R III 1996 Role of acid tolerance response in virulence of *Salmonella typhimurium*. Infect Immun 64:1085–1092

Yamada M, Sedgwick B, Sofuni T, Nohmi T 1995 Construction and characterization of mutants of *Salmonella typhimurium* deficient in DNA repair of O^6-methylguanine. J Bacteriol 177:1511–1519

DISCUSSION

Padan: Have you looked for suppressors of the H144L mutant?

Foster: Not yet. We know that this histidine residue is important though because Saito et al (1991) said that it is most affected by changes in pH.

Padan: You mentioned that PhoQ functions at low magnesium concentrations. If you are correct then increasing magnesium concentration will rescue the pH effect.

Foster: Eduardo Groisman and colleagues have shown that magnesium is sensed by PhoQ (Soncini et al 1996), but in terms of pH I don't know. Eduardo's magnesium concentrations are higher, but neither of us have tried to modulate the magnesium concentration further.

Quivey: Are there any differences in the proteins expressed immediately following acid shock and later during long-term acid tolerance?

Foster: We have not looked in detail at the proteins expressed following long-term adaptation at pH 5.8. We see a small difference in the protein pattern produced at pH 5.8 but it's not significant. At pH 4.4 following acid shock for 20 min, we saw the dramatic protein array I presented. Joan Slonczewski may have looked in more detail at long-term acid protein composition.

Slonczewski: I haven't looked at *Salmonella*. There are differences between proteins expressed at steady-state levels over the long term and proteins that are induced following a sudden shock. Proteins induced suddenly tend to be universal stress responses, whereas proteins more specific to pH stress are expressed in the long term.

Quivey: RpoS has a short half life, which begs the question of what is happening downstream. MviA is orchestrating some of the long-term effects downstream, because σ^S is present for longer, but I wonder whether there are organisms that have acid adaptation but not RpoS. I'm not saying that they don't exist, I'm saying that they haven't been identified yet. We know that the F-ATPase in *Streptococcus mutans* and the family of chaperones are induced, but these are short-term phenomena.

Slonczewski: RpoS does regulate many interesting proteins. It seems to switch on proteins that may be useful at some future time when the cell is in a growth phase.

Quivey: Do you mean proteins that have a longer half-life than RpoS itself?

Slonczewski: Exactly. And it's interesting that RpoS has been found in such a wide range of organisms. It is even present in mycobacteria.

Matin: I agree. All the starvation proteins, for example, that are regulated by σ^S are synthesized before the end of stationary phase. If their synthesis is inhibited, an effect on survival is not observed immediately, but rather in the long term.

Cook: Can an rpoS mutant regulate its intracellular pH (pH_i)?

Foster: Yes.

Cook: You showed that the *rpoS* mutant survived in the presence of organic acids and when they were removed. Are there any differences in ΔpH under those two conditions, i.e. in the presence and absence of organic acids?

Foster: We didn't look at ΔpH.

Slonczewski: We showed, in collaboration with Pam Small, that RpoS is a component of acid resistance in *Escherichia coli* (Small et al 1994). We found that *rpoS* mutant cells that are just surviving (but not growing) in extreme acid have impaired pH homeostasis.

Bennett: Have you looked at the RpoS acid tolerance effect with different organic acids and different sugars as carbon sources?

Foster: I have not looked at different carbon sources, although I have used different organic acids (propionate, butyrate and acetate) and RpoS is required for protection against those acids.

Booth: I prefer not to say that RpoS has something to do with stationary phase. We have coined the term 'safety net' instead. If the exponential growth rate of *E. coli* is decreased to below $\mu = 1.1$ per h, which is relatively rapid and is equivalent to growing in an amino acid/glucose/phosphate buffer with all the bases, then RpoS is a dominant character in the survival of any stress. If you start varying the sugar, and you observe that the growth rate changes, you will get RpoS effects that come about just through growth rate modulation. The term 'safety net' covers it beautifully because RpoS is preparing the cell for the eventuality that it will encounter some stress that will kill it. It doesn't know what it is, it just knows there's a problem ahead.

Matin: We published a paper in *Molecular Biology* showing that in the chemostat there's an inverse correlation between growth rate and RpoS concentration (Zgurskaya et al 1997).

Booth: But the growth rates that you studied were lower than 1.1 per h.

Stock: One could imagine that σ^S exists in several states. When RpoS is in a complex with polymerase, is it susceptible to degradation, or does it have to be released? Does the level of RpoS utilization control its degradation?

Foster: We have been able to manipulate the cells, such that we allow RpoS and MviA to accumulate, and proteolysis of RpoS is switched on after MviA has accumulated. We find that RpoS still turns over, but we don't know whether this is at the same rate. If σ^S was associated with RNA polymerase and this was preventing degradation, one would expect that the production of MviA would not have an effect on turnover.

Stock: It would stimulate the cycle when it is being utilized.

Matin: I agree because the site on σ^S that is attacked by ClpXP appears to be the one with which it binds to the RNA core enzyme.

Booth: I have a problem with that experiment, because the crystal structure of ClpXP was generated at the end of last year (Wang et al 1997). It has a barrel structure, and the polypeptide chain, which is fed in like a fishing reel in reverse, winds round the active site. It's an incredible structure because it raises the protease active sites to 400 mM in that local environment, and it clips every 10th residue. So the idea that this is where the protease attacks has to be revisited.

Foster: The site you speak of may be where MviA interacts with RpoS and facilitates interaction with ClpXP. It may not be the site of proteolytic attack.

Booth: It is possible that the problem with RpoS is not proteolysis, but is two potential folds. The protein may fold into two complete configurations, one of which is recognized as abnormal and is fed through to ClpXP, and the other of which is recognized by MviA in its phosphorylated form so it goes down the correct folding pathway which stabilizes it. In a sense ClpXP is a distraction because it's what happens to unfolded proteins, whereas what we should be concentrating on is how the protein is stabilized.

Moir: In the MviA mutant are other ClpXP targets degraded?

Foster: Most of the known ClpXP targets are viral targets.

Matin: Zhou & Göttesman (1998) showed that if RssB (MviA) is overexpressed σ^S is affected but not other targets of ClpXP, such as λO.

Foster: That's true, but in terms of proteins that are normally in *E. coli* or *Salmonella*, as far as I know σ^S is the only target.

Poole: We've heard about a number of specific molecular reactions, some of which have been studied *in vitro*, e.g. the turnover of RpoS, the autophosphorylation of CheA and the regulation of PhoP/PhoQ interactions. What studies have been done on the dependence of those reactions *in vitro* on pH_i? For example, the turnover of RpoS can be catalysed by certain proteins, but is that degradative reaction directly affected by pH *in vitro* and could it therefore have anything to do with pH sensing *in vivo*?

Foster: Susan Göttesman has started to look at this *in vitro* using isolated proteins, but as far as I know the answer is not known yet.

Matin: Schellhorn & Stones (1992) showed that weak acids can induce σ^S-dependent gene fusions, which may indicate that affecting internal pH can influence σ^S levels.

Foster: But it probably also affected growth.

Booth: If you drop *E. coli* Frag1 down to pH 6.0, it produces large amounts of RpoS within about 5 min. Acid tolerance in Frag1 is virtually wholly dependent on RpoS. But if you acidify with acetate at pH 7, then it takes about 15–30 min for the levels of RpoS to increase to the same level. Therefore, it is possible that different signals are being seen. We know that osmotic stress, for example, also controls RpoS turnover and doesn't affect internal pH too much.

Moir: I would like to change the subject and ask anyone to comment on the response to pH stress and virulence.

Foster: PhoP and PhoQ are essential for virulence in *Salmonella*. The macrophage phagolysosome, in which *Salmonella* often finds itself, is acidic. Therefore, PhoP and PhoQ need to be operative to survive in that condition. Everyone has been working on the assumption that the magnesium concentration in phagolysosomes is low and that's why the PhoP/PhoQ system is turned on. However, our results question whether the magnesium and/or pH are involved. We've knocked out several genes that affect acid tolerance and tested them in mice, and found that most of them are avirulent, but we cannot say that the avirulence is due to the inability to survive pH because some of these mutations affect other things.

Slonczewski: Several well-known virulence systems are acid regulated; for example, ToxR in *Vibrio*. ToxR is homologous to CadC in *E. coli*, and it looks as though this protein started out evolutionarily as a metabolic regulator. ToxR regulates proteins requiring ToxT, as well as ToxT-independent proteins. ToxT-independent proteins function in fermentation and other metabolic activities. ToxR appears to have acquired the ToxT subset of functions, such as cholera toxin and other virulence factors. Therefore, this looks like an acid-regulated system that acquired virulence functions later.

Foster: Most acid-regulated proteins are also regulated by other things.

Booth: In a survey of *E. coli* O157 isolates, Waterman & Small (1996) observed that roughly 30% of the isolates that cause disease in humans were RpoS$^-$, and theoretically ought to be acid sensitive. Also, we divide RpoS$^-$ strains into two: the so-called genetically created Rpos$^-$, which we create; and the evolutionary RpoS$^-$, which are isolated from the field. We found that acid-tolerant regimes in the latter group are RpoS$^-$ independent. They have been RpoS$^-$ for a long time and have found a different way of switching on tolerance genes.

Foster: In stationary phase there are also RpoS-independent acid tolerance response (ATR) systems.

Slonczewski: We managed to knock out one of the RpoS-independent acid resistance systems, and we're now trying to figure out what it is. In addition, anaerobically grown cells have a reduced need for RpoS systems; they have other systems that are turned on at low pH.

Hill: We have been working on *Listeria monocytogenes*, which also has to survive in a macrophage phagolysosome. We identified a mutant that cannot induce an ATR, although it does have stationary phase tolerance (O'Driscoll et al 1996). When we infect mice with this mutant, we find it is also almost completely avirulent. We do get some survivors in the spleen, about one in every 10^5 that we introduce, but we found that every single survivor has restored its ability to switch on the ATR.

Booth: One interesting point is that people seem to record RpoS$^-$ variants in natural strains but they don't seem to record ATR-minus variants. I don't know whether anyone has even found a natural ATR-minus variant.

Hill: The corollary of my earlier remark is that we have identified mutants which are constitutively acid tolerant all through the growth phase and are up to 1000-fold more virulent than the wild-type (Marron et al 1997). We've also shown that if you recover wild-type *Listeria* that hasn't been induced from the macrophage phagolysosome, you find that they have switched on the ATR, but this might be expected, given that the phagosomal pH is about 5.0, which is an inducing pH for *Listeria*.

References

Marron L, Emerson N, Gahan C, Hill C 1997 A mutant of *Listeria* monocytogenes L028 unable to induce an acid tolerance response displays diminished virulence in a murine model. Appl Environ Microbiol 63:4945–4947

O'Driscoll B, Gahan C, Hill C 1996 Adaptive acid tolerance response in *Listeria monocytogenes*: isolation of an acid tolerant mutant which demonstrates increased virulence. Appl Environ Microbiol 62:1693–1698

Saito T, Duly D, Williams RJ 1991 The histidines of the iron-uptake regulation protein, Fur. Eur J Biochem 197:39–42

Schellhorn HE, Stones VL 1992 Regulation of *katF* and *katE* in *Escherichia coli* K-12 by weak acids. J Bacteriol 174:4769–4776

Small P, Blankenhorn D, Welty D, Zinser E, Slonczewski JL 1994 Acid and base resistance in *Escherichia coli* and *Shigella flexneri*: the role of *rpoS* and growth pH. J Bacteriol 176:1729–1737

Soncini FC, García Véscovi E, Solomon F, Groisman EA 1996 Molecular basis of the magnesium deprivation response in *Salmonella typhimurium*: identification of PhoP-regulated genes. J Bacteriol 178:5092–5099

Wang J, Hartling JA, Flanagan JM 1997 The structure of ClpP at 2.3 Å resolution suggests a model for ATP-dependent proteolysis. Cell 91:447–456

Waterman SR, Small PL 1996 Characterization of the acid resistance phenotype and *rpoS* alleles of shiga-like toxin-producing *Escherichia coli*. Infect Immun 64:2808–2811

Zgurskaya H, Keyhan M, Matin A 1997 The sigma S levels in starving *Escherichia coli* cells increase solely as a result of its increased stability, despite decreased synthesis. Mol Microbiol 24:643–651

Zhou M, Göttesman S 1998 Regulation of proteolysis of stationary phase sigma factor RpoS. J Bacteriol 180:1154–1158

Acid and base regulation in the proteome of *Escherichia coli*

Joan L. Slonczewski and Darcy Blankenhorn

Department of Biology, Kenyon College, Gambier, OH 43022, USA

Abstract. Acid and base conditions have many significant effects on the growth of *Escherichia coli*. External and internal pH perturbations induce different classes of genes. pH-dependent regulation of genes intersects with other regulatory responses, e.g. oxygen level or osmolarity. 2D electrophoretic gels were used to compare global patterns of protein induction in *Escherichia coli* grown in complex media buffered at the acid or alkaline ends of the pH range for growth (pH 4.4 vs. pH 9.1). Preliminary results indicate new classes of acid- and base-dependent regulation, in some cases highly dependent on oxygen level. Other proteins are induced strongly at both extremes of pH, compared to pH 7. Current work continues to dissect the relationship between effects of pH, oxygen level and osmolarity.

1999 Bacterial responses to pH. Wiley, Chichester (Novartis Foundation Symposium 221) p 75–92

Escherichia coli maintains its pH within a narrow range, near pH 7.6 (Slonczewski et al 1981, Zilberstein et al 1984) during growth over a relatively wide range of external pH (pH$_o$; pH 4.4–9.0). Furthermore, many *E. coli* strains can survive for several hours in a resistant, non-growing state at even greater extremes of pH, as low as pH 1.5 or as high as pH 10 (Gorden & Small 1993, Small et al 1994, Lin et al 1995). A number of comprehensive reviews on acid and base response and survival have been published (Foster 1995, Olson 1993, Padan & Schuldiner 1987, Slonczewski & Foster 1996). This chapter highlights some particularly interesting features of pH stress, and includes some new directions of recent research approached from the viewpoint of the proteome.

Acid effects on growth

When *E. coli* grows at moderately low pH (pH 4.5–7.0), its transmembrane pH gradient, ΔpH, makes a substantial contribution to the proton potential. As external pH decreases, the ΔpH increases, actually increasing the energy available to the cell. A problem, however, is that growth, particularly in dense cultures, usually produces large quantities of end products that act as membrane-permeant weak acids (or 'permeant acids'). Permeant acids are taken up primarily in the electroneutral protonated form, but their total concentration increases along the pH gradient, in proportion to the hydronium ion concentration. This has serious effects on the cell: (1) the acid loading depresses internal pH (pH_i), and decreases proton motive force; and (2) the dissociated anion may build up to toxic concentrations. If the anion also is sufficiently hydrophobic to cross the membrane, then cyclic transfer of protons may partly or totally collapse the proton gradient, the 'uncoupler' effect (Salmond et al 1984).

Thus, *E. coli* faces two distinct, although related, kinds of acid stress: external acidity and permeant acids that depress pH_i. The two forms of stress are related in that: (1) extreme external acidification can overload the pH homeostasis and substantially depress pH_i; and (2) the degree of internal acidification caused by permeant acids increases in proportion to the ΔpH. Nevertheless, it is interesting that most of the pH-dependent genetic systems discovered so far respond specifically either to external acid or to permeant acids, not both.

External acid

The best characterized system of response to external acid is that of the amino acid decarboxylases, particularly lysine decarboxylase, encoded with its product transporter in the operon *cadBA* (Fig. 1; Neely et al 1994, Meng & Bennett 1992, Watson et al 1992). Lysine decarboxylase removes carbon dioxide from lysine, producing the basic amine cadaverine. Cadaverine is then transported from the cell by CadB, eventually alkalinizing the growth media. Expression of *cadBA* requires lysine, and it is subject to extremely tight regulation by external acid. Acid regulation is mediated by the pH sensor protein CadC, a homologue of ToxR in *Vibrio cholerae*. Several other decarboxylases show a similar pattern of regulation, especially arginine and ornithine decarboxylases. Regulation of glutamate decarboxylase appears to be more complex (see below).

Induction of decarboxylases is increased several-fold by anaerobiosis. Anaerobic enhancement is also found in a number of other important acid-regulated pathways, particularly fermentation components such as lactate dehydrogenase (Mat-Jan et al 1989) and formate hydrogenlyase (Rossmann et al 1991). Both these cause preferential production of mildly acidic or neutral products, instead of strongly acidic products; thus, their induction during acidification is advantageous.

Permeant acids

A few genes have been shown to respond to the acid-driven uptake of permeant acids. The *mar* system, including unlinked loci such as *inaA*, is induced by benzoate derivatives (Podglajen et al 1998, Rosner & Slonczewski 1994).

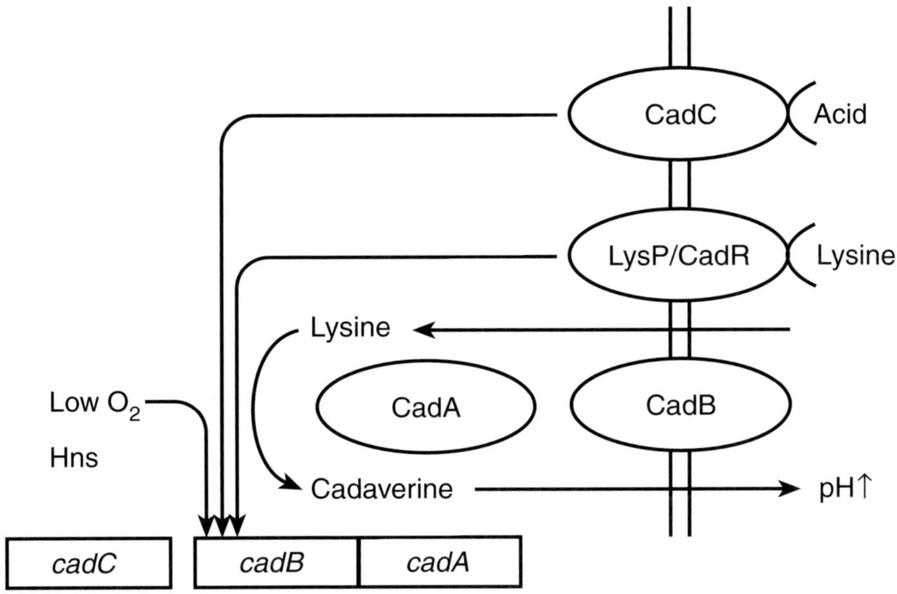

FIG. 1. Regulation of lysine decarboxylase and cadaverine transport: a mechanism to reverse external acidification. Reprinted with permission from Slonczewski & Foster (1996).

Mutants constitutive for *marA* show multiple drug resistance; the mechanism for this resistance is unknown. The mechanism of the response appears to involve both internal acidification and the accumulation of the anion, but the details remain unclear.

Benzoate has often been used to investigate perturbation of pH_i because it crosses the membrane predominantly in the protonated form, and it is not metabolized by *E. coli*; but at higher pH_o, permeation by the anion becomes significant (Salmond et al 1984). To distinguish between acid-dependent and non-acid-dependent effects of benzoate, we used 2D gel electrophoresis to compare the proteins induced by benzoate at pH_o 6.5 vs. pH 8.0 (Lambert et al 1997). The VanBogelen–Neidhardt protocols were used (VanBogelen et al 1996). Thirty-three proteins were observed to be induced by addition of 20 mM benzoate at pH 6.5; of these, 18 appeared only at lower pH. The proteins identified based on VanBogelen's reference gel are summarized in Table 1.

Of the proteins induced by benzoate both at high and low pH_o, most are also induced by the uncoupler dinitrophenol (Gage & Neidhardt 1993). This is consistent with previous evidence that benzoate acts as a weak uncoupling agent above pH 7. At pH 6.5, where benzoic acid crosses the membrane primarily in the protonated form, several other proteins are induced. The most interesting of these

TABLE 1 Proteins induced in *Escherichia coli* W3110 by 20 mM benzoate at pH 6.5

Protein	*Induced by benzoate at pH 8.0 as well as pH 6.5*	*Other published inducing conditions*[a]
Fur	No	
HslU	No	42C
Hns	No	CSI, PSI, 10C
IbpA	No	42C
AhpC	Yes	DNP, PSI, OXY
ClpB	Yes	DNP, 42C, CSI
DnaK	Yes	DNP, 42C, CSI, NSI, PSI
GroLS	Yes	DNP, 42C, CSI, PSI
GrpE	Yes	DNP, 42C, CSI, PSI
HtpG	Yes	DNP, 42C
TyrS	Yes	
UspA	Yes	DNP, 42C, CSI, NSI, PSI

[a]Inducing conditions are based on those compiled in VanBogelen et al (1996) and Gage & Neidhardt (1993). Abbreviations: CSI, induced by carbon starvation; DNP, proteins induced by 0.5 mM dinitrophenol; NSI, induced by nitrogen starvation; OXY, member of OxyR regulon; PSI, induced by phosphate starvation; 10C, shift from 28 °C to 10 °C; 42C, shift from 28 °C to 42 °C. Modified with permission, from Lambert et al (1997).

is Fur, the ferric ion regulator, which has also been implicated in acid regulation in *Salmonella typhimurium* (Foster & Hall 1992). Further testing will reveal whether Fur in *E. coli* shows dependence on pH_i.

Most of the proteins induced by benzoate shock, however, either at low or high pH, were general stress proteins, known to respond to many different kinds of shock (Table 1). This finding suggests that the sudden shock of benzoate reveals little about the cell's mechanisms of steady-state adaptation to growth in the presence of a permeant acid. For this reason, in subsequent work we have adjusted our experimental design to observe steady-state protein expression in cultures grown for several generations in a pH stress condition (see below).

External base

Growth at pH values higher than the pH_i causes several problems for *E. coli.* Most seriously, the ΔpH is inverted, subtracting from the total proton motive force; thus the energy available for growth is significantly decreased. In addition, sodium ion becomes limiting, requiring the Na^+/H^+ antiporter to exclude sodium ions (Karpel et al 1991). Expression of the antiporter NhaA is induced by external base, and pH-dependent residues have been defined.

Alkaline growth appears to induce several porins known for osmotic regulation (Heyde & Portalier 1987). In our work, we have preliminary evidence of similar alkaline induction of periplasmic transporters (J. L. Slonczewski, unpublished results 1998). We are interested in further investigation of the links between pH stress and osmolarity.

Just as acid effects are often co-regulated by anaerobiosis, effects of alkalinization are often co-induced by aerobic growth. An example is expression of cytochrome o (*cyoABCDE*) in preference to cytochrome d (*cydAB*) at higher pH (Cotter et al 1990). Since cytochrome o produces considerably more hydrogen ions per electron, the resulting acidification is more favourable to the cell at higher pH_o.

Finally, it should be noted that permeant amines such as methylamine can become concentrated in the cell by the inverted pH gradient; the pH_i rises, causing a chemotactic response (Repaske & Adler 1981). Some commonly used buffers, such as triethanolamine and tris[hydroxymethyl]aminomethane, are subject to this effect. The physiological effects of permeant bases on cell growth have been little studied.

The proteome in extreme acid or base

Some of the more interesting questions about pH-regulated genes are best addressed by a global approach such as that of 2D gels. Are particular functional classes of genes induced at the extreme ends of the pH range? Do acid and base induce any proteins in common, compared to growth at pH 7? How do related stress conditions, such as anaerobiosis or high osmolarity, interact with pH stress?

There have been surprisingly few studies of global response to acid or base. The one major 2D gel study, by Hickey & Hirshfield (1990), was done before direct protein sequencing techniques were available. Recently we decided to investigate the induction of proteins at the extreme ends of the pH range of growth, using the 2D gels and N-terminal sequencing of the protein spots (J. L. Slonczewski & D. Blankenhorn, unpublished work 1998). Preliminary initial findings are described here.

The 2D protein separation system was based on that of VanBogelen et al (1996), but for culture growth we chose a complex medium, Luria broth (LB), instead of the defined glucose-based medium generally used. Growth in complex media reveals a wider range of pH responses than is likely to appear in minimal media; for example the repression of the SOS response by acid (Dri & Moreau 1994) and the anaerobic/acid-regulated acid resistance factor *aniG* in *Salmonella* (Aliabadi et al 1988). More dramatically, intracellular growth of *Salmonella* cells within macrophages reveals protein induction patterns different from those seen under relevant stress conditions (such as low pH or anaerobiosis) in cells grown in standard defined media (Abshire & Neidhardt 1993).

Another advantage of buffered LB is that it supports growth of *E. coli* at greater extremes of pH than in minimal media (as low as pH 4.4 and as high as pH 9.2 in aerated cultures.) We defined the pH limits of growth as the values of pH requiring growth for approximately 24 h at 37 °C to reach mid-log phase. Buffers of appropriate pK_a were used at 100 mM concentration to ensure maintenance of the pH for each growth condition (Slonczewski & Foster 1996).

In our first exploratory experiments, we have compared the protein patterns for mid-log phase cultures grown at extreme acid or base, either aerobically or anaerobically. Proteins were identified by N-terminal sequencing. A number of proteins were induced by acid, either aerobically or anaerobically; for example, GatY, a component of galactitol fermentation (Lin 1996). Other acid-induced proteins (not yet identified) appeared only in aerobiosis, or only in anaerobiosis, and not in both. One protein that showed substantial induction at high pH was tryptophanase (TnaA). TnaA removes ammonia from several different amino acids, producing pyruvate, which can be further broken down to other acids (McFall & Newman 1996). TnaA is considered to function mainly to enable the consumption of amino acids. Yet it clearly can produce acidity for the cell; it would make sense for this system to be most active under alkaline stress, just as decarboxylases are induced under acid stress.

A more surprising and complex result was the induction pattern for GadA, one of two homologous genes encoding glutamate decarboxylase (Smith et al 1992). In aerobic cultures, our preliminary analysis is consistent with previous findings that the glutamate systems are induced by acid (McFall & Newman 1996) and that glutamate metabolism helps respond to acid and osmotic shock (McLaggan et al 1994, Roe et al 1998). Anaerobically, however, we find just the opposite—that GadA is induced strongly at high pH. This result, if confirmed, would be of interest because our studies of acid survival (Hersh et al 1996, Waterman & Small 1996) show that glutamate decarboxylase enables cultures grown at pH 7–8 to survive subsequent exposure at pH 2.5 (Fig. 2). Unlike other decarboxylases, glutamate decarboxylase is only active below pH 6, and therefore is particularly suited to neutralize pH_i lowered to this range when exposed to extreme acid (Small & Waterman 1998). Perhaps GadA is induced anaerobically because anaerobic metabolism usually acidifies the cell.

Another interesting induction pattern was that of isocitrate lyase, AceA (Cronan & LaPorte 1996). The glyoxylate bypass is known to be induced by aerobic growth, as well as by acetate. In our gels, however, we found much greater induction by extreme acid or base, compared with growth at neutrality. For unknown reasons, AceA appears to show an extreme pH response, either acid or base.

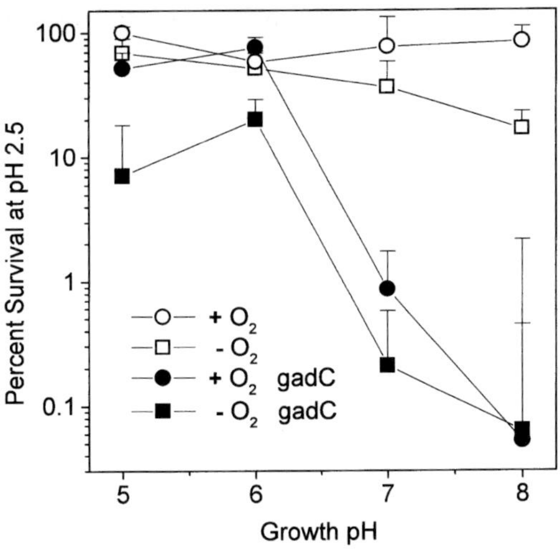

FIG. 2. Extreme acid survival of *Esherichia coli* MC4100 and a *gadC* mutant. Survival in acid was measured as the percentage of colony forming units on Luria broth (LB), following incubation of cells for 2 h in LB at pH 2.5. Modified with permission from Hersh et al (1996).

In summary, our current work suggests that many complex features of pH-regulated gene expression remain to be revealed. We are resuming the study of benzoate and other weak acids to determine their effects on the patterns of *E. coli* proteins at steady state. In addition, it will be most interesting to see how pH-dependent mechanisms intersect with regulation by oxygen level and by osmolarity.

Acknowledgements

This work was supported by grant MCB 960963 from the National Science Foundation, and by a faculty development grant from Kenyon College. We thank Jack Presley at the Protein Structure Laboratory, University of California at Davis, for peptide sequence analyses.

References

Abshire KZ, Neidhardt FC 1993 Analysis of proteins synthesized by *Salmonella typhimurium* during growth within a host macrophage. J Bacteriol 175:3734–3743

Aliabadi Z, Park YK, Slonczewski JL, Foster JW 1988 Novel regulatory loci controlling oxygen- and pH-regulated gene expression in *Salmonella typhimurium*. J Bacteriol 170:842–851

Cotter PA, Chepuri V, Gennis RB, Gunsalus RP 1990 Cytochrome *o* (*cyoABCDE*) and *d* (*cydAB*) oxidase gene expression in *Escherichia coli* is regulated by oxygen, pH, and the *fnr* gene product. J Bacteriol 172:6333–6338

Cronan JE, LaPorte D 1996 Tricarboxylic acid cycle and glyoxalate bypass. In: Neidhardt FC, Curtiss III R, Ingraham JL et al (eds) *Escherichia coli* and *Salmonella*: cellular and molecular biology, 2nd edn. ASM Press, Washington, DC, p 206–215

Dri AM, Moreau PL 1994 Control of the LexA regulon by pH: evidence for a reversible inactivation of the LexA repressor during the growth cycle of *Escherichia coli*. Mol Microbiol 12:621–629

Foster JW 1995 Low pH adaptation and the acid tolerance response of *Salmonella typhimurium*. Crit Rev Microbiol 21:215–237

Foster JW, Hall HK 1992 Effect of *Salmonella typhimurium* ferric uptake regulator (*fur*) mutations on iron- and pH-regulated protein synthesis. J Bacteriol 174:4317–4323

Gage DJ, Neidhardt FC 1993 Adaptation of *Escherichia coli* to the uncoupler of oxidative phosphorylation 2,4-dinitrophenol. J Bacteriol 175:7105–7108

Gorden J, Small PLC 1993 Acid resistance in enteric bacteria. Infect Immun 61:364–367

Hersh BM, Farooq FT, Barstad DN, Blankenhorn DL, Slonczewski JL 1996 A glutamate-dependent acid resistance gene in *Escherichia coli*. J Bacteriol 178:3978–3981

Heyde M, Portalier R 1987 Regulation of major outer membrane porin proteins of *Escherichia coli* K-12 by pH. Molec Gen Genet 208:511–517

Hickey EW, Hirshfield IN 1990 Low-pH-induced effects on patterns of protein synthesis and on internal pH in *Escherichia coli* and *Salmonella typhimurium*. Appl Environ Microbiol 56: 1038–1045

Karpel R, Alon T, Glaser G, Schuldiner S, Padan E 1991 Expression of a sodium proton antiporter (NhaA) in *Escherichia coli* is induced by Na^+ and Li^+ ions. J Biol Chem 266:21753–21759

Lambert L, Abshire K, Blankenhorn D, Slonczewski JL 1997 Proteins induced in *Escherichia coli* by benzoic acid. J Bacteriol 179:7595–7599

Lin ECC 1996 Dissimilatory pathways for sugars, polyols, and carboxylates. In: Neidhardt FC, Curtiss III R, Ingraham JL et al (eds) *Escherichia coli* and *Salmonella*: cellular and molecular biology, 2nd edn. ASM Press, Washington, DC, p 307–327

Lin J, Lee IS, Frey J, Slonczewski JL, Foster JW 1995 Comparative analysis of acid survival in *Salmonella typhimurium*, *Shigella flexneri*, and *Escherichia coli*. J Bacteriol 177:4097–4104

Mat-Jan F, Alam KY, Clark DP 1989 Mutants of *Escherichia coli* deficient in the fermentative lactate dehydrogenase. J Bacteriol 171:342–348

McFall E, Newman EB 1996 Amino acids as carbon sources. In: Neidhardt FC, Curtiss III R, Ingraham JL et al (eds) *Escherichia coli* and *Salmonella*: cellular and molecular biology, 2nd edn. ASM Press, Washington, DC, p 358–375

McLaggan D, Naprstek J, Buurman ET, Epstein W 1994 Interdependence of K^+ and glutamate accumulation during osmotic adaptation of *Escherichia coli*. J Biol Chem 269:1911–1917

Meng S-Y, Bennett GN 1992 Nucleotide sequence of the *Escherichia coli* cad operon: a system for neutralization of low extracellular pH. J Bacteriol 174:2659–2669

Neely MN, Dell CL, Olson E 1994 Roles of LysP and CadC in mediating the lysine requirement for acid induction of the *Escherichia coli cad* operon. J Bacteriol 176:3278–3285

Olson ER 1993 Influence of pH on bacterial gene expression. Mol Microbiol 8:5–14

Padan E, Schuldiner S 1987 Intracellular pH and membrane potential as regulators in the prokaryote cell. J Membr Biol 95:189–198

Podglajen I, McDermott PF, Draper MP, Levy SB 1998 Regulation of *mar*C and *ychE* expression, and exploration of their role in the *mar*-dependent resistance phenotype. American Society for Microbiology General Meeting, Atlanta, USA, May 1998, p 529 (abstr V-94)

Repaske DR, Adler J 1981 Change in intracellular pH of *Escherichia coli* mediates the chemotactic response to certain attractants and repellents. J Bacteriol 145:1196–1208

Roe AJ, McLaggan D, Davidson I, O'Byrne C, Booth IR 1998 Perturbation of anion balance during inhibition of growth of *Escherichia coli* by weak acids. J Bacteriol 180:767–772

Rosner JL, Slonczewski JL 1994 Dual regulation of *inaA* by the multiple antibiotic resistance (Mar) and superoxide (SoxRS) stress response systems of *Escherichia coli*. J Bacteriol 176: 6262–6269

Rossmann R, Sawers G, Bock A 1991 Mechanism of the regulation of the formate-hydrogenlyase pathway by oxygen, nitrate and pH: definition of the formate regulon. Mol Microbiol 5:2807–2814

Salmond CV, Kroll RG, Booth IR 1984 The effect of food preservatives on pH homeostasis in *Escherichia coli*. J Gen Microbiol 130:2845–2850

Slonczewski JL, Foster JW 1996 pH-regulated genes and survival at extreme pH. In: Neidhardt FC, Curtiss III R, Ingraham JL et al (eds) *Escherichia coli* and *Salmonella*: cellular and molecular biology, 2nd edn. ASM Press, Washington, DC, p 1539–1552

Slonczewski JL, Rosen BP, Alger JR, Macnab RM 1981 pH homeostasis in *Escherichia coli*: measurement by ^{31}P nuclear magnetic resonance of methylphosphonate and phosphate. Proc Natl Acad Sci USA 78:6271–6275

Small PL, Waterman SR 1998 Acid stress, anaerobiosis and gadCB: lessons from *Lactococcus lactis* and *Escherichia coli*. Trends Microbiol 6:214–216

Small PL, Blankenhorn D, Welty D, Zinser E, Slonczewski JL 1994 Acid and base resistance in *Escherichia coli* and *Shigella flexneri*: the role of *rpoS* and growth pH. J Bacteriol 176:1729–1737

Smith DK, Kassam T, Singh B, Elliott JF 1992 *Escherichia coli* has two homologous glutamate decarboxylase genes that map to distinct loci. J Bacteriol 174:5820–5826

VanBogelen RA, Abshire KZ, Pertsemlidis A, Clark RL, Neidhardt FC 1996 Geneprotein database of *Escherichia coli* K-12, edition 6. In: Neidhardt FC, Curtiss III R, Ingraham JL et al (eds) *Escherichia coli* and *Salmonella*: cellular and molecular biology, 2nd edn. ASM Press, Washington, DC, p 2067–2117

Waterman SR, Small PL 1996 Identification of sigma S-dependent genes associated with the stationary-phase acid-resistance phenotype of *Shigella flexneri*. Mol Microbiol 21:925–940

Watson N, Dunyak DS, Rosey EL, Slonczewski JL, Olson ER 1992 Identification of elements involved in transcriptional regulation of the *Escherichia coli cad* operon by external pH. J Bacteriol 174:530–540

Zilberstein D, Agmon V, Schuldiner S, Padan E 1984 *Escherichia coli* intracellular pH, membrane potential, and cell growth. J Bacteriol 158:246–252

DISCUSSION

Foster: You showed a slide indicating pI values of proteins in *Escherichia coli* with no proteins evident that possessed pI values in the neutral range. Is it possible that one of the reasons why cells die in acid conditions is because proteins are precipitating inside the cell as intracellular pH (pH_i) decreases?

Slonczewski: This slide contained data from Ruth Van Bogelen's observations of the distribution of protein pI values (A. Pertsemlidis & R. Van Bogelen, unpublished data 1998). It certainly makes sense that there is a limited pH_i range over which the proteins are stable, and perhaps pH 5.0 is the critical limit below which proteins precipitate.

Matin: Tomb et al (1997) report the sequence of *Heliobacter pylori* and show that in this species about 70% of proteins are positively charged below pH 7.0, as opposed to *Escherichia coli* and *Haemophilus influenzae*, which are neutrophilic.

Booth: Before we discuss this I must say that this paper has been withdrawn because there was an error in the way they had done the work. However, one interesting point is that *Saccharomyces cerevisiae* doesn't have an optimal pH_i of between 7.0 and 8.0, but fits your graph. Its optimum pH_i for growth is in the range of 5.8–6.8, and yet it has the same pH_i profile.

Slonczewski: But what is its actual pH_i when it is growing normally? I would have to know more about how pH_i was measured in that experiment. The slide actually demonstrated that the mimimum of the pH_i distribution was lower in *S. cerevisiae* than in *E. coli*.

Booth: A number of people have done these studies using a number of different methods — e.g. the recently developed carboxyfluorescein-based methods — that give accurate pH_i values and they all end up with the same answer. Even the NMR method, which is one of the ones you use, gives the same answer, so there is no reason to doubt that the pH_i of *S. cerevisiae* under growth conditions is between 5.8 and 6.8.

Stock: It would be interesting to generate random protein sequences with typical amino acid compositions and estimate their pIs using a computer. This would allow you to determine whether or not random fluctuations in arginine, lysine, glutamate and aspartate could explain the kind of bimodal distribution you observe.

Skulachev: You mentioned that at high pH cytochrome o is induced and cytochrome d is repressed. In our hands, however, the opposite occurs, i.e. at high pH cytochrome d is induced and cytochrome o is repressed. You also mentioned that cytochrome o produces more acid than cytochrome d. What do you mean by this?

Slonczewski: I was referring to the proton to oxygen ratio.

Skulachev: But the situation inside the cell is the reverse because cytochrome o pumps four protons per two electrons. This results in a more alkaline cytosol than in the case of cytochrome d, which pumps two protons per two electrons.

Slonczewski: It is interesting to consider which aspect the cell is most concerned about. Is it most concerned about maintaining its pH_i or changing what's going on outside?

Booth: Joan Slonczewski's point was relating more to ATP generation potential, rather than pH. This pathway may be preferable because it generates more ATP. Most of the protons pumped out are taken back in by an energy-transducing machinery, otherwise the cell would grind to a halt. And in this case there is no change in pH.

Poole: In a case like this where the two systems have such different properties anyway, trying to estimate their effects is fraught with difficulties. For example, those two oxidases have different affinities for oxygen, so it's difficult to predict to what extent each of them will be working at full capacity under any set of growth conditions.

Rowbury: I would like to raise a point about GadA. We've shown L-glutamate-induced acid tolerance at neutral pH and at alkaline pH, but we haven't looked at acidic pH. I wondered whether that might be related to what you're seeing. The tolerance requires glutamate decarboxylase, but I was surprised that you could get induction at pH 7.0 under the conditions you described.

Slonczewski: I wasn't looking at induction, I was looking at survival. The situation for glutamate is complicated because there are several systems. There are two decarboxylases, and the one that was induced in our experiments at high pH was GadA, whereas GadBC was essential for acid survival of cells grown at pH 8.0. I had originally assumed that the Gad system may function at other pHs, but there are several different systems available to cells grown in mixed media. If the cells are first grown at pH 8.0 a shift to acidic conditions is a more extreme stress, and so knocking out GadA may knock out resistance completely, whereas if GadA is knocked out in cells originally grown in acid they may survive because they have other alternatives.

Rowbury: Have you thought about the possibility that the system goes further than decarboxylation? Because in our inducible tolerance system, there is also induction by amino butyrate, so it's possible that you get deamination of the decarboxylase product.

Slonczewski: Yes, that's possible.

Konings: There is evidence that some of these decarboxylases are coupled to precursor–product exchangers, which is an effective way of generating a proton motive force. Protons are consumed inside in the decarboxylation reaction; net negative charges are transported to the inside, and as a result $\Delta\Psi$ and ΔpH are generated.

Slonczewski: It may be true that another function of the decarboxylases is to generate a proton gradient. But it's also true that at least three or four of the decarboxylase systems are extremely acid inducible, indicating a role for the neutralization of the acid.

Konings: The changes are pH sensitive; they only occur within a narrow pH range. This is why I am puzzled by the induction of GadA at high pH.

Slonczewski: You are questioning why, if it is only going to work under acid conditions, it is made at high pH. You don't want it to work at high pH. You want it to be available to work when the pH is lowered.

Bennett: Fermentation products such as acetate enter the cell and then glutamate is released. Is there any connection between this exit of glutamate and its decarboxylation?

Slonczewski: This is not known at present, but it would be interesting to look at.

Foster: Although we haven't looked at the proteins specifically, we find that in *E. coli* the Gad acid protection system is switched on no matter what the pH is. The only requirement is that the cells have to be in complex medium. It is possible, however, that GadA might operate under alkaline conditions and GadB might work in acid conditions.

Slonczewski: We wondered about that too. We were puzzled by this effect, and little is known about how GadA is regulated.

Booth: You say that the Gad system operates only in enriched medium. I always think of the glutamate pool as was being relatively tightly regulated by potassium balance in the cell. Do these cells stock up on glutamine? Have you looked at the amino acid pools, for instance, prior to and subsequent to the acid shock? Is the glutamate consumed?

Foster: No, we haven't looked.

Slonczewski: Neither have we.

Kobayashi: One of the roles of decarboxylation is to supply carbon dioxide or carbonate, which are essential for the growth of *E. coli*. When oxygen is limited, it is possible that amino acid decarboxylation is induced because the supply of carbon dioxide from the TCA cycle linked to respiration is limited. This is one of the physiological roles of the decarboxylation enzyme. I am not surprised, therefore, that glutamine decarboxylase is induced at high pH, because carbon dioxide is essential at high pH as well.

Booth: I thought that this was an elegant paper that demonstrated the importance of the carbon dioxide at low pH (Takayama et al 1994).

Fillingame: Joan Slonczewski mentioned growing cells at pH 9.1, and showed that the pH_i increases under those conditions. What was the maximum pH_i?

Slonczewski: The pH_i increases to about 8.0.

Fillingame: Is there agreement between this and Mugikura et al (1990), which describes the growth of *E. coli* in Luria–Bertani (LB) medium in the presence of uncouplers at pH 9 with a pH_i of 8.5? Do you agree that *E. coli* can grow with a pH_i that high?

Kobayashi: I will talk in detail about this in my presentation, but *E. coli* can grow even if the cytoplasm is alkalinized beyond pH 8.0.

Slonczewski: For the growth of *E. coli* the upper pH_i limit is about 8.0.

Skulachev: In spite of the presence of an uncoupler, pH_i will be lower than extracellular pH (pH_o) because of the electrophoretic uptake of protons to compensate for the electrogenic efflux of sodium. If there are only proton pumps, then if you add an uncoupler the pH_i will be equal to pH_o. However, other pumps are obviously involved.

Fillingame: My recollection of the experiments of Mugikura et al (1990) are that the concentration of carbonyl cyanide m-chlorophenylhydrazone (CCCP) they were using (in the range of 50–100 μM) would completely inhibit growth on succinate, but cells would continue to grow on LB. We have grown *E. coli* on succinate at alkaline pH, and our conclusion is that the growth on succinate at alkaline pH with low concentrations of CCCP is due to $\Delta\bar{\mu}_{H}+$ and not to a sodium potential. We have not measured pH_i, but the simplest interpretation of your published paper is that *E. coli* can grow, albeit somewhat slower, with an alkaline pH_i.

Skulachev: What was the concentration of sodium ions in your experiments?

Fillingame: 170 mM.

Matin: With respect to the use of CCCP, we should keep in mind that there is a multidrug efflux system that can extrude this protonophore.

Schäfer: What was the membrane potential when the pH was inverted?

Fillingame: In our case at pH 8.7 with or without CCCP the $\Delta\bar{\mu}_{H}+$, as measured indirectly by lactose accumulation, was about 150 mV, versus 190 mV at pH 8.0. There was a significant slowing of growth but there was still a substantial $\Delta\bar{\mu}_{H}+$, and growth was sensitive to higher concentrations of CCCP. Therefore, we believe that growth under those conditions was due to $\Delta\bar{\mu}_{H}+$ and to the F_1F_0 ATP synthase.

Booth: Could you exclude the presence of an MDR-type pump that pumped out the CCCP, so the CCCP doesn't actually have an effect?

Fillingame: That is possible at lower concentrations of CCCP. I'm saying that the alkaline pH by itself is enough to slow growth significantly and reduces $\Delta\bar{\mu}_{H}+$, which suggests that an MDR-type pump is not operating.

Poole: Have you done those experiments with cytochrome o (*cyoABCDE*) and d (*cydAB*) oxidase mutants to distinguish the roles of the two systems?

Fillingame: We did not. All we did was to ensure that the ATP synthesis we saw under these conditions was due to the F_1 F_0 ATP synthase. Also, only one set of genes in the *E. coli* genome codes for those subunits, so we know this has to be due to the normal type of F_1F_0 ATP synthase. One of the questions that I will address in my presentation is whether the cation specificity of this enzyme is modified under these conditions.

Poole: In *E. coli* cytochrome o is repressible at low pH under aerobic conditions: at pH 5.5, the level of expression is fourfold lower than at pH 7.5 (Cotter et al 1990). Under anaerobic conditions, however, *cyo* expression is completely repressed at all pH values tested. Interestingly, in a *fnr* mutant derepression is observed anaerobically above about pH 6.5. This strongly implicates a Fnr-independent mechanism for repression that operates at low pH values. Does anyone know what such a repressor might be?

Slonczewski: Cytochrome d is also induced in the *fnr* mutant, supporting the idea that there is an underlying system.

Krulwich: I have a comment not about *E. coli*, but about *Bacillus subtilis*. In this organism there seems to be an emerging thought that the laboratory strains may be different with respect to which terminal oxidases they are using compared to wild-type strains that have not been cultivated for long periods under laboratory conditions. The caa$_3$ pathway has at most a minor pathway that can be mutated with little or no phenotypic result.

Booth: In *E. coli* some of the so-called stationary phase genes are actually catabolite-repressed genes (Schultz et al 1988). The point I wanted to make is that the strength of catabolite repression in *E. coli* is pH dependent, so changes in catabolite repression occur when the external pH changes from 6.1 to 5.9, because glucose loses its catabolite-repressing potential at about 5.9. Thomas & Booth (1992) also pointed out that the low pH induction of OmpC is purely catabolite repression.

Matin: Joan Slonczewski, you said that when you suspended *E. coli* cells at between pH 2.5 and 4.0 they remained viable, and the pH$_i$ was in the range of 5.0 to 6.5. Did you look at the membrane potential under these conditions?

Slonczewski: No.

Matin: When you then added acid to those cells, the cells lost viability.

Slonczewski: Yes. If the cells are at pH 2.5–3.0 and we add benzoate, if we add enough benzoate to decrease the pH$_i$ down to 4.5 then the cells rapidly lose viability.

Matin: Did you measure the pH$_i$ when you added the weak acids?

Slonczewski: Yes, but I have to say that it's extremely difficult to measure pH$_i$ accurately under these conditions. We tried to use both the benzoate reporter and the salicylate reporter.

Schäfer: Wouldn't you expect an uncoupling effect on oxidative phosphorylation at this pH?

Slonczewski: At the lower pH there is enough protonated salicylate to ensure that uncoupler effects (permeation of the anion) are minimal. We would have to calculate the pK_a effects, but at least we were able to see a ΔpH, and we used low concentrations of salicylate, in the micromolar range.

Matin: But salicylate won't be a significant uncoupler at such low concentrations. The problem may be that you already had excess weak acid to bring the pH down.

Slonczewski: But we started with pH 2.5 and the cells were diluted by 1000-fold into fresh LB broth so the weak acid should not affect the cells to a large extent. We were surprised to see that the cells maintained a significant ΔpH under those conditions, however.

Matin: This is a thermodynamically stable situation in which the residual ΔpH is balanced by a positive intracellular $\Delta\Psi$.

Slonczewski: This is possible.

Cook: I would like to change the subject. In your 2D gels you showed that MalE, which is a periplasmic binding protein, was strongly induced at alkaline pH. You seemed to think it was an artefact. Would you like to comment on this?

Slonczewski: I was speculating whether osmolarity was involved because our initial view was that some of the things induced by growth in alkaline media are also induced by growth at high osmolarity. I'm concerned that the cells may be perceiving a higher pH as higher osmolarity. There could be other explanations for why these things are induced specifically at high pH.

Booth: You have to be careful when you use the word 'specifically'. Because if I understood your experiment correctly, the growth rate decreased quite considerably. Therefore, many regulatory circuits are probably induced, such as RpoS-dependent systems, other slow growth systems and catabolite repression systems, because catabolite repression is dependent upon the rate of glucose flux through the cell. So when you use the word 'specifically', you can only say that it is specific to the condition that you chose, and not that it is specifically induced by alkaline pH.

Slonczewski: I agree, and that's why we would like to use this experiment as a starting point to then test the range of conditions for systems of interest.

Poole: The question of how growth rate changes when you do almost anything else to a culture is age old, and is something which we all try to be aware of. One could argue that one always ought to do experiments at defined growth rates and change any one variable at a time, but that is only advantageous if what you are trying to do is to dissect several interacting overlapping systems. However, I can't think of any system or any environment *in vivo* where an organism's growth rate is maintained as in a chemostat, so what we observe in laboratory experiments with fluctuating conditions in many ways is a life-like situation and it is valuable, but it's difficult to work out what's going on in terms of growth control methods.

Slonczewski: One advantage of the 2D gel technique is that it has become automated enough so that you can do many gels under reproducible conditions. In this way you can separate two factors at a time to look at.

Quivey: Having seen the effects of GadA upregulation, are you considering expanding your 2D gel studies into the basic region to look for potential DNA-binding proteins, which are almost always basic?

Slonczewski: I would be interested in doing that. The gel technique required to look at the basic region is more difficult, and we haven't worked it out yet.

Poole: I have a technical question that several people might find interesting. Are your gel analyses done using professional software packages or do you simply visualize them?

Slonczewski: I haven't yet seen a gel analysis package that I consider worth the investment. I scan the gels on a standard scanner and then just manipulate the axes and the contrast, which makes comparisons easier. I am wary of software packages because of the problems of defining the baseline when you are trying to find out whether something is being induced. How do you define the total protein level?

Quivey: I would like to address that because we are doing similar experiments with *Streptococcus* grown in steady-state conditions. I find that my own eye is reasonable as a starting point. We then use NIH Image (which is available at http://rsb.info.nih.gov/nih-image) to plot the values, so that we can do statistics on the intensity numbers. We average the intensities of the spots on three gels and obtain each S.D. using the intensity numbers in the *z* coordinate. To address the issue of total protein, we subtract the background value from all the dots, and then we sum all the dots and analyse the dot versus the background-subtracted dots. This generates a more robust number, although the cut-off point is entirely arbitrary.

Slonczewski: How do you decide the baseline when you compare one culture and another? Suppose overall there's twice as much protein made at pH 7 than at pH 8, do you look at the same amount of protein under these conditions? Or do you look at half the amount of protein per gel for pH 8 compared to pH 7?

Quivey: Y_{max} question is true for *Streptoccoccus*. At pH 5.0 it uses most glucose to produce ATP for removing protons. Therefore, the maximum yield at pH 5 is half that at pH 7.0. However, we always load the same amount of protein on each gel.

Slonczewski: But that choice is arbitrary.

Booth: Ingraham et al (1983) produced a concept that is important for all physiologists. They suggested that there are three protein fractions in the cell. (1) A fraction that is transcribed and translated proportional to growth rate, so the amount of that protein doesn't change as the growth rate changes. (2) A fraction, such as ribosomes and the polymerization assembly machinery for protein synthesis, that is transcribed and translated at a rate greater than μ, so that the fraction of the total cell proteins increases with the growth rate. In other words, if you make more protein in a shorter period, you need more ribosomes. (3) A fraction that results from genes which are constitutively expressed; they are not subject to transcriptional or translational control. The consequence is that this fraction decreases as the growth rate increases.

The important point is that if the growth rate decreases, this fraction of proteins increases. This is just a response to having spare protein synthesis capacity, and it is not a response to a specific stress. This is when people get into problems of defining what is a stress response and what is purely an artefact of physiology. *E. coli* can sacrifice 30% of its non-protein translational machinery without changing its growth rate. It is making 30% more protein than it actually needs at all times. The reason it does that is because it has to be able to adapt. If it only made enough of each protein, then each time the growth rate increased it would have to sacrifice some protein. When the growth rate increases, they can sacrifice a fraction of that material to make more ribosomes, with the result that they continue to grow at the same rate and have the same biochemical flux.

Matin: The concentration of a number of substrate-capturing enzymes increases as the growth rate decreases. In the light of what you have just said, would this have no physiological consequence?

Booth: That's a difficult question. *E. coli* has been around for a long time, and the fact that it has survived means that some of these things that don't appear to confer a strong selective advantage have an advantage under a particular condition. A classic example is that when you look at the lysine or glutamate requirement for survival, it's not only the amount of lysine or glutamate you have to take into consideration, it's also the population size, because if you go from 10^7 to 10^6 organisms then you need 10-fold less material. As long as you're in the K_m range for the transporter, although it's actually the K_s that's important rather than the K_m, the cells can pick up the stuff they need even though they are in a dilute medium. We showed that you can get osmo-protection with 1 nM betaine in *Salmonella typhimurium*, but you have to dilute the cells to 10^5 cells per ml, because at 10^7 cell per ml, which is in the normal physiological range, there isn't enough material to give rise to protection. So, my answer to your question is that by having these things you can scavenge, which may be advantageous rather than disadvantageous.

Bennett: I would like to change the subject and ask Joan Slonczewski if some of the proteins on the 2D gels are transaminases, and whether the tryptophanase was metabolizing enough substrate to contribute to the pH response, even at those high levels?

Slonczewski: We're interested in the deaminases. We have looked for them for a number of years. These data are recent, and we have just started generating mutants to pursue such studies.

Rowbury: You mentioned AhpC, the small subunit of alkyl hydroperoxide reductase. Was this induced at acidic pH but not at alkaline pH?

Slonczewski: It is induced by many different stress conditions, so I don't know how significant that is.

Rowbury: Presumably, it is related to peroxides being produced under those stress conditions.

Slonczewski: Either that or AhpC may be induced by many stress conditions, and pH just happens to be one of them.

Rowbury: We showed that AhpC was induced at alkaline pH, which made me wonder whether any condition that caused damage-inducing peroxide expression indirectly leads to AhpC formation.

References

Cotter PA, Chepuri V, Gennis RB, Gunsalus RP 1990 Cytochrome o (*cyoABCDE*) and d (*cydAB*) oxidase gene expression in *Escherichia coli* is regulated by oxygen, pH, and the *fnr* gene product. J Bacteriol 172:6333–6338

Ingraham JL, Maaloe O, Neidhardt FC 1983 Growth of the bacterial cell. Sinauer, Sunderland, MA

Mugikura S, Nishikawa M, Igarashi K, Kobaashi H 1990 Maintenance of a neutral cytoplasmic pH is not obligatory for growth of *Escherichia coli* and *Streptococcus faecalis* at an alkaline pH. J Biochem (Tokyo) 108:86–91

Schultz JE, Latter GI, Matin A 1988 Differential regulation by cyclic AMP of starvation protein synthesis in *Escherichia coli*. J Bacteriol 170:3903–3909

Takayama M, Ohyama T, Igarashi K, Kobayashi H 1994 *Escherichia coli cad* operon functions as a supplier of carbon dioxide. Mol Microbiol 11:913–918

Thomas AD, Booth IR 1992 The regulation of expression of the porin gene *ompC* by acid pH. J Gen Microbiol 138:1829–1835

Tomb JF, White O, Kerlavage AR et al 1997 The complete genome sequence of the gastric pathogen *Helicobacter pylori*. Nature 388:539–547

Acid tolerance induced by metabolites and secreted proteins, and how tolerance can be counteracted

Robin J. Rowbury

Biology Department, University College London, Gower Street, London WC1E 6BT, UK

Abstract. Several metabolites and salts including glucose, L-glutamate, L-aspartate, $FeCl_3$, KCl and L-proline induce acid tolerance at neutral external pH (pH_o) in log phase *Escherichia coli*. For induction by glucose and L-glutamate, the processes are independent of integration host factor (IHF), H-NS, CysB, ferric uptake regulator (Fur) and RelA. For most of the above, tolerance does not appear if induction occurs and NaCl, sucrose, SDS or DOC are present. For several responses, cAMP inhibits induction. For many established acid tolerance and sensitization processes, including those tolerance responses switched on at pH 5.0 and by glucose, glutamate or aspartate, induction is associated with secretion of extracellular induction proteins. These proteins bring about the response if added to organisms under normally non-inducing conditions. Secreted components also influence inherent acid tolerances and sensitivities. Analysis of some established tolerance responses indicates that induction is a two-stage process, secreted extracellular proteins playing an obligate role in induction. For example, the functioning of the acid-induced medium protein(s) is essential for acid habituation at pH_o 5.0. It seems likely that such two-stage mechanisms are essential for many inducible processes in bacteria.

1999 Bacterial responses to pH. Wiley, Chichester (Novartis Foundation Symposium 221) p 93–111

It is now clear that inducible stress tolerance influences whether bacteria resist stresses assailing them in the environment, in foodstuffs and in the body, and inducible tolerance to lethal acidity is important because there are many locations where high proton concentrations occur. It is clear that bacteria which cause disease after ingestion of small numbers of organisms do so because acid tolerance allows them to resist gastric acidity.

Inducible acid tolerance was discovered in *Escherichia coli* and *Salmonella typhimurium* (Rowbury et al 1989, Foster & Hall 1990) but has now been demonstrated in many other bacteria including *Listeria monocytogenes* (Kroll & Patchett 1992), root nodule bacteria (O'Hara & Glenn 1994) and *Aeromonas hydrophila* (Karem et al 1994). Several inducible acid tolerances have been reported, e.g. the acid tolerance response (ATR; or acid habituation) appearing at external pH (pH_o) *c.* 5.0 (Rowbury et al 1989, Foster & Hall 1990), stationary-phase acid tolerance (Cooper & Rowbury 1986, Lee et al 1994) and heat-induced acid tolerance (Humphrey et al 1993). Three types of study are urgently required: (1) studies that attempt to investigate mechanisms of ATRs; (2) those aimed at discovering novel inducible ATRs; and (3) those attempting to counteract, reverse or subvert such responses. We are researching in each of these areas.

Although this chapter relates mainly to acid tolerance, it is relevant to other aspects of this Novartis Foundation symposium and to wider aspects of stress tolerance and mechanisms of physiological response induction. The extracellular components involved in tolerance induction described here, which influence every nuance of inherent and inducible acid tolerance, exemplify secreted agents (usually proteins) that have been recently discovered at University College London, and that also affect alkali tolerance, thermotolerance and other stress tolerances. There are almost certainly similar secreted components that induce responses not related to stress, and this may transform our ideas on inducible processes.

Log phase acid tolerance responses

The intention is to consider new responses, with respect to conditions for induction and properties of induced cells, and compare these with those of cells acid habituated after exposure to pH_o 5.0.

The acid habituation response

Log phase *E. coli* transferred to pH 5.0 broth becomes acid tolerant within 10–15 min at 37 °C. Protons cross the outer membranes via PhoE, inducing tolerance by activating a component that has a periplasmic sensor (Rowbury et al 1992). Tolerant cells are less damaged by acid and are better able to repair acid damage (Raja et al 1991). Tolerance induction is dependent on CysB but not on any of 16 other regulatory components (Rowbury 1997, Rowbury & Goodson 1997).

Novel acid tolerance responses in Escherichia coli

Many agents allow log phase cells at pH_o 7.0 to become acid tolerant (Rowbury & Goodson 1997), and we consider here mainly tolerance induction at 37 °C by glucose, L-glutamate, L-aspartate, L-proline and KCl. At neutral pH_o, log phase organisms also rapidly become acid tolerant if pre-exposed to $FeCl_3$ (1 mM), NH_4Cl (200 mM), glycerol (300 mM), L-glutamine (250 mg/l) or glucosamine (250 mg/l).

With 0.03–1% glucose, log phase cells at pH_o 7.0 became tolerant within 30 min, without any change in medium pH or growth rate. Table 1 shows the acid tolerance of glucose-exposed cells compared to that of control cells. Glucose-exposed cells were also more resistant to acid-induced DNA and protein damage than uninduced cells. Induction by glucose was most active at pH_o 7.0, but it also occurred at pH_o 7.5, 8.0 and 8.5. There are similar tolerance responses with glutamate, aspartate, KCl or L-proline (Table 1), with no pH_o change during induction. Therefore, it is not medium acidification that induces tolerance, and for the glutamate response, at least, induction also occurs at pH_o 7.5, 8.0 and 8.5. Our limited study of regulation of glucose- and L-glutamate-induced tolerance shows that CysB, H-NS and integration host factor (IHF), which function in many pH responses (Shi et al 1993, Shi & Bennett 1994, Hassani et al 1995), are not involved in induction, nor are ferric uptake regulator (Fur) or RelA. Glucose, glutamate, aspartate and KCl probably induce new pH homeostatic processes. Thus, tolerance in each type of induced cell was abolished by adding weak acids. In connection with L-glutamate-induced acid tolerance, we note that a mutant lacking L-glutamate decarboxylase does not induce the response.

Subversion of acid habituation and other acid tolerance responses

Several metabolites, e.g. NaCl, sucrose, SDS and DOC, prevent organisms becoming acid tolerant at pH_o 5.0 (Rowbury & Goodson 1997), or at pH_o 7.0 in the presence of glucose or other inducers (Table 2). Also, cAMP, which reduces

TABLE 1 Effect of cAMP on induction of novel acid tolerance responses at pH_o 7.0

	Acid tolerance for cells induced:[a]	
Inducer of acid tolerance at pH_o 7.0	*Minus cAMP*	*Plus 3 mM cAMP*
None	0.5 ± 0.06	0.77 ± 0.14
Glucose (1%)	41.3 ± 4.3	6.5 ± 1.8
L-glutamate (250 mg/l)	29.9 ± 1.8	2.2 ± 0.31
L-aspartate (250 mg/l)	11.7 ± 1.6	0.66 ± 0.24
L-proline (250 mg/l)	17.7 ± 1.9	3.6 ± 0.33
$FeCl_3$ (1 mM)	28.1 ± 1.1	13.8 ± 2.8
NH_4Cl (200 mM)	13.5 ± 1.6	9.0 ± 1.6
KCl (200 mM)	13.9 ± 0.7	3.1 ± 0.11

[a]Log phase organisms of strain 1829 ColV were diluted into broth with the stated inducer (± cAMP) and incubated for 60 min at 37 °C. After washing, organisms were acid challenged at pH_o 3.0 for 7 min; results are given as % survival ± S.E.M. for three experiments.

habituation (Rowbury & Goodson 1997), markedly inhibited tolerance induction by glucose, glutamate, aspartate, proline or KCl (Table 1), but it had little effect on the responses induced by NH_4Cl, $FeCl_3$ (Table 1) or glucosamine.

Inducible acid sensitization responses

Acid sensitivity is induced by transfer to alkaline pH_o, addition of high [NaCl] or addition of L-leucine (Rowbury 1997). Sensitization by L-leucine is absent from *ompA* mutants lacking the OmpA protein (OmpAp) but occurs in some that make abnormal OmpAp. In these, we find that the response is apparently lost only if acidic amino acids at the surface (surface loops are at *c.* 25, 70, 110 and 155) are replaced by neutral or basic ones, or are deleted. For example, Glu68→Gly, Glu68→Lys or deletion of Asp105 abolishes acid sensitization caused by L-leucine, whereas Gly28→Val, Gly70→Asp and Val110→Asp do not affect such sensitization. OmpAp appears to form a pore in the presence of L-leucine and loss of acidic amino acids at the surface may abolish pore activity.

TABLE 2 Agents that counteract acid tolerance induced by glucose and other metabolites

	Effect of potential inhibitor on acid tolerance induction by:			
Potential induction inhibitor	*Glucose*	*L-glutamate*	*L-aspartate*	*$FeCl_3$*
NaCl	Tolerance abolished	Tolerance abolished	Tolerance abolished	Tolerance abolished
Sucrose	Tolerance abolished	Tolerance abolished	Tolerance abolished	Tolerance abolished
Ethanol	Slight inhibition	Slight inhibition	No effect	Slight inhibition
Urea	No effect	No effect	Marked inhibition	Slight inhibition
Phosphate	Marked inhibition	Marked inhibition	Slight inhibition	Slight inhibition
SDS	Tolerance abolished	Tolerance abolished	Marked inhibition	Tolerance abolished
DOC	Tolerance abolished	Tolerance abolished	No effect	Slight inhibition

Organisms of strain 1829 ColV were grown to log phase in pH 7.0 broth and then transferred to fresh broth with the stated inducer of acid tolerance and, where appropriate, the potential induction inhibitor. After 60 min at 37 °C, organisms were washed and then acid challenged at pH_o 3.0 for 7 min.

Extracellular proteins and other components involved in acid tolerance

Inducible bacterial processes, although controlled by numerous mechanisms, have in common that external stimuli interact with intracellular sensors which, by a series of intracellular reactions, lead to enhancement of transcription and/or translation, thus producing the response. An exception occurs for some Gram-negative bacteria in association with higher organisms. These bacteria initially produce extracellular agents, usually N-acyl-L-homoserine lactones (AHLs), which subsequently induce the responses. These are not normal inducible responses because there is no clear initial stimulus, except for a build-up of cell mass that leads to a build-up of autoinducer. Here we describe inducible responses with defined stimuli where some induction components are clearly extracellular.

Inducible ATRs have been considered to behave like other inductions with all stages and regulatory components being intracellular. However, because of how some stress exposures occur, we examined whether two-stage inductions might be involved. In water, effluent releases of acidity are periodic with acidification periods being followed by periods of higher pH before the cycle is repeated. It might be advantageous if organisms surviving acidity produced extracellular agents that converted unchallenged organisms to tolerance, allowing all the latter to survive the next acidification cycle.

For several pH responses, we have established that induction is associated with secretion of agents (usually proteins) which themselves induce the response under normally non-inducing conditions (Hussain et al 1998, Rowbury & Goodson 1998). It is likely that some or all of the original responses are dependent on these extracellular agents, i.e. are two-stage inductions.

Novel acid tolerance induction by a protein or proteins accumulated in acidic media

Neutralized filtrates from *E. coli* grown at pH_o 5.0 induced acid tolerance if added to log phase organisms at pH_o 7.0; filtrates from pH_o 7.0-grown cultures were ineffective. Filtrates from cultures incubated for 60–90 min at pH_o 5.0 induced tolerance in pH_o 7.0-grown cells after incubation for 30 min at 37 °C (Fig. 1).

We have studied: (1) the stability, properties and nature of active filtrate component(s); (2) the conditions for synthesis of active agent(s); and (3) the conditions for acid tolerance induction by the active agent(s). The agent(s) in medium filtrates is stable, i.e. can be stored overnight at 4 °C or frozen for a week without losing activity. It is also heat stable, retaining appreciable activity at 75 °C but losing all activity on boiling. The efficacy of pH_o 5.0 filtrates was lost after protease treatment, indicating that a heat-stable protein (or proteins) is essential for tolerance induction (Table 3). Other large molecules could be involved but

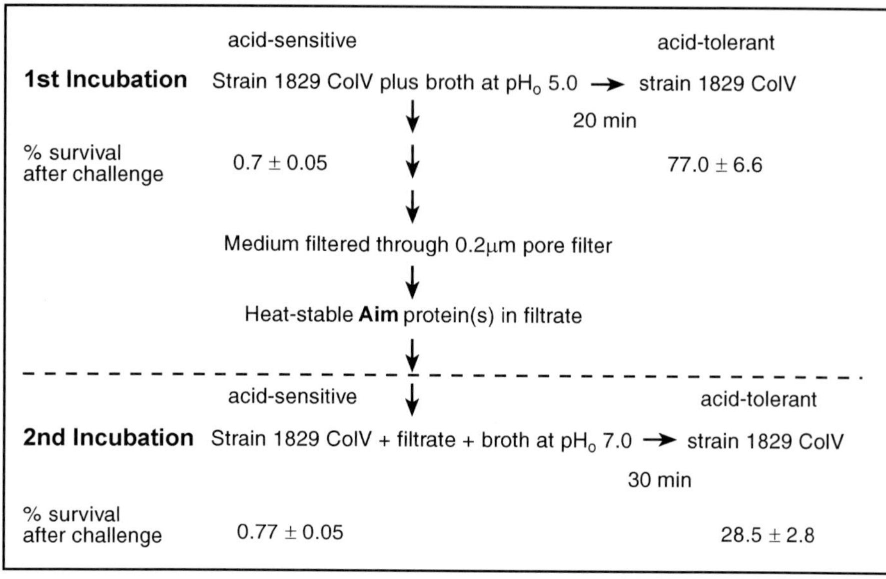

FIG. 1. Acid tolerance induced by filtrates from organisms habituating at pH_o 5.0. Strain 1829 ColV was transferred to pH_o 5.0 (at 37°C) and incubated further (1st incubation). After 20 min, a sample was removed and after washing, acid tolerance was assessed (using challenge at pH_o 3.0 for 7 min). Incubation was continued at pH_o 5.0 and after 90 min, a filtrate was prepared by filtration through a 0.2 μm pore size filter. The filtrate was neutralized and then added (one part filtrate to three parts culture, 2nd incubation) to log phase pH_o 7.0-grown strain 1829 ColV, with subsequent incubation at 37°C for 30 min at pH_o 7.0. Cells were then washed and challenged (pH_o 3.0 for 7 min). Results are shown as the mean per cent survival $\pm$ S.E.M. for three experiments.

small ones could not, as dialysis of filtrates had no effect. We have termed the filtrate induction proteins, the acid-induced medium protein(s), or Aim protein(s).

Phosphate and cAMP inhibit acid habituation (Rowbury et al 1992, Rowbury & Goodson 1997), so we tested whether either inhibits synthesis of active component(s); both were markedly inhibitory. Phosphate and cAMP were also tested on tolerance induction in pH_o 7.0 cells by filtrate component(s), as was bicarbonate. The cyclic nucleotide and bicarbonate prevented tolerance induction; phosphate did not. We also found that the synthesis of the Aim protein(s) was CysB dependent.

Dissecting the pH_o 5.0 acid habituation process

Acid habituation is dependent on CysB and inhibited by phosphate and cAMP. The synthesis/utilization of the extracellular components showed the same

TABLE 3 Extracellular components inducing acid tolerance

Source of medium filtrate inducing acid tolerance	*Filtrate protease treated prior to incubation with sensitive cells*	*Acid tolerance of pH_o 7.0-grown cells after exposure to medium filtrate*[a]
None	NA	0.5
pH_o 7.0[b]	No	0.1
pH_o 5.0[b]	No	28.8
pH_o 5.0[b]	Yes	0.8
pH_o 7.0[c]	No	0.5
pH_o 7.0 plus glucose[c]	No	21.6
pH_o 7.0 plus glucose[c]	Yes	11.1
pH_o 7.0 plus L-glutamate[c]	No	36.1
pH_o 7.0 plus L-aspartate[c]	No	13.1
pH_o 7.0 plus L-aspartate[c]	Yes	2.5
pH_o 7.0 plus $FeCl_3$[c]	No	12.0
pH_o 7.0 plus $FeCl_3$[c]	Yes	14.2

[a]Per cent survival after 7 min at pH_o 3.0.
[b]Log phase organisms of strain 1829 ColV were incubated at 37°C in pH 7.0 or 5.0 broth for 90 min, and cultures were filtered through 0.2 μm pore size filters. Filtrates were neutralized and stored at 4°C. Where stated, filtrates were incubated for 1 h at 30°C with 2 U/ml Sigma P4531 protease, followed by removal of excess protease by filtration through glass fibre filters. Filtrates were added to log phase pH_o 7.0-grown 1829 ColV cultures (1 part filtrate to three parts culture) and mixtures incubated for 30 min at 37°C prior to washing cultures and acid challenge.
[c]Log phase organisms of strain 1829 ColV were incubated at 37°C in pH 7.0 broth plus or minus inducer for 120 min, and cultures were filtered through 0.2 μm pore size filters. Filtrates were neutralized and dialysed (against 0.065 M phosphate buffer pH 7.0) overnight at 4°C. Where stated, filtrates were incubated for 1 h at 30°C with 2 U/ml Sigma P4531 protease, followed by removal of excess protease by filtration through glass fibre filters. Filtrates were added to log phase pH_o 7.0-grown 1829 ColV cultures at one part filtrate to one part culture, and mixtures incubated for 45 min at 37°C prior to washing cultures and acid challenge.
NA, not applicable.

dependencies and inhibitions, so it is likely that habituation (which is shown in Fig. 1 as a single stage induction process under '1st incubation', with induction of tolerance by the Aim protein(s) shown as the '2nd incubation') is actually a two-stage induction involving the synthesis of the Aim protein(s) at stage one and the interaction of Aim protein(s) with cells producing acid tolerance at stage two. This is supported by several lines of evidence (Table 4), e.g. the response is inhibited if extracellular component(s) are destroyed with protease during habituation, if they are removed by conducting habituation in a vessel that allows proteins to diffuse out (Table 5) or if both procedures are conducted together.

TABLE 4 Comparison of acid habituation at pH_0 5.0 and the two-stage acid tolerance induction process involving the extracellular components that are formed at pH_0 5.0

	Effect of inhibitory condition on:	
Inhibitory condition affecting acid tolerance induction	*Acid habituation at pH_0 5.0*	*Two-stage induction of acid tolerance*
Presence of phosphate	Inhibits	Inhibits stage one
Presence of cAMP	Inhibits	Inhibits both stages
Presence of protease	Inhibits	Destroys component(s) formed in stage one
Absence of CysB	Inhibits	Inhibits stage one
Lesion in strain 1157	Slows down habituation	Slow extracellular component synthesis in stage one
Filtration through 0.2 μm pore membrane filter	Inhibits	Removes components formed in stage one
Filtration and presence of protease	Strongly inhibits	Removes/destroys components formed in stage one

Can medium component(s) formed at acidic pH_0 induce alkali sensitivity?

Organisms shifted to acidic pH_0 become acid tolerant but also, at pH_0 5.5–6.0, become alkali sensitive by a distinct process (Rowbury 1997). We now find that neutralized filtrates from pH_0 5.5 cells sensitize pH_0 7.0-grown organisms to alkali. The component(s) needed for alkali sensitization is probably different to that involved in acid tolerance induction, because activity in filtrates was protease sensitive but resistant to boiling, implying that it is a heat-stable protein. Also, low molecular weight components are needed for induction because dialysis inactivated filtrates. Protease prevents alkali sensitization at pH_0 5.5, suggesting that switching on of the original sensitization response is dependent on functioning of the heat-stable extracellular protein.

Acid tolerance induced by glucose and other agents: dependence on extracellular components

Log phase organisms become tolerant at neutral pH_0 plus glucose, etc. We now find that filtrates prepared from organisms incubated at pH_0 7.0 with glucose, L-glutamate, L-aspartate or $FeCl_3$ (followed by dialysis of filtrate to remove glucose, etc.) induce acid tolerance on addition to log phase pH_0 7.0-grown organisms. Proteins in the filtrates are needed for the effects of these inducers;

TABLE 5 Effects of protease and filtration on acid habituation

Incubation pH[a]	*Protease*	*Vessel type*	*Significance of incubation*	*Per cent survival after challenge*[b]
7.0	No	Single compartment	Negative control. Non-habituated	0.5
5.0	No	Single compartment	Positive control. Acid habituated	62.8
5.0	Yes	Single compartment	Protease added to destroy extracellular induction proteins	6.0
5.0	No	Two compartments. 1829 ColV separated from pH 5.0 broth by a 0.2 μm pore size filter	Second compartment to filter off extracellular induction components	6.3

[a]Log phase strain 1829 ColV was incubated at 37°C under the stated conditions for 20 min at pH_o 7.0 or pH_o 5.0 (with or without 2 U/ml Sigma P4531 protease) and organisms were then washed and acid challenged.
[b]pH_o 3.0 for 7 min.

preliminary studies suggest that $FeCl_3$ may be an exception (Tables 3 & 6). The original responses appearing at pH_o 7.0 in the presence of glucose, L-glutamate, etc. need these filtrate components for induction.

Medium filtrates modify inherently acid-tolerant or acid-sensitive strains

Some *E. coli* derivatives show exceptional acid tolerance or sensitivity. We now find that pH_o 7.0 filtrates from sensitive strain 1829 sensitize pH_o 7.0 cultures of tolerant strain 1157. Protease did not affect such filtrates but they might contain protease inhibitor. Filtrates of another acid-tolerant strain NK1 (*cysB*) induced tolerance in *cysB*$^+$ strains (Table 6).

Induction of medium components that enhance acid sensitivity

E. coli undergoes acid sensitivity induction (ASI) on transfer to alkaline media ($pH_o > 8.4$), with some strains (e.g. strain 1157) showing intense responses (Rowbury 1997). ASI involves two sensitization components, one is probably independent of protein synthesis (component one), whereas the other is protein synthesis dependent. Here we note that strain 1157 accumulates extracellular components at pH_o 9.0 which sensitize pH_o 7.0 cultures to acid. Such sensitization was partially dependent on a protein(s) in filtrates, i.e. sensitization

TABLE 6 Stress tolerance and sensitization responses induced by extracellular components contained in medium filtrates

Conditions for synthesis of extracellular (filtrate) induction components	*Response induced by extracellular component on log phase organisms at pH_o 7.0*	*Properties and/or nature of extracellular components*
Filtrates inducing acid tolerance/sensitization		
pH_o 5.0	Acid tolerance	Heat-stable protein
Glucose at pH_o 7.0	Acid tolerance	Protein[a]
L-glutamate at pH_o 7.0	Acid tolerance	Protein[a]
L-aspartate at pH_o 7.0	Acid tolerance	Protein[a]
$FeCl_3$	Acid tolerance	Large component. May not be a protein
pH_o 9.0	Acid sensitization[b]	Non-protein component
pH_o 9.0	Acid sensitization[c]	Protein[a]
Filtrates inducing alkali tolerance/sensitization		
pH_o 5.5	Alkali sensitization	Heat-stable protein plus low molecular weight compound
pH_o 9.0	Alkali tolerance	Protein[a] plus another component
Filtrates inducing inherent acid tolerance/sensitivity		
Strain 1829 at pH_o 7.0	Acid sensitivity in inherently acid-tolerant strain (1157)	Not determined
Strain NK1 (*cysB*) at pH_o 7.0	Acid tolerance in inherently acid-sensitive strain (JA199)	Not determined

[a]Properties of extracellular protein not yet tested.
[b]Acid sensitization component one (probably analogous to acid sensitivity induction component one).
[c]Acid sensitization component two (probably analogous to acid sensitivity induction component two).
Log phase organisms of various strains — 1829 ColV (for filtrates from organisms grown at pH_o 5.0 or at pH_o 7.0 plus metabolites); 1829 (for filtrates from pH_o 5.5-grown organisms being examined for alkali sensitization and pH_o 9.0-grown organisms being studied for alkali tolerance); 1157 (for studies of acid sensitivity induction); and NK1 grown as indicated above—were used to prepare filtrates that were neutralized and stored overnight or, where necessary, neutralized and then dialyzed overnight (see Table 3). Filtrates were then added to pH_o 7.0-grown strain 1829 ColV (for all acid tolerance studies), to pH_o 7.0-grown 1157 (for studies of acid sensitization and of inherent acid sensitivity), to pH_o 7.0-grown strain 1829 (for examining alkali tolerance changes) or to pH_o 7.0-grown JA199 (see above), with incubation in all cases at pH_o 7.0, followed by washing of cells and acid or alkali tolerance assessment. In all cases it has been assumed that a single protein induction component or other filtrate component functions; multiple components cannot yet be ruled out, however.

was partially prevented by protease treatment of filtrates. Accordingly, the second ASI sensitization component may appear due to the functioning of an extracellular protein, in contrast to the first component which depends on a non-protein extracellular component (Fig. 2). Interestingly, neutralized filtrates from pH_o 9.0-grown organisms (strain 1829) induced alkali tolerance in pH_o 7.0-grown cells and protease partially reversed induction (Table 6).

Why do stressed organisms produce extracellular induction components?

Extracellular induction components are probably produced to allow organisms surviving stress exposures to convert to stress tolerance others nearby that have not been challenged, permitting all the latter to survive if subsequently challenged. This would benefit populations facing periodic exposures to extremes of pH, heat or cold, or to extreme chemical challenges.

Conclusions and future work

This study revealed new acid tolerance responses induced at pH_o 7.0 in log phase; other novel acid tolerance responses may occur under other environmental

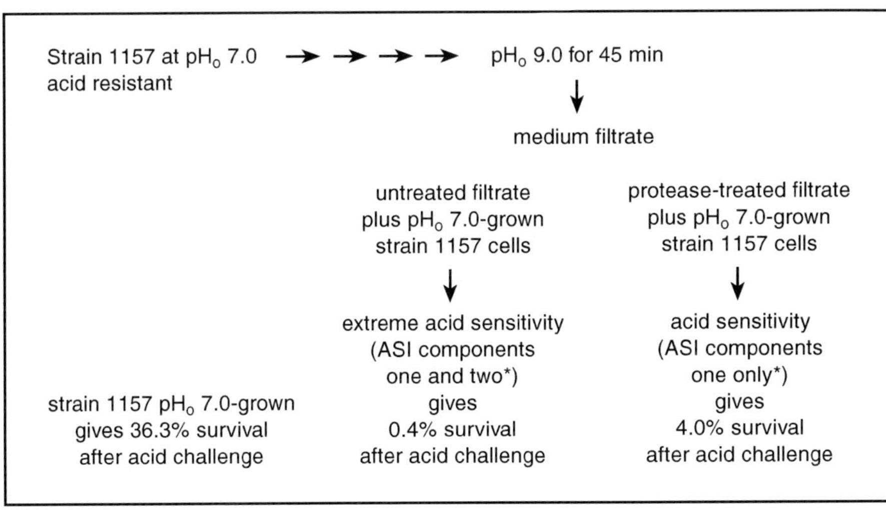

FIG. 2. Two filtrate components involved in acid sensitization. Log phase strain 1157 was transferred from pH_o 7.0 to pH_o 9.0 and incubated for 45 min prior to preparing medium filtrates (see Fig. 1 legend). Filtrates were neutralized and, where stated, treated with protease (2 U/ml Sigma P4531 protease for 1 h at 30°C). Filtrates were then incubated (in a 1:1 ratio) with pH_o 7.0-grown log phase 1157 cells at pH_o 7.0 for 45 min at 37°C, prior to washing and acid challenge (pH_o 3.0 for 5 min). ASI, acid sensitivity induction.

conditions. Encouragingly, it proved possible to counteract acid tolerance induced as above by adding agents such as NaCl, sucrose, SDS and DOC. We note also that NH_4Cl induces novel alkali tolerance in log phase organisms at pH_o 7.0.

We have found that many stress responses are associated with secretion of extracellular components (generally proteins, Table 6) which induce the corresponding responses, these agents inducing only phenotypic changes in exposed organisms. We propose a detailed examination of: (a) how these components are synthesized; (b) how they interact with uninduced organisms; and (c) whether all established stress responses need extracellular agents for induction. These stress responses are regulated quite differently than supposedly classical inducible responses (Fig. 3) or quorum-induced responses. We propose to

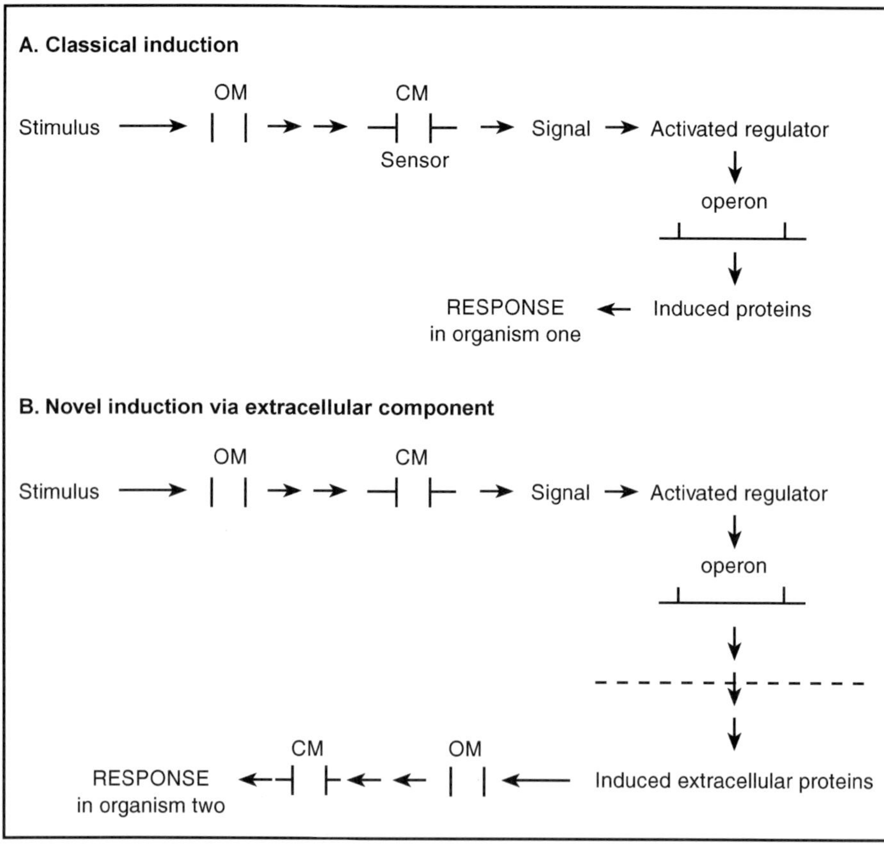

FIG. 3. Novel inducible processes involving extracellular components. This shows, in general terms, the widely held view of how inducible processes take place (A) and the novel mechanism, involving obligate extracellular intermediate components, now established in this work for many responses (B). CM, cytoplasmic membrane; OM, outer membrane.

test whether other inductions, e.g. some inducible enzyme syntheses, involve functioning of extracellular induction components; if so, this could transform our views on regulation of inducible processes.

Other proposals:

(1) We will: (a) examine whether thermotolerance and alkylhydroperoxide reductase synthesis induced at alkaline pH_o need extracellular induction components; and (b) test whether responses to metal ions, u.v. irradiation, cold, starvation and osmotic stress involve extracellular induction components.
(2) We will study synthesis and functioning of extracellular components which induce alkali tolerance in pH_o 7.0 cells.
(3) Strains lacking RpoS show stress sensitivity; we will investigate whether tolerance of *rpoS*$^+$ strains involves extracellular stress induction components.
(4) Plasmid-encoded properties, e.g. ColV$^+$ strains, are more inhibitor sensitive; we will examine whether extracellular components produced by p^+ or p^- organisms influence sensitivities. Low temperature abolishes some ColV properties; we will test whether filtrates from 25 °C-grown organisms influence properties of 37 °C-grown ones or vice versa. Other temperature responses (e.g. flagellar synthesis in *Salmonella* spp) may be associated with secretion of extracellular components.
(5) We propose to study for inducible antibiotic resistance whether extracellular components are secreted during induction which confer antibiotic resistance on organisms in absence of antibiotics.

Acknowledgements

The author is grateful to N. H. Hussain and Margaret Goodson for dedicated experimental work and to T. J. Humphrey for advice and encouragement.

References

Cooper GC, Rowbury RJ 1986 Virulence plasmid-associated sensitivity to acid in *Escherichia coli* and its possible significance in human infections. J Med Microbiol 22:231–236

Foster JW, Hall HK 1990 Adaptive acidification tolerance response of *Salmonella typhimurium*. J Bacteriol 172:771–778

Hassani M, Bennett GN, Hirshfield IN 1995 The effects of *cysB*, *himA* and *himD* mutations on the induction of *lysU* and other acid-inducible genes in *Escherichia coli*. Abstr Am Soc Microbiol H158

Humphrey TJ, Richardson NP, Statton, KM, Rowbury RJ 1993 Effects of temperature shifts on acid and heat tolerance in *Salmonella enteritidis* phage type 4. App Environ Microbiol 59:3120–3122

Hussain NH, Goodson M, Rowbury RJ 1998 Intercellular communication and quorum sensing in microorganisms. Sci Prog 81:69–80

Karem KL, Foster JW, Bej AK 1994 Adaptive acid tolerance response (ATR) in *Aeromonas hydrophila*. Microbiology 140:1731–1736

Kroll RG, Patchett RA 1992 Induced acid tolerance in *Listeria monocytogenes*. Letts Appl Microbiol 14:224–227

Lee IS, Slonczewski JL, Foster JW 1994 A low pH-inducible, stationary-phase acid tolerance response in *Salmonella typhimurium*. J Bacteriol 176:1422–1426

O'Hara GW, Glenn AR 1994 The adaptive acid tolerance response in root nodule bacteria and *Escherichia coli*. Arch Microbiol 161:286–292

Raja N, Goodson M, Smith DG, Rowbury RJ 1991 Decreased DNA damage and increased repair of acid-damaged DNA in acid-habituated *Escherichia coli*. J Appl Bacteriol 70:507–511

Rowbury RJ 1997 Regulatory components, including integration host factor, CysB and H-NS, that influence pH responses in *Escherichia coli*. Letts Appl Microbiol 24:319–328

Rowbury RJ, Goodson M 1997 Metabolites and other agents which abolish the CysB-regulated acid tolerance induced by low pH in log-phase *Escherichia coli*. Recent Res Dev Microbiol 1: 1–12

Rowbury RJ, Goodson M 1998 Induction of acid tolerance at neutral pH in log-phase *Escherichia coli* by medium filtrates from organisms grown at acidic pH. Letts Appl Microbiol 26:447–451

Rowbury RJ, Goodson M, Whiting GC 1989 Habituation of *Escherichia coli* to acid and alkaline pH and its relevance for bacterial survival in chemically polluted natural waters. Chem Ind 1989:685–686

Rowbury RJ, Goodson M, Wallace AD 1992 The PhoE porin and transmission of the chemical stimulus for induction of acid resistance (acid habituation) in *Escherichia coli*. J Appl Bacteriol 72:233–243

Shi X, Bennett GN 1994 Effects of *rpoA* and *cysB* mutations on acid induction of biodegradative arginine decarboxylase in *Escherichia coli*. J Bacteriol 176:7017–7023

Shi X, Waasdorp BC, Bennett GN 1993 Modulation of acid-induced amino acid decarboxylase gene expression by *hns* in *Escherichia coli*. J Bacteriol 175:1182–1186

DISCUSSION

Foster: Have you tried taking these extracts and mixing them with other strains of *Escherichia coli* or *Salmonella* to see if you can get cross-feeding?

Rowbury: So far, we have only cross-fed the *hns* mutant. The *hns* mutant doesn't produce the pH 5.5 extracellular induction component (EIC), but it is possible to cross-feed it with the *hns*$^+$ strain by placing the *hns* mutant in one compartment and the *hns*$^+$ strain in another with pH 5.5 broth, and separating the two by a filter. We haven't yet done the experiment with other strains of *E. coli*, but it would be interesting because in the natural environment many organisms won't be related to the one producing the EIC.

Foster: Have you purified the protein?

Rowbury: No. We've been concentrating on observing the responses to prove whether the EICs occur for only a few unusual responses or whether these extracellular components are commonly involved in inducible response mechanisms; the latter is clearly the case. We will now purify the EICs. It will be interesting to know how many proteins synthesized at acidic pH are in the medium, because people have tended to look at whole-cell proteins.

Konings: Do you have any idea how those proteins can exert their action?

Rowbury: No. If you compare acid-tolerant *E. coli* cells with cells that are not tolerant, you find they are less damaged by acid and they're better able to repair acid damage to DNA. We assume that the protein complex which interacts with the DNA switches on both of these properties.

Matin: I find this phenomenon intriguing. A certain concentration of this protein in the external medium has to be reached before the culture can develop resistance, so why is it necessary for the cell to have a quorum-sensing mechanism before it develops resistance?

Rowbury: The argument we have been following is that in the natural environment the organisms which survive lethal acidity from effluents, for example, may produce a compound which may confer tolerance on other organisms in the vicinity, so next time they are exposed to acid they will be able to survive. Alternatively, some organisms may sense the stress conditions at low levels (i.e. act as 'watchmen' for the population) and produce EICs, switching on a response in the other organisms in anticipation of the stress levels rising.

Poole: Perhaps we shouldn't confuse this too quickly with quorum sensing. I understood from Robin Rowbury's presentation that there wasn't a firm relationship between the amount of any particular component and a particular number of cells, i.e. this is a semi-quantitative phenomenon at present.

Matin: But is there a threshold concentration?

Rowbury: One of the reasons why fewer of the cells become tolerant when the filtrate is used is that when we induce tolerance at pH 5.5 to all intents and purposes we use 100% filtrate, i.e. 100% medium. When we make the filtrate and add it to cells, we normally use one part filtrate to three parts cells, and we have titrated it and shown that one part in 20 gives no effect, whereas one in six has a dramatic effect.

Matin: Is more than one protein involved?

Rowbury: We don't know this yet.

Schäfer: Is the protein newly synthesized or just released?

Rowbury: It's newly synthesized because the synthesis of the EICs and the acid tolerance response (ATR) at pH 5.5 can be blocked by inhibitors such as cAMP and phosphate. In addition, CysB, the rather intriguing regulatory component, is needed both for acid tolerance induction and for the production of the extracellular components.

Matin: You said the phenomenon was inhibited by cAMP. *E. coli* excretes 99.9% of the cAMP that it makes out into the external medium. Wouldn't this result in an antagonistic situation?

Rowbury: It could, although I was using high concentrations of cAMP, in the millimolar range.

Matin: But high concentrations would accumulate in the medium. In a natural environment, unless you're thinking of a biofilm, the secreted products of an isolated cell will be removed, and that would be true of your protein as well as cAMP.

Rowbury: I agree. The phenomenon may occur in biofilms and in the natural environment; this could be important.

Glenn: Does the protein have to be secreted in order for it to have its effect?

Rowbury: It is possible to destroy it with protease at pH 5, so a large proportion of it must be secreted or else the cells would induce tolerance. However, it's difficult to say whether it all has to be secreted.

Glenn: There are *E. coli* mutants defective in secretion, which might give you a handle on trying to address this.

Booth: E. coli does not generally secrete proteins across the outer membrane. You used a ColV strain, and most of these strains produce an immunity protein that binds to a receptor and prevents colicin from re-entering and killing the cell. It is possible, therefore, that you are dealing with a perturbation of the association between the immunity protein and the receptor, i.e. when the pH is shifted, the balance between colicin, available receptor and immunity protein changes. If the immunity protein did dissociate, and it is a fairly heat-stable protein, when you add it back to the filtrate you shift the equilibrium back, and you may be looking at a phenomenon specific to strains that secrete colicins. Have you done this experiment in other strains?

Rowbury: Yes, we have. The CysB work was done with a strain that isn't ColV$^+$ and the 1829 strain, which also doesn't have ColV, produces EICs as well. The second phenomenon I talked about involved work done with 1829, and all the mutants were isolated from 1829. Therefore, it is unlikely that ColV is involved.

Poole: Have you looked at the possibility that these proteins are not so much secreted but rather released? Have you looked for the presence of marker enzymes of the periplasm and the cytoplasm in these media to confirm that the cells are not lysing and releasing a factor?

Rowbury: This is a possibility, except that it is required for the original response at pH 5.5, which suggests to me that this isn't the case.

Park: Does the filtrate confer cross-resistance to other environmental stresses? The reason I ask this is because we see a similar phenomenon during the inactivation of *E. coli* by high hydrostatic pressure. If we spin down stationary phase cells, and re-suspend the exponential cells in the stationary phase supernatant we can confer resistance to high hydrostatic pressure.

Rowbury: There are some cross-responses. For example, at pH 9 both alkali tolerance and acid sensitivity are induced. As for the basis for the responses, I should note that after we preincubate the cells with the filtrate, we wash the cells to remove the filtrate. Sometimes we add protease to make sure we have removed everything from the filtrate. I emphasize that we are not looking at protection against stress by the extracellular components, rather an induction of tolerance or sensitivity. We also know that some EICs are formed as precursors by non-stressed cells, and the precursors act as stress sensors, being converted to EICs by the stress in the absence of organisms. Such extracellular stress sensors may be common to all stress responses, and such sensing in the growth medium may also occur for other inducible organisms.

Park: We also find that supernatants from *E. coli* O157 strains have more significant effects than supernatants from any other *E. coli* strains.

Rowbury: That's fascinating. Tom Humphrey and myself also looked at the inherent differences in acid tolerance, heat tolerance, tolerance to hydrogen peroxide and in the ability to survive on surfaces between various strains of *Salmonella*. We found that there were large differences between the strains, and the ones which survived best on the surfaces and were more inherently tolerant were also more virulent. We didn't show that EICs were involved in that, although I suspect they were, so we should look at that. I suspect that EICs from inherently virulent strains will be able to confer increased virulence on other organisms, the effect being only phenotypic of course.

Poole: How do you define per cent survival, and why is it not much closer to 100% in any of the situations that you showed?

Rowbury: It depends on how long we incubate the cells for tolerance or for sensitization, and it also depends on what challenge we use. In the case of acid tolerance, if we incubate for 30 min at pH 5.0 or 5.5 we get about 90% survival on subsequent challenge, but we generally only incubate for 15 min, which gives 60–80% survival. We have done various pH/time combinations in various strains over the years, and we find that tolerant strains are better able to resist every type of acid challenge.

Poole: Of what figure is the survival a percentage? What is 100% survival?

Rowbury: 100% refers to all the cells that are alive at the start of the experiment.

Slonczewski: For the purposes of comparing data, it's helpful to understand the differences between how people define the acid survival response. What we are really talking about is death rates, i.e. there is a negative exponential function of death rate no matter what system you're looking at. For example, if you're talking about 43% survival after 15 min, after 2 h, it would not come out as significantly resistant by the definition of several other researchers in this field (10% survival after 2 h). Were all the exposure times in the work you presented today the same?

Rowbury: Yes. For the acid tolerance experiments they were pH 3.0 for 7 min, and for the alkali tolerance experiments they were pH 9.75 for 30 min.

Glenn: I was wondering whether behind Joan Slonczewski's question was a plea to agree on some standard approaches to the assays, so that we can compare data from lab to lab.

Slonczewski: Equally one could make the opposite argument that people should look at different exposures because it's difficult to know what's right. My point is that exposure times should always be stated clearly. Also, one needs to indicate the range of error in one's observations of survival, which can be large.

Epstein: If under stress the cells are dying at an exponential rate, it doesn't matter at what time you measure survival because the log of the fractional survival will be a linear function of time.

Cook: I have a question on the alkali sensitization response induced by the filtrate. You showed that it is a low molecular weight component which is not a protein. Could it be a siderophore, and if so what is the effect of a *fur* mutant on the production of this component?

Rowbury: The *fur* mutation knocks out the production of the pH 5.5-induced extracellular component, but I'm not sure whether we have distinguished between the protein and the non-protein component. This is why we're doing the cross-feeding experiments. In any case, it is still possible that there is only one component produced as pH 5.5, namely a low molecular weight protein (i.e. rather than a low molecular weight non-protein component and a larger protein).

Schäfer: If I understand you correctly, the cells are not growing in these experiments. If they acquire tolerance, would you expect them to grow and continue to produce these factors?

Rowbury: It takes about 15–20 min to induce acid tolerance, so they would be growing in that situation. It would be interesting to start them off at pH 5.0 until they start secreting the EIC and then put them back to pH 7 to see if they still go on making it.

Konings: I would like to ask Robin Rowbury about the proteins involved in both acid and alkali tolerance in *E. coli*. Have you managed to obtain any information on those proteins from the genomic database?

Rowbury: Not yet. We don't yet know how many of these proteins are secreted to the outside by *E. coli*. *E. coli* doesn't usually secrete many proteins, so it should be quite easy to determine which secreted proteins are involved in tolerance induction and sequence them so that we can map them on the genome. We hope to do this in the future. We have focused on establishing this principle for as many systems as we can before we start to look in detail at individual systems.

Konings: Are all the proteins secreted?

Rowbury: It is just possible that some could be produced by lysis of the organism. However, when the organisms are inducing the response they are also dependent on the EIC, so I would think that lysis is unlikely because in that case the organisms that didn't lyse would be able to induce tolerance even in the presence of protease, and they don't.

Poole: How many proteins does *E. coli* secrete?

Rowbury: It depends on the laboratory conditions. It doesn't generally secrete large numbers, unlike some pseudomonads and *Proteus* strains, but if it is carrying plasmids it will probably secrete some, so our $ColV^+$ will secrete some.

Acid tolerance in root nodule bacteria

A. R. Glenn[1], W. G. Reeve, R. P. Tiwari and M. J. Dilworth

Centre for Rhizobium Studies, School of Biological Sciences and Biotechnology, Division of Science, Murdoch University, Murdoch, Perth, Western Australia, 6150, Australia

Abstract. Biological nitrogen fixation, especially via the legume–*Rhizobium* symbiosis, is important for world agriculture. The productivity of legume crops and pastures is significantly affected by soil acidity; in some cases it is the prokaryotic partner that is pH sensitive. Growth of *Rhizobium* is adversely affected by low pH, especially in the 'acid stress zone'. Rhizobia exhibit an adaptive acid tolerance response (ATR) that is influenced by calcium concentration. Using Tn*5*-mutagenesis, *gusA* fusions and 'proteome' analysis, we have identified a range of genes that are essential for growth at low pH (such as *actA*, *actP*, *exoR*, *actR* and *actS*). At least three regulatory systems exist. The two-component sensor–regulator system, *actSR*, is essential for induction of the adaptive ATR. Two other regulatory circuits exist that are independent of ActR. One system involves the low pH-induced regulator gene, *phrR*, which may control other low pH-regulated genes. The other circuit, involving a regulator that is yet unidentified, controls the expression of a pH-regulated structural gene (*lpiA*). We have used pH-responsive *gusA* fusions to identify acid-inducible genes (such as *lpiA*), and then attempted to identify the regulators of these genes. The emerging picture is of a relatively complex set of systems that respond to external pH.

1999 Bacterial responses to pH. Wiley, Chichester (Novartis Foundation Symposium 221) p 112–130

The nitrogen-fixing symbiosis between legumes and root nodule bacteria (*Rhizobium*, *Sinorhizobium* and *Bradyrhizobium*) plays a vital role in world agriculture because it is often uneconomic to use large inputs of nitrogenous fertilizers. Biological nitrogen fixation via the legume–*Rhizobium* symbiosis is cost-effective, and avoids the leaching of nitrogen fertilizer into water bodies and their subsequent eutrophication.

Soil acidity and acidification significantly decrease the productivity of legume crops and pastures (Robson & Bottomley 1991, von Uexküll & Mutert 1995). The total area of naturally acidic soils is increasing in large areas of the world due to industrial pollution and agricultural practice.

[1]Current address: Office of the Pro Vice Chancellor, University of Tasmania, Hobart, Tasmania, 7001, Australia.

The soil acidity problem can be solved: (i) by the management of changes to industrial and agricultural practice to reduce the acidification rate; (ii) chemically, by the use of lime and gypsum to ameliorate soil pH; and (iii) biologically, by developing legume–*Rhizobium* systems that perform well under acidic conditions. This third approach, which requires the identification of acid-tolerant strains of root nodule bacteria that nodulate and fix N_2 at high rates, has been used successfully in Western Australia on large areas of acid soils with a climate suitable for medic pastures. The medic root nodule bacteria traditionally used in Australia are acid sensitive and cannot persist in such soils. This problem was overcome by isolating strains of root nodule bacteria from *Medicago* sp. growing on acid soils of the southern and eastern Mediterranean rim. Strains such as WSM419 have allowed the successful development of medic pastures on over 500 000 ha of land in Western Australia, posing two important questions for our work:

(1) how do agriculturally important root nodule bacteria respond to low pH; and
(2) how do acid-tolerant strains differ from their acid-sensitive counterparts? Do they have unique genes or more efficient homologues of genes also found in acid-sensitive strains?

The effect of pH on root nodule bacteria

As the pH falls below 7.0 there is initially little or no effect on the mean generation time of root nodule bacteria, but further decreases in pH lead eventually to rapid increases over a narrow range of 0.2–0.3 pH units (Reeve et al 1993), which is now designated the 'acid stress zone' (Glenn & Dilworth 1994, Watkin et al 1997). At still lower pH the cells fail to grow but maintain viability for prolonged periods— the 'survival zone'—and further decreases in pH causes cells to die. The more extreme the pH the more rapid the rate of cell death (Fig. 1; Y. Benggu & G.W. O'Hara, unpublished data 1998). In physiological studies of cells in the acid stress zone, therefore, precise control and maintenance of pH are required to prevent cell metabolism modifying the external pH and moving cells out of the stress zone into a rapid growth zone, or into the death zone.

Studies using ^{14}C-5,5-dimethyloxazolidine-2,4-dione, ^{14}C-butyrate or dihydrophthalonitrile show that root nodule bacteria such as WSM419 maintain a slightly alkaline intracellular pH (pH_i) of 7.2–7.5 (O'Hara et al 1989, Graham et al 1994). Acid-sensitive strains of *Sinorhizobium meliloti*, such as WSM244 or WSM301, show a significant decrease in pH_i as the external pH (pH_o) becomes acidic (O'Hara et al 1989); a pH_i of 6.5–7.0 seems to stop cell growth (O'Hara et al 1989, Perez-Galdona & Kahn 1994).

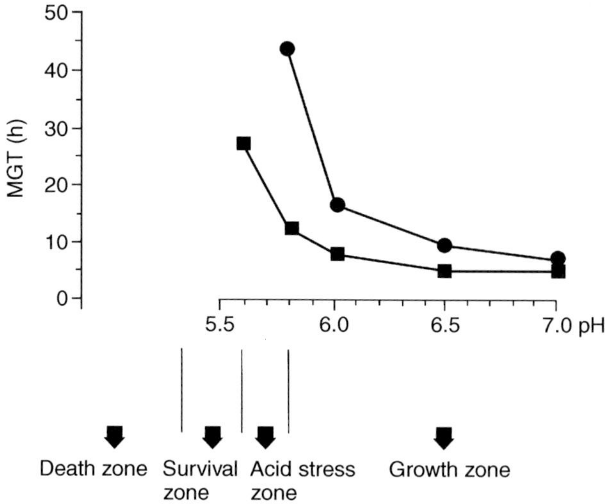

FIG. 1. Effect of calcium concentration on mean generation time (MGT) for growth of *Sinorhizobium meliloti* WSM419. Solid circles, 0.3 mM Ca^{2+}; solid squares, 3.0 mM Ca^{2+}. Higher calcium concentration extends the growth and acid stress zones to lower pH values.

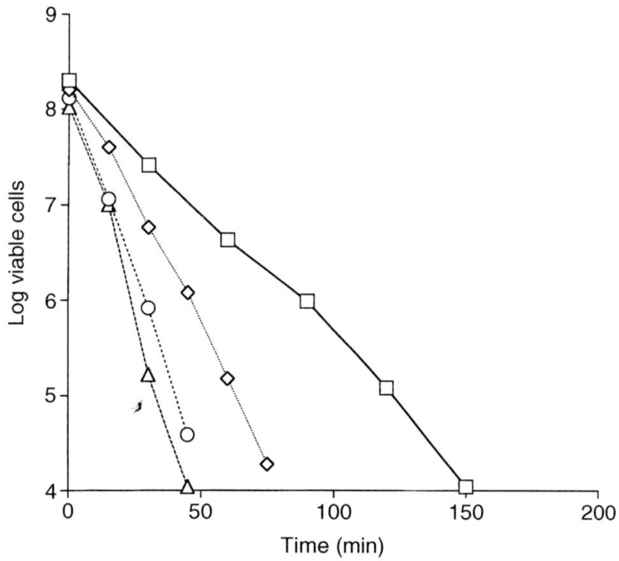

FIG. 2. Adaptive acid-tolerance response for *Sinorhizobium meliloti* WSM419 and its mutant derivatives. Cells were grown at pH 5.8 or pH 7.0 and subjected to acid shock in minimal medium at pH 3.5; the times required for numbers to fall by 90% (decimal reduction time, D) were derived from the line slopes. Squares, WSM419 (wild-type) grown at pH 5.8, D = 30 min; diamonds, WSM419 grown at pH 7.0, D = 20 min; circles, RT295 (*actS*$^-$) grown at either pH 5.8 or 7.0, D = 13 min; triangles, TG5-46 (*act*R$^-$) grown at either pH 5.8 or 7.0, D = 11 min.

The concentration of Ca^{2+} in the medium influences the effects of acidity on root nodule bacteria, stimulating the growth rate within the growth zone and shifting the acid stress zone to a more acidic region (Fig.2; Glenn & Dilworth 1994).

The effects of pH and Ca^{2+} on growth and exopolysaccharide synthesis are complex (Table 1). While increasing Ca^{2+} at pH 7.0 did not stimulate the rate of exopolysaccharide production in *S. meliloti* WSM419, it markedly increased the rate of synthesis at pH 5.8. Proton NMR studies have shown (A. Vivas, unpublished results 1998) that acidic pH decreases overall acetyl, pyruvyl and succinyl substitution of the exopolysaccharide and interacts with Ca^{2+} to affect the ratios between them. There is currently no adequate explanation for the effect(s) of Ca^{2+} on root nodule bacterial growth at low pH.

When rhizobial cells grown under mildly acidic conditions are exposed subsequently to a more severe acid shock, they survive significantly better than cells previously grown at neutral pH (O'Hara & Glenn 1994). This phenomenon, termed 'acid habituation' (Goodson & Rowbury 1989) or the acid tolerance response (ATR; Foster & Hall 1990, Foster 1991), requires new protein synthesis and has been described in a range of micro-organisms (Glenn et al 1997).

Besides influencing growth, pH and calcium affect survival of root nodule bacteria. The calcium concentration in the growth medium prior to acid shock has a small but significant effect on the subsequent rate of cell death under low pH conditions. The main decrease in death rate, however, is due to the concentration of Ca^{2+} in the acid shock medium itself (Table 2).

TABLE 1 Effects of calcium concentration and pH on rates of exopolysaccharide synthesis and on its composition

Growth conditions	*Mean generation time (h)*	*Rate of exopolysaccharide synthesis (mg/mg protein per h)*	*Degree of exopolysaccharide substitution*[a]	*Ratio of succinyl/acetyl substitution*[b]	*Ratio of acetyl/pyruvyl substitution*[b]
pH 7.0					
0.3 mM Ca^{2+}	4.0	0.139	64.1	0.23	0.94
3.0 mM Ca^{a+}	4.1	0.150	54.8	0.66	0.78
pH 5.8					
0.3 mM Ca^{2+}	42.5	0.025	28.2	1.18	0.90
3.0 mM Ca^{2+}	9.8	0.225	28.4	0.66	0.90

[a]Arbitrary ratio between integrated ^1H NMR signals for acetyl, pyruvyl and succinyl protons compared to sugar protons.

[b]Arbitrary ratios between integrated ^1H NMR signals for acetyl, pyruvyl and succinyl protons.

TABLE 2 Effects of growth pH, $CaCl_2$ concentration in the growth medium, and $CaCl_2$ concentration in the exposure medium at pH 4.0 on decimal reduction time (h) for WSM419

	Decimal reduction time[a]		
Growth medium (JMM)	*No added Ca^{2+}*	*0.3 mM Ca^{2+}*	*3 mM Ca^{2+}*
pH 7.0, 0.3 mM Ca^{2+}	0.98 ± 0.22	1.63 ± 0.40	4.67 ± 1.73
pH 7.0, 3 mM Ca^{2+}	0.93 ± 0.17	1.16 ± 0.04	6.20 ± 0.60
pH 5.8, 0.3 mM Ca^{2+}	3.15 ± 1.26	5.46 ± 1.31	9.69 ± 0.70
pH 5.8, 3 mM Ca^{2+}	1.75 ± 0.21	2.45 ± 0.20	5.97 ± 0.85

[a]The time (in hours) taken to reduce the viable number by 90%.
The effects are significant at the levels indicated: growth pH, $p<0.01$; calcium concentration in growth medium, $p<0.05$; calcium concentration in exposure medium, $p<0.001$; growth pH X calcium concentration in growth medium, $p<0.05$. Values given are means of four observations ± S.D.

Approaches to the problem

How can we develop an understanding of the effect of acidity on root nodule bacteria? In this chapter we outline approaches to analyse the responses of *S. meliloti* and *Rhizobium leguminosarum* bv. *viciae* to low pH. Genes in these bacteria associated with low pH are listed in Table 3.

Mutagenesis

Transposon Tn*5* mutagenesis of wild-type strains such as *S. meliloti* WSM419 and R. *leguminosarum* bv. *viciae* WSM710 has been used to generate acid-sensitive derivatives (Goss et al 1990, Tiwari et al 1992). For these mutants to be useful in subsequent studies a number of criteria must be satisfied. Firstly, the growth rates

TABLE 3 Genes in root nodule bacteria involved with response to low pH

Gene	*Phenotype*	*Function*
actA	Acid sensitive	Apoprotein N-acyltransferase
actP	Acid sensitive	P-type ATPase
*act*R	Acid sensitive	Regulator protein
actS	Acid sensitive	Sensor protein
exoH	Acid sensitive	Succinylation of exopolysaccharide
*exo*R	Mildly acid sensitive	Regulator of exopolysaccharide synthesis
lpiA	Low pH induced	Sulfate permease
*phr*R	Low pH induced	Regulator protein

of such mutants at neutral pH must be essentially identical to those of the wild-type. Secondly, the acid sensitivity must be shown by transduction, complementation or reconstruction to be due to the insertion of Tn*5* at a particular locus. Cloning and sequencing can then identify the gene essential for growth at low pH.

Acid-inducible/repressible genes

The use of *lacZ*, or preferably *gusA*, fusions in root nodule bacteria enables the identification of promoters that are induced or repressed by low pH_o. Such genes may not be required for growth at low pH but may be required for survival at still lower pH.

Proteome analysis

Contemporary 2D PAGE techniques and subsequent computer analysis of the separated proteins (proteome analysis) can be applied to protein preparations derived from cells grown under neutral and acidic conditions (Hall et al 1995).

Genes essential for growth under acidic conditions

Structural genes

Transposon Tn*5* mutagenesis of *S. meliloti* WSM419 or *R. leguminosarum* bv. *trifolii* WSM710 has generated a number of acid-sensitive mutants that grow like the wild-type at pH 7.0. Based on the frequency with which such mutants can be isolated there may be 20–50 genes essential for growth of root nodule bacteria at low pH.

actA. *S. meliloti* TG2-6 is a Tn*5*-induced mutant of WSM419 that fails to grow below pH 6.0 and is defective in the capacity to maintain pH_i above pH 7.0 when the pH_o is below pH 6.5 (O'Hara et al 1989). That Tn*5* insertion into *actA* was the cause of its acid sensitivity was confirmed using marker exchange, reversion and complementation studies (Tiwari et al 1996a).

The *actA* gene codes for a basic protein with a hydrophobic NH_2-terminal portion consistent with a membrane location, and a hydrophilic COOH-terminal. The ActA protein has 69% similarity to the *Escherichia coli* CutE protein (Rogers et al 1991), a homologue of an apoprotein N-acyltransferase in *Salmonella* (Gupta et al 1993). Lack of this protein results in copper sensitivity in *E. coli*; *S. meliloti* TG2-6 is also sensitive to copper and zinc. The plasmid pRT206-6, which complemented the acid-sensitive phenotype, also restored copper and zinc tolerance to TG2-6. The *CutE* mutant of *E. coli* is also zinc sensitive (data not shown). Removal of Cu^{2+} and Zn^{2+} from the rhizobial minimal salts medium failed to restore growth of TG2-6 at pH 5.6, suggesting that the sensitivity to low pH is not due to Cu^{2+} or Zn^{2+} toxicity. Introduction of a plasmid-borne *actA* gene from *S. meliloti* WSM419 did not restore copper tolerance to a *cutE*-defective mutant of *E. coli*. Constitutive expression of *actA* has been shown using *actA-lacZ* fusions and northern analysis with an intragenic probe.

actP. The need to exercise caution about the impact of metal ions such as Cu^{2+} and Zn^{2+} at low pH during studies with TG2-6 was reinforced in subsequent work on strains RT3-27 and WR1-14, acid-sensitive mutants of *S. meliloti* and *R. leguminosarum* bv. *viciae*, respectively. These fail to grow below pH 5.8 and pH 5.5, respectively, on a minimal salts medium containing 2.5 μM Cu^{2+}; if copper is removed RT3-27 and WR1-14 grow normally at low pH, thus displaying pH-dependent copper sensitivity. The disrupted genes show homology to P-type ATPases involved with Cu^{2+} transport, and have been designated *actP*. Both ActP proteins show clear ATP-binding motifs, DGINDA in *S. meliloti* and DGVNDA in *R. leguminosarum* bv. *viciae* (Glenn et al 1997). Extracts of mutant RT3-27 lack a Cu^{2+} ATPase activity that is found in the wild-type (N. Kale, unpublished data 1997).

Downstream from the *actP* gene (160 bp in *S. meliloti* WSM419 and 4 bp in *R. leguminosarum* bv. *viciae* WSM710) there is a small gene with considerable similarity to heavy-metal resistance regulators. That pH-dependent sensitivity to Cu^{2+} in the RT3-27 mutant is due to inactivation of *actP* and not to a polar effect on expression of this small gene has been established by complementation studies.

exoH/exoR. Recent work (A. Vivas, R.P. Tiwari & W.G. Reeve, unpublished data 1998) has shown that a mutant of WSM419 carrying a single Tn*5* which results in the inactivation of its *exoH* gene was both acid and Zn^{2+} sensitive, and produced non-succinylated exopolysaccharide. Complementation with a plasmid carrying 20 kb of *exo* and other genes from *S. meliloti* Rm1021 restored wild-type exopolysaccharide synthesis, as well as acid and Zn^{2+} tolerance (A. Vivas & R. P. Tiwari, unpublished data 1998). Subcloning and complementation will be needed to identify unequivocally the gene responsible for the acid tolerance. Mild acid sensitivity in *R. leguminosarum* bv. *viciae* WR6-35 is associated with over-production of exopolysaccharide caused by a defective regulatory gene (*exoR*) (Reeve et al 1997a).

Regulatory genes

In prokaryotes gene expression in response to environmental signals is generally regulated by two-component sensor–regulator systems.

The actS-actR system. A contiguous sensor (*actS*)–regulator (*actR*) gene pair essential for growth at low pH has been described in *S. meliloti* WSM419 (Tiwari et al 1996b). Strains TG5-46 (*actR*$^-$) and RT295 (*actS*$^-$) are acid-sensitive mutants of WSM419 unable to grow below pH 6.0. ActS is a putative membrane-bound protein, the hydrophobicity profile of which, from the Kyte–Doolittle algorithm, predicts five strongly hydrophobic domains in the NH_2-terminal region and at least three hydrophilic regions that potentially protrude from the membrane and play a role in sensory input. The COOH-terminal portion of ActS has homology to the protein kinase portion of known sensor proteins; three COOH-terminal sequences of ActS show the invariant His and Asn residues and the glycine-rich region of two parts separated by 20–50 amino acids typical of the histidine kinase family (Tiwari et al 1996b).

ActS shows a 71% similarity to PrrB and RegB of *Rhodobacter sphaeroides*, 65.6% similarity to RegA of *Rhodobacter capsulatus* (Tiwari et al 1996b) and about 60% similarity to RegS from *Bradyrhizobium japonicum* (Bauer et al 1998). The *regB-regA* and *prrB-prrA* pairs are involved in the regulation of photosynthesis (Eraso & Kaplan 1994, 1995) in response to O_2 or redox state, while the *regS-regR* pair is similarly required for *nifA* regulation for N_2 fixation (Bauer et al 1998). ActS lacks the S boxes of PAS domains present in many, but not all, proteins involved in O_2/redox state sensing (Rebbapragada et al 1997, Zhulin et al 1997).

Adjacent to *actS* is the regulator gene *actR*, which codes for a protein of deduced molecular mass 21 463 and which contains the invariant aspartate and lysine residues found in this group of regulator proteins; Asp30, Asp73 and Lys123 residues of ActR appear to correspond to Asp12, Asp57 and Lys109, respectively, of the CheY protein. Asp57 in CheY is phosphorylated (Stock et al 1989), so Asp73 in ActR may be the residue phosphorylated by ActS, although this will require direct confirmation using γ^{32}P-ATP.

The ActR protein shows 87% similarity to the RegA and PrrA proteins of *R. capsulatus* and *R. sphaeroides* (Tiwari et al 1996b), and about 80% similarity to the RegR protein of *B. japonicum* (Bauer et al 1998). Inactivation of *regR* in *B. japonicum* results in an inability to fix N_2 symbiotically, but *actR*$^-$ mutants of *S. meliloti* nodulate and fix N_2 with *Medicago murex* at rates comparable to those of the wild-type. If the RegA and PrrA proteins are to be considered as a special subgroup of the response regulator proteins on the basis of their C-terminal sequences (Eraso & Kaplan 1994), ActS and RegS should probably be placed there also.

Recent analysis of the ActR sequence shows it contains a helix-turn-helix motif NVSETARRLNMNRRTLQRILAK (a score of 1690 using the weight matrix of Dodd & Egan 1987), consistent with DNA binding involved in a role in controlling gene expression.

Complementation studies show that multiple copies of *actS* on pRT546-6 restores acid tolerance to RT295 (*actS*$^-$) but not TG5-46 (*act*R$^-$). By contrast, multiple copies of *act*R on pRT546-11 complement both these mutants (Tiwari et al 1996b), suggesting that multiple copies of the *act*R gene can either bypass the need for phosphorylation, or allow the 'cross-talk' phosphorylation known for other sensor–regulator systems (Stock et al 1989).

Studies using *lacZ* fusions have shown that expression of both the *actS* and *act*R genes is constitutive (Tiwari et al 1996b); this behaviour is logical for a system needing a rapid response to change in pH, although others (e.g. the *kdp* system for K^+ transport in *E. coli*) have a basal level of expression that is up-regulated in response to K^+ concentration.

A working model of the *actRS* system (Fig. 3) is of a sensor–regulator pair responding to changes in pH and controlling gene expression, although the significant task now is to identify the genes controlled.

To identify genes controlled by *actRS* we have first identified pH-controlled *gusA* reporter gene fusions. A strain with an insertionally inactivated *actS*

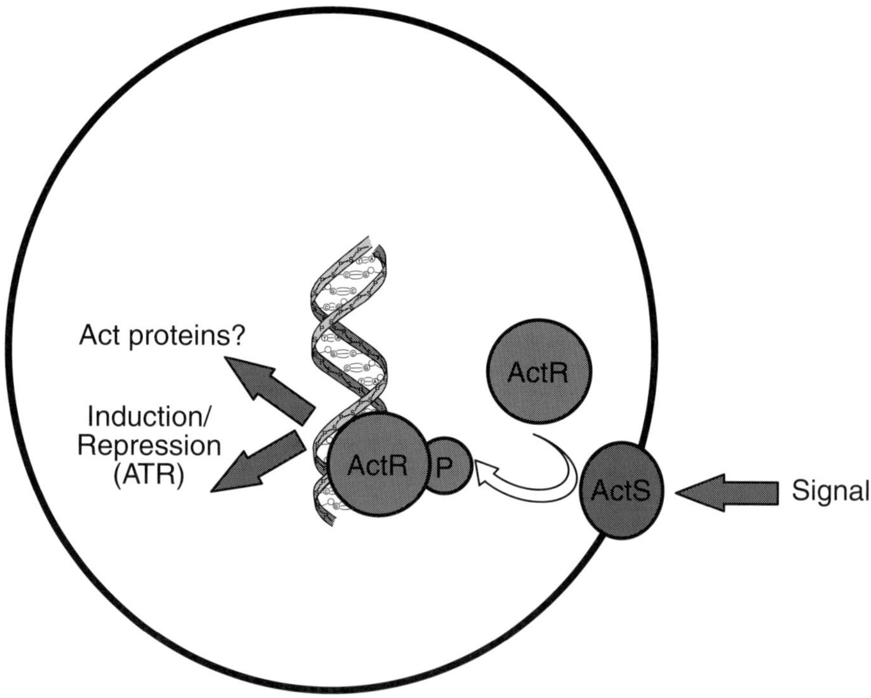

FIG. 3. Proposed mechanism for gene regulation for acid-tolerance via the ActS-ActR system in *Sinorhizobium meliloti* WSM419. ATR, acid tolerance response.

complemented with a plasmid-borne *actS* was randomly mutated with mTn*5*-GNm (containing a promoterless *gusA*). Low pH-regulated clones identified by screening on X-Glc plates were confirmed by broth assays, after which the plasmid-borne *actS* was eliminated by selecting for the introduction of an incompatible plasmid. Clones showing differential expression of β-glucuronidase in the presence and absence of *actS* were further studied. What has emerged from these experiments is that the major role of *actS* may be in gene repression rather than in gene induction.

Because genes regulated by two-component systems are sometimes located adjacent to the sensor–regulator genes, the region upstream of *act*R*S* has been explored. A DEAH RNA helicase gene, *helO*, directly upstream of *actS*, appears to have no role in acid tolerance (Reeve et al 1997b) and no pH-responsive gene has been located in this region.

The adaptive ATR in WSM419 is induced in cells grown at pH 6.0 (Y. Benggu & G. W. O'Hara, unpublished data 1998). TG5-46 (*act*R$^-$) and RT295 (*actS*$^-$) both lack the ATR, implying that a functional *actS*R system is essential for its induction.

pH regulated genes

*phr*R

During sequence analysis of the *actA* gene in *S. meliloti* WSM419 a gene was identified downstream that codes for a protein with similarities to regulator

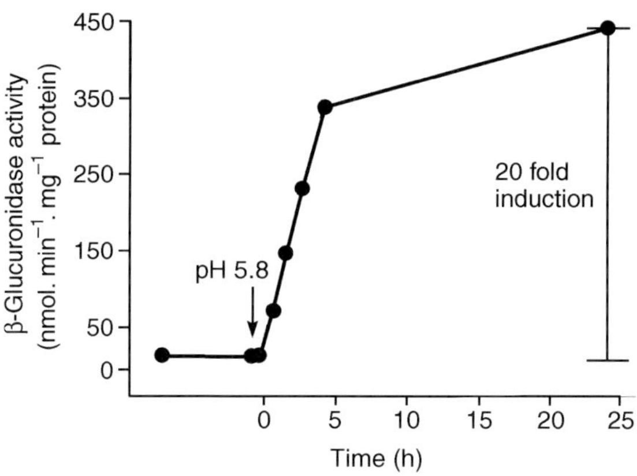

FIG. 4. Time course of induction of β-glucuronidase activity in *Sinorhizobium meliloti* WR101 (mTn*5*-GNm) on transfer of cells grown at pH 7.0 in minimal medium to pH 5.8 (arrow).

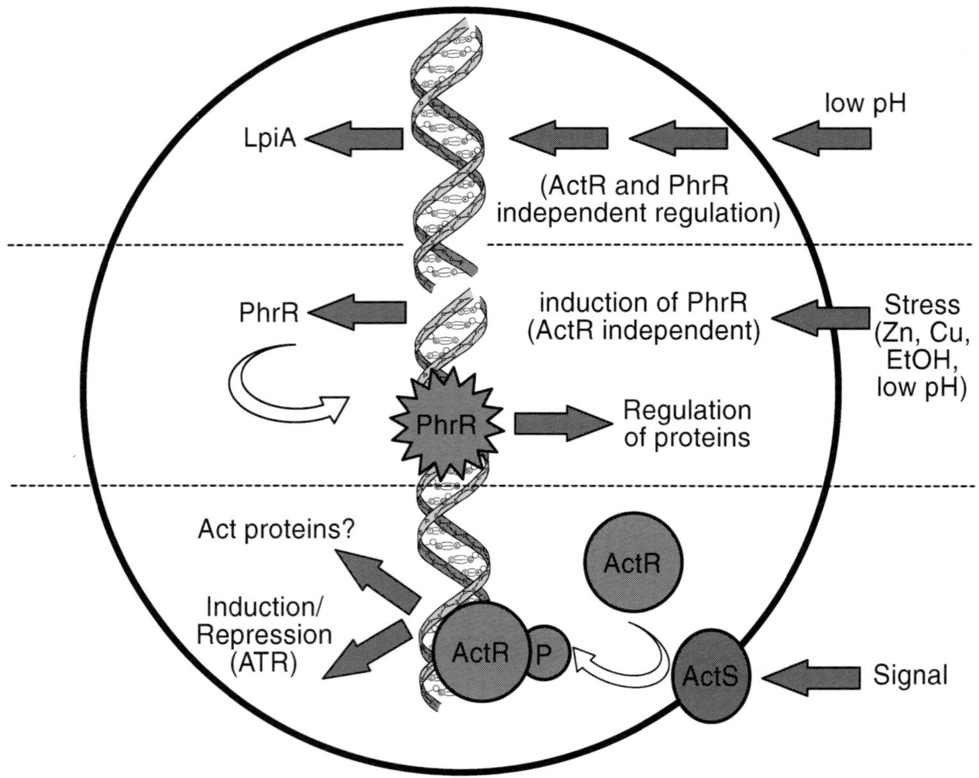

FIG. 5. Current model for acid-controlled gene regulation in *Sinorhizobium meliloti* WSM419. The *actS-act*R system regulates some genes for acid tolerance, including proteins involved in the acid tolerance response (ATR). The *phr*R system responds to an unknown pH-induced signal (not the *actS-act*R system) and presumably also controls other pH-responsive genes. *lpiA* responds to acidic pH but not via either *actS-act*R or *phr*R.

proteins from the y4WC and y4aM genes of *Rhizobium* sp. NGR234 (Freiberg et al 1997). All of these proteins contain a DNA-binding helix-turn-helix motif similar to those reported in URF4 in *Rhodospirillum rubrum* and repressors in coliphage. Because *lacZ* or *gusA* fusions showed that the gene adjacent to *actA* is induced fivefold at low pH, it has been termed *phr*R (pH-regulated regulator).

Insertional inactivation of *phr*R at codons 10 or 126 using ΩKm affects neither the ability of cells to grow at low pH nor the capacity to nodulate and fix N_2 in symbiosis with *M. murex*. The *phr*R gene is also induced by exposure to stressful concentrations of Cu^{2+}, Zn^{2+}, ethanol or hydrogen peroxide, but not to high temperature, high sucrose concentrations, phosphate starvation or in stationary phase. When the *phr*R-*lacZ* fusion (pRT206-8) was mobilized into TG5-46 (*act*R$^-$), β-galactosidase expression was comparable to the wild-type suggesting that the *act*R*S* system is not involved in the induction of *phr*R.

lpiA

A pH-responsive promoter library created by mutagenesis of *S. meliloti* with mTn*5*-GNm has helped to identify genes whose expression responds to changes in pH_o. Strain WR101 shows a 20–25-fold induction of β-glucuronidase activity when the pH falls from pH 7.0 to pH 5.7; the gene disrupted has been provisionally designated *lpiA* (low pH inducible A). In contrast to *phr*R, *lpiA* is induced only by low pH (Fig. 4) and by no other stress. The exact physiological function of *lpiA* in *S. meliloti* remains to be determined.

Acid-inducible genes such as *phr*R and *lpiA* must be controlled by regulatory mechanisms responding directly or indirectly to changes in pH or ΔpH. Separate experiments have established that neither *phr*R nor *lpiA* is under the control of the *act*R*S* sensor–regulator control system, and that *lpiA* is not under the control of *phr*R.

Expression of the pH-inducible gene *lpiA* is affected by Ca^{2+}, increasing concentrations of which shift the pH for initial and maximal induction to lower values. Taken together (Fig. 5) these data indicate that there must be:

(1) genes controlled by *act*R*S*;
(2) a gene to control *phr*R; and
(3) a separate gene to control *lpiA*.

Conclusions

(1) There may be 20–50 genes that are essential for growth and survival of *S. meliloti* at low pH; a number of these appear to be constitutively expressed.

(2) Where tested, alleles of genes essential for growth at acidic pH occur in both acid-tolerant and acid-sensitive strains of *S. meliloti*, suggesting that there may not be unique genes conferring acid tolerance.
(3) Ca^{2+} plays an important role in acid tolerance, affecting growth, the cut-off pH for growth, survival at low pH, expression of pH-inducible genes, and the quantity and quality of exopolysaccharide synthesized.
(4) There is an interaction between ability to tolerate acidic pH and the ability to tolerate stress from heavy metals such as Cu^{2+} and Zn^{2+}.
(5) In *S. meliloti* WSM419 there are likely to be at least three mechanisms for regulating response to low pH: the *actS-act*R system; the *phr*R system; and the system controlling *lpiA*.
(6) The *actS-act*R system appears mainly to be involved in gene repression rather than induction, and it is essential for the adaptive ATR.
(7) It is likely that low pH-inducible genes will include some (e.g. *phr*R) that also respond to other stresses, and others (e.g. *lpiA*) that are specific to low pH.
(8) There may be practical application of the ATR to the preparation of inoculants for soils of low pH.
(9) Although the complexity of the genetic background to acid tolerance makes it a difficult target for *in vitro* construction of more acid-tolerant strains of root nodule bacteria by genetic engineering, any hope of success will depend on our understanding the mechanisms of acid tolerance and their control. Selection of naturally acid-tolerant variants and evaluation of their symbiotic properties (nodulation, N_2 fixation), perhaps with subsequent introduction of an appropriate *Sym* plasmid (Chen et al 1991) offer alternative strategies for selecting acid-tolerant inoculant strains of root nodule bacteria.

Acknowledgements

We thank the Australian Wool Corporation, the Australian Research Council and the Grains Research and Development Corporation for financial support for various phases of this work.

References

Bauer E, Kaspar T, Panglungtshang K, Emmerich R, Fischer H-M, Hennecke H 1998 A two component regulatory system ensures *nifA* expression in *Bradyrhizobium japonicum*. In: Elmerich C, Kondorosi A, Newton WE (eds) Biological nitrogen fixation for the 21st century. Kluwer Academic, Dordrecht, p 135–138

Chen J, Richardson AE, Gartner E, Djordjevic MA, Roughley RJ, Rolfe BG 1991 Construction of an acid-tolerant *Rhizobium leguminosarum* biovar *trifolii* strain with enhanced capacity for nitrogen fixation. Appl Environ Microbiol 57:2005–2012

Dodd IB, Egan JB 1987 Systematic method for the detection of potential λ Cro-like DNA-binding regions in proteins. J Mol Biol 194:557–564

Eraso JM, Kaplan S 1994 PrrA, a putative response regulator involved in oxygen regulation of photosynthesis gene expression in *Rhodobacter sphaeroides*. J Bacteriol 176:32–43

Eraso JM, Kaplan S 1995 Oxygen-insensitive synthesis of the photosynthetic membranes of *Rhodobacter sphaeroides*—a mutant histidine kinase. J Bacteriol 177:2695–2706

Foster JW 1991 *Salmonella* acid-shock proteins are required for the acid-tolerance response. J Bacteriol 173:6896–6902

Foster JW, Hall HK 1990 Adaptive acidification tolerance response of *Salmonella typhimurium*. J Bacteriol 172:771–778

Freiberg C, Fellay R, Balroch A, Broughton WJ, Rosenthal A, Perret X 1997 Molecular basis of symbiosis between *Rhizobium* and legumes. Nature 387:394–401

Glenn AR, Dilworth MJ 1994 The life of root nodule bacteria in the acidic underground. FEMS Microbiol Lett 123:1–10

Glenn AR, Tiwari RP, Reeve WG, Dilworth MJ 1997 The response of root nodule bacteria to acid stress. In: Moniz AC, Furlani AMC, Schaffert RE, Fageria NK, Rosolen CA, Cantarella H (eds) Plant–soil interactions at low pH. Brazilian Soil Science Society, Brazil, p 123–138

Goodson M, Rowbury RJ 1989 Habituation to normally lethal acidity by prior growth of *Escherichia coli* at a sub-lethal acid pH value. Lett Appl Microbiol 8:77–79

Goss TJ, O'Hara GW, Dilworth MJ, Glenn AR 1990 Cloning, characterization and complementation of lesions causing acid-sensitivity in Tn*5*-induced mutants of *Rhizobium meliloti* WSM419. J Bacteriol 172:5173–5179

Graham PH, Draeger KJ, Ferrey ML 1994 Acid pH tolerance in strains of *Rhizobium* and *Bradyrhizobium*, and initial studies on the basis for acid tolerance of *Rhizobium tropici* UMR1899. Canad J Microbiol 40:198–207

Gupta SD, Gan K, Schmid MB, Wu SC 1993 Characterization of a temperature sensitive mutant of *Salmonella typhimurium* defective in apolipoprotein acyltransferase. J Biol Chem 268:16551–16556

Hall HK, Karem KL, Foster JW 1995 Molecular responses of microbes to environmental pH stress. Adv Microb Physiol 37:229–272

Howieson JG, Ewing MA 1986 Acid tolerance in the *Rhizobium meliloti–Medicago* symbiosis. Aust J Agric Res 37:55–64

O'Hara GW, Glenn AR 1994 The adaptive tolerance response in root nodule bacteria and *Escherichia coli*. Arch Microbiol 161:286–292

O'Hara GW, Goss TJ, Dilworth MJ, Glenn AR 1989 Maintenance of intracellular pH and acid tolerance in *Rhizobium meliloti*. Appl Environ Microbiol 55:1870–1876

Perez-Galdona R, Kahn ML 1994 Effects of organic acids and low pH on *Rhizobium meliloti* 104A14. J Gen Microbiol 140:1231–1235

Rebbapragada A, Johnson MS, Harding GP 1997 The Aer protein and the serine chemoreceptor Tsr independently sense intracellular energy levels and transduce oxygen, redox, and energy signals for *Escherichia coli* behavior. Proc Natl Acad Sci USA 94:10541–10546

Reeve WG, Tiwari RP, Dilworth MJ, Glenn AR 1993 Calcium affects the growth and survival of *Rhizobium meliloti*. Soil Biol Biochem 25:581–586

Reeve WG, Dilworth, MJ, Tiwari RP, Glenn AR 1997a Regulation of exopolysaccharide production in *Rhizobium leguminosarum* biovar *viciae* WSM710 involves *exo*R. Microbiology 143:1951–1958

Reeve WG, Tiwari RP, Dilworth MJ, Glenn AR 1997b A helicase gene (*helO*) in *Rhizobium meliloti* WSM419. FEMS Microbiol Lett 153:43–49

Robson AD, Bottomley PJ 1991 Limitations in the use of legumes in agriculture and forestry. In: Dilworth MJ, Glenn AR (eds) Biology and biochemistry of nitrogen fixation. Elsevier Science, Amsterdam, p 320–349

Rogers SD, Bhave MR, Mercer JFB, Camakaris J, Lee BTO 1991 Cloning and characterization of *cutE*, a gene involved in copper transport in *Escherichia coli*. J Bacteriol 173:6742–6748

Stock JB, Ninfa AJA, Stock AM 1989 Protein phosphorylation and regulation of adaptive responses in bacteria. Microbiol Rev 53:450–490

Tiwari RP, Reeve WG, Glenn AR 1992 Mutations conferring acid sensitivity in the acid-tolerant strains *Rhizobium meliloti* WSM419 and *Rhizobium leguminosarum* biovar *viciae* WSM710. FEMS Microbiol Lett 100:107–112

Tiwari RP, Reeve WG, Dilworth MJ, Glenn AR 1996a An essential role for *actA* in acid tolerance of *Rhizobium meliloti*. Microbiology 142:601–610

Tiwari RP, Reeve WG, Dilworth MJ, Glenn AR 1996b Acid tolerance in *Rhizobium meliloti* strain WSM419 involves a two-component sensor–regulator system. Microbiology 142:1693–1704

von Uexküll HR, Mutert E 1995 Global extent, development and economic impact of acid soils. In: Date RA, Grundon NJ, Rayment GE, Probert ME (eds) Plant–soil interactions at low pH: principles and management. Kluwer Academic, Dordrecht, p 5–19

Watkin ELJ, O'Hara GW, Glenn AR 1997 Calcium and acid stress interact to affect the growth of *Rhizobium leguminosarum* bv. *trifolii*. Soil Biol Biochem 29:1427–1432

Zhulin IB, Taylor BL, Dixon R 1997 PAS domain S-boxes in archaea, bacteria and sensors for oxygen and redox. Trends Biochem Sci 22:331–333

DISCUSSION

Cook: I have a question regarding the intracellular pH (pH_i) measurements you made on the pH-sensitive mutants. You showed that these mutants were unable to maintain a ΔpH. What is the mechanism of pH_i regulation in wild-type strains?

Glenn: We haven't yet done much work on the RT327 mutant in terms of pH_i.

Cook: You mentioned that some of your acid-sensitive mutants were calcium repairable. Does this mean that they are now able to maintain their pH_i?

Glenn: Yes. We did some work on TG2-6 and found that if we added external calcium, it helped pH_i maintenance.

Booth: Is it possible that calcium is blocking copper entry? Because copper is toxic in *Escherichia coli*: it causes many disastrous events, including the release of almost all the potassium from the cell, and we don't know the mechanism by which this occurs.

Glenn: It is possible in the case of RT327 because if copper is removed from the medium, acid tolerance is restored. TG2-6 is also copper sensitive; however, if trace amounts of copper in the medium are removed, TG2-6 is still acid sensitive, so that's not the answer.

Poole: It is surprising that such high concentrations of calcium are required for this repair. Most microbial cells maintain a low intracellular concentrations, so 10 mM extracellular calcium is unusual to say the least. Is this repair phenomenon restricted to the rhizobia? Is there something special about the natural environment of rhizobia that makes these high concentrations of calcium physiologically significant?

Glenn: It is possible that these high millimolar concentrations of calcium can arise in soils by the continued application of superphosphate. I'm not aware that people have looked in other organisms. At neutral pH rhizobia have only a small requirement for calcium for growth.

Foster: I was struck by how similar your *lpiA* results were to the PhoP-PhoQ system. Have you looked at the effect of magnesium?

Glenn: Yes. Magnesium does not replace calcium in terms of the growth, although we haven't looked at the effect of magnesium on the induction of *lpiA* at different pHs. That could be an interesting experiment.

Cook: You have shown that calcium has an effect on exopolysaccharide production in *Rhizobium*. When you added calcium to the medium, the cells seemed to make more exopolysaccharide on plates. Does this protect them from acidic pH?

Glenn: This is a complicated story. We know that calcium has some profound effects on the amounts of exopolysaccharides that are made by rhizobia. There are also changes in the organic acid substitution on the exopolysaccharides. An insertional mutant of *exoH* makes a non-succinylated exopolysaccharide, so the exopolysaccharide just has acetyl and pyruvyl residues on it, and this mutant is extremely acid sensitive. If we do complementation analyses with a suite of *exo* genes, we can restore both normal succinylation of the exopolysaccharide and normal acid tolerance. We don't yet know whether this is directly due to *exoH* or a polar effect. All we can say is that there are some significant effects on exopolysaccharides, but we cannot say they are direct effects. A mad-cap experiment that we should do is to take isolated wild-type exopolysaccharide, which is succinylated, and add it to the cells of the *exoH*$^-$ mutant to see if this restores acid tolerance. It's the kind of bizarre experiment that we don't expect to work, but we should try it and see what happens.

Slonczewski: Did calcium repair only the copper deficiency or did it also repair defects that did not appear to involve copper?

Glenn: There does not appear to be a simple relationship between copper-related systems and the calcium effect. For example, calcium repairs strain TG2-6 (a lipoacyl transferase-defective strain) which is copper sensitive, but not RT3-27 (the Cu^{2+} ATPase-defective strain). However, it appears to be true that strains which are acid sensitive and copper sensitive are repaired by calcium.

Dilworth: But it does repair the sensor–regulator mutation.

Padan: Perhaps you should look for mutants that are not responsive to calcium.

Glenn: We have been trying to do this.

Slonczewski: Is 10 mM Ca^{2+} required, or have you looked at a range of concentrations?

Glenn: Effects are observed at lower concentrations. We see differences in both growth and in *lpiA* expression when we use 1 mM as opposed to 0.1 mM.

Epstein: Have you looked at genes that are induced by calcium?

Glenn: This is something we are currently doing by looking at the effect of calcium on a *gusA* promoter library of WSM419. There are a few clones that may be responding to calcium.

Poole: Are there any calcium-responsive genes in other systems?

Booth: There is an emerging story that calcium regulates gene expression and protein phosphorylation in *E. coli.* It is possible that there is a protein kinase C homologue in *E. coli* which is regulated by calcium transients.

Poole: It has been suggested for a long time that calcium is an ideal signalling molecule in all cell types because the concentration is normally kept so low. Tiny spikes in calcium concentration can result in huge differences in the steady-state situation.

Glenn: The literature suggests that micromolar concentrations of calcium are adequate for growth. Most people do these experiments at around neutral pH, so you may end up with different results at low pH.

Poole: When you do these experiments with relatively high calcium concentrations, like 10 mM, did you have to be careful about precipitation of other inorganic components of the medium?

Dilworth: At low pH that's not a problem.

Booth: Tisa et al (1993) found a calcium channel in *E. coli* on the basis of inhibition by ω-conotoxin. There is also a channel in *E. coli* that is homologous to the Shaker channel in humans. It's closest relative is the *Drosophila* slow Shaker channel, which is regulated by calcium. Therefore, there may be a precedent for calcium-regulated circuits that haven't yet been explored.

Skulachev: It is known that nitrogen fixation is sensitive to oxygen and oxidative stress, and that copper is a catalyst for the Fenton reaction, when hydrogen peroxide is converted to the hydroxyl radical OH. You mentioned that the effects of ethanol and hydrogen peroxide are similar, but have you tried using antioxidants? Because it is possible that this story is related to oxidative stress.

Glenn: All we know is that all of these mutants nodulate their host and they fix nitrogen.

Matin: You mentioned that copper is toxic at low pH. Is an uptake mechanism involved?

Glenn: It's possible. The toxic effect could occur at the cytoplasmic membrane, but one could equally well argue that the effect is at the cell surface.

Matin: Your experiments were performed at pH 5.8, and at that pH a positive-inside membrane potential may be required. Do you know what the potential is?

Glenn: No.

Booth: Why would you need a positive-inside membrane potential at 5.8?

Matin: If there is a large pH gradient, one way to deal with the force with which the protons try to get into the cell and the energy required to extrude them is to generate a positive-inside membrane potential. We know that a positive potential is generated, but we do not know whether cationic uptake is involved.

Booth: I understand your argument for pH 1 and pH 2, but at pH 5.8 the pH_i is 8 and the pH gradient is two units, which is well within the normal range.

Matin: It depends on the pumping strength.

Fillingame: If I understand you correctly, you have a Tn*5* insertion in a gene that you think is a copper P-type ATPase, and yet the phenotype is that it grows at pH 7 but not at pH 5.5. Have you looked at the effects of copper at pH 7 to see if excess copper at this pH has an effect on growth?

Glenn: At pH 7 the effect of copper on the mutant and the wild-type appears to be indistinguishable. By contrast, at pH 5.8 the mutant is much more sensitive to copper than the wild-type.

Fillingame: Is this typical for other copper P-type ATPase mutants in other organisms that are selected on the basis of their copper sensitivity?

Glenn: My understanding is that there is usually a difference in sensitivity to copper, so we have an interesting situation.

Poole: We have identified a zinc P-type ATPase mutant that has an increased sensitivity to zinc (Beard et al 1997), but we haven't looked at its pH sensitivity.

Fillingame: Unless the copper uptake system does not work at pH 7, one would expect sensitivity at higher copper concentrations.

Foster: Do you have any mutants that are sensitive to acid that are not sensitive to copper?

Glenn: No, they are all sensitive.

Poole: You mentioned that the *actSR* mutants are sensitive to cadmium and/or zinc. The resistance and sensitivity to that pair of metal ions is precisely what one would expect, e.g. the zinc P-type ATPase in *E. coli* has phenotypic consequences for cadmium tolerance (Beard et al 1997). Do you understand the basis for this?

Glenn: Not at all. But it's a convenient screening process.

Booth: Barry Rosen published a paper last year showing that mutants which lack disulfide isomerase become sensitive to zinc and cadmium (Rensing et al 1997).

Poole: But the basis for this sensitivity is not understood.

Booth: I'm just wondering if some of these effects are mediated by changes in expression of periplasmic protein-folding enzymes, so that stress sensitivity may be induced by the loss of those proteins.

Poole: Only in the case of copper has a family of proteins being implicated in homeostasis. For example, there are copper chaperones, copper import systems and copper export systems. For all the other metals that Andrew Glenn mentioned that sort of information isn't available.

Dilworth: What is the pH profile for copper import systems? Because we would guess that there is increased copper availability at low pH, and if the organism still has the same level of copper transport then it's going to be in trouble at low pH.

Poole: I'm not aware that anyone has looked at this. An important message from Andrew Glenn's presentation is that there are indirect and unexpected effects of changing pH; for example, the greater availability of toxic metal ions at low pH. Another example is the way in which cells respond to nitrite as the pH is reduced. This turns out to be a purely chemical phenomenon due to the increased availability of nitric oxide in acidic solutions of nitrite. There must be dozens more examples of situations we need to be aware of as the pH changes.

References

Beard SJ, Hashim R, Membrillo-Hernández J, Hughes MN, Poole RK 1997 Zinc(II) tolerance in *Escherichia coli* K-12: evidence that the *zntA* gene (*o732*) encodes a cation transport ATPase. Mol Microbiol 25:883–891

Rensing C, Mitra B, Rosen BP 1997 Insertional inactivation of *dsbA* produces sensitivity to cadmium and zinc in *Escherichia coli*. J Bacteriol 179:2769–2771

Tisa LS, Olivera BM, Adler J 1993 Inhibition of *Escherichia coli* chemotaxis by ω-conotoxin, a calcium ion channel blocker. J Bacteriol 175:1235–1238

How can archaea cope with extreme acidity?

G. Schäfer

Institute of Biochemistry, Medical University of Luebeck, 23538 Luebeck, Germany

Abstract. Archaea are the most extremophilic of the acidophilic microbes, combining, in many cases, acidophilicity with hyperthermophilicity. They form one of the three branches of the phylogenetic tree, and they are specifically found within the so-called crenarchaeota, typical members of which thrive at pH 1–3 and at temperatures of 75 °C to nearly 100 °C. Despite this, these cells can maintain a near neutral cytosol, and they use H^+ for chemiosmotic coupling of ADP phosphorylation. These phenomena require efficient exclusion and disposal of protons. This is achieved by multiple synergistic mechanisms that act in parallel. One strategy is to use bipolar tetraether lipids as a matrix of their plasma membranes, providing low ion permeabilities, even at high temperatures. Additionally, an inverted membrane potential can help to balance a large pH gradient of up to 4 at a proton motive force of $\Delta p = 140–180$ mV. This is not a general rule, because in several species the membrane potential contributes only minimally. Also, local buffering capacity and charge profiles across the membrane may significantly influence adaptation to bulk phase acidity. Neither complex I nor complex III electron transport-coupled proton pump equivalents have been found in aerobic archaea. Only terminal oxidases seem to provide either H^+ pumping or the generation of a proton gradient by chemical charge separation. Organization, redox centres and primary structures of some archaeal terminal quinol oxidase complexes are known and will be discussed. Much less is known about anaerobic sulfur reducers. For those a possible mechanism for proton exclusion is proposed.

1999 Bacterial responses to pH. Wiley, Chichester (Novartis Foundation Symposium 221) p 131–151

Many prokaryotic organisms grow under moderate or extreme acidic conditions, but keep a near neutral cytosolic pH, a phenomenon that is of great interest for various reasons. On the one hand, their membrane-spanning proteins face completely opposite environments with respect to ionic strength and polarity. The rupture of such cells, and subsequent exposure of both surfaces to homogeneous solvent conditions, may cause significant or even denaturating changes that could impose experimental limits to the successful study of isolated membrane components. On the other hand, the strong proton motive force provided by a large inward-directed pH gradient is an ideal precondition for the operation of ATP synthases using protons as coupling ions. However, the structures and mechanisms that function as molecular machines that continuously balance the influx of protons at expense of metabolic energy may be essentially different and in many cases are only partially understood.

Archaea form a distinct evolutionary domain (Woese 1993) and have unusual physiological adaptations that combine acidophilicity with hyperthermophilic properties. These conditions range from near neutrality to pH values as low as <1 (Van de Vossenberg et al 1998), and temperatures from about 60 °C to near or even above 100 °C. Interestingly, acidophilicity is mainly restricted to aerobes, such as in the genus *Sulfolobales* or *Thermoplasmales* (Stetter 1996), although anaerobic growth at pH 1–2 and 75–90 °C is observed in the genus *Acidianus. Acidianus ambivalens* is a strictly litho-autotrophic organism and is capable of switching between both conditions, either oxidizing or reducing elemental sulfur as sole energy source, whereas *Sulfolobales* and *Thermoplasmales* grow organotrophically. Both *Sulfolobales* and *Acidianus* have been specifically investigated in our laboratory.

Adaptation strategies

Four principles for coping with extremely acidic environments may be distinguished. (1) The surface properties of the cell envelope and of the exposed domains of integral proteins of the plasma membranes may provide strong buffering power (not discussed here). (2) The plasma membrane may demonstrate low proton conductivity due to its lipid composition. (3) An inverted membrane potential may be maintained (inside positive) to prevent positive charge influx by proton currents. (4) Highly efficient proton pumps driven by membrane-residing redox systems may operate, or vectorial charge separation may occur due to the topological organization of metabolic reactions that involve charge release and consumption on opposite sides of the plasma membrane (termed 'chemical charge separation').

Archaeal membranes

Thermoacidophilic archaea are characterized by an extremely well-organized, rigid and quasi-crystalline surface layer that is, in general, firmly attached to the plasma membrane, leaving practically no periplasmic space (Baumeister & Lembcke 1992). This envelope mainly serves as barrier against osmotic swelling, whereas the lipid structure provides extreme thermostability. Exceptions exist, however, such as *Thermoplasma acidophilum*, which is devoid of such a surface layer. The rigid monolayer structure of the membrane-spanning bipolar lipid matrix presumably plays the major role, as has been shown with liposomes from

Sulfolobus acidocaldarius (Komatsu & Chong 1998). In a comparative study it has been shown that vesicles formed of membrane-spanning tetraether lipids from thermoacidophiles are extremely tight. Only at high temperatures (75 °C) do they achieve a proton conductivity equivalent to those measured at about 30 °C with bilayer-forming glycerol lipids from mesophiles (Elferink et al 1993). Thus, these lipids contribute to the ability of archaea to withstand a large proton motive force, even at high temperatures, without leakage. Thermophilic organisms such as *Bacillus stearothermophilus* possess bilayer glycerol phospholipids and have to compensate for temperature-induced ion leakage by a high turnover of ion pumps.

Polarity of potentials

According to the chemiosmotic principle (Mitchell 1961, 1963), the electrical potential ($\Delta\Psi$) and the pH difference across the membrane imposed by a proton gradient (ΔpH) contribute to the proton motive force according to equations (1) and (2).

$$\Delta\mu H^+ \,[kJ/mol] = -F\Delta\Psi + 2.3RT\Delta pH \tag{1}$$

$$\Delta p[mV] = -\Delta\mu H^+/F = \Delta\Psi - 2.3RT/F\Delta pH \tag{2}$$

$\Delta\Psi$ and ΔpH may mutually replace each other. For example, in halo-alkaliphilic archaea the membrane potential can represent the predominant driving force (Wittenberg 1996).

For a number of acidophilic eubacteria an inverse relation of $-2.3RT\Delta pH$ and $\Delta\Psi$ is established when the external pH is changed from neutral to acidic conditions. $\Delta\Psi$ changes from inside negative to inside positive at $pH<3$, thereby preventing a free influx of H^+. Values of $\Delta\Psi$ between +10 to +100 mV have been reported in, for example, *Bacillus acidocaldarius*, *T. acidophilum* and some *Thiobacilli* (Bakker 1991). *Heliobacter pylori* and *Picrophilus oshimae* also adapt to low pH by the same mechanism (Matin et al 1996, Van de Vossenberg et al 1998).

Table 1 shows the components of the proton motive force of the hyperthermophile *S. acidocaldarius* under steady-state respiration at an outside pH of 3.5, the upper limit of growth. Its growth optimum is pH 2.5, but this is experimentally unfavourable for ΔpH determinations. A positive-inside potential is found in respiring and in partially respiration-inhibited cells. N_3 inhibition of terminal oxidases induces an inside positive shift of membrane potential that is further enhanced via the action of uncouplers (4,5,6,7-tetrachloro-2′-trifluormethyl benzimidazole, TTFB) or of general ionophores (gramicidin) concomitant with an internal acidification and a collapse of ΔpH (Moll & Schäfer 1988, Lübben & Schäfer 1989).

TABLE 1 Bioenergetic parameters of *Sulfolobus* cells at moderate acidity

Conditions	$\Delta\Psi$ *(mV)*	$-Z\Delta pH$ *(mV)*	Δp	pH_i	pH_o
Control (5 mM K_o^+)	+30	199	170	6.5	3.5
Plus 0.75 mM N_3^-	+65	174	109	6.2	3.5
Plus 6.8 μM TTFB	+74	160	86	5.8	3.5
Plus 12 μM gramicidin	+110	129	19	5.3	3.5

Data taken from Moll et al (1987) and Schäfer et al (1990). Experimental procedures and details of analytical techniques are described in Moll (1987). TTFB, 4,5,6,7-tetrachloro-2′-trifluoromethyl benzimidazole.

TABLE 2 Relationship between internal potassium concentration and membrane potential of respiring *Sulfolobus acidocaldarius* cells

Conditions	K_i^+	K_i^+/K_o^+	ΨK^+ *(mV)*	$\Delta\Psi app$ *(mV)*	Δp *(mV)*
Control (11.5 mM K_o^+, pH 3.5)	18–21	1.7	+12 to +16	+30	170
Plus uncoupler (21 μM SF6847)	10.5	0.78	−7	+74	77
Plus 25 μM nigericin	3.2	0.27	−37	n.d.	n.d.
Plus 33 μM gramicidin	13.6	1.15	+4.8	+110	18

Sulfolobus acidocaldarius cells (0.6 mg/ml) were incubated under aeration containing 9.8 mM ammonium sulfate and 11.5 mM K_o^+. The pH was adjusted to 3.5 at the beginning of the incubation and equilibrium allowed for about 15 min before addition of ionophores. Final $\Delta\Psi$, Δp and K_i^+ values were determined after another equilibration time of about 20 min. $\Delta\Psi K^+$ is a fractional potential calculated from the potassium distribution. Inside-positive potentials are indicated by +. $\Delta\Psi$app and Δp are determined by probe distribution methods. Experimental details and analytical methods are given in Moll (1987) and (Schäfer et al (1990). n.d., not determined; SF6847, 3,5-di (tert-butyl)-4-hydroxybenzyliden malononitril.

$[K^+]$ plays an essential role in maintaining the normal potential, which is close to zero. This is supported by the experiments shown in Table 2. At an outside concentration of $[K_o^+] = 11$ mM $[K_i^+]$ was almost twice as high and declined immediately on addition of uncoupler or ionophore with a parallel drastic increase in membrane potential ($\Delta\Psi$app) due to proton influx and/or K^+/H^+ exchange. In all circumstances the proton motive force decreased. Even at 2.5 mM $[K_o^+]$ average inside concentrations of 12–21 mM $[K_i^+]$ are observed in cells harvested in the exponential growth phase. From this and the experiments in Table 2, we would expect the existence of a high affinity K^+ uptake system. As yet, however, our attempts to identify such systems protein chemically or by isotope labelling have failed.

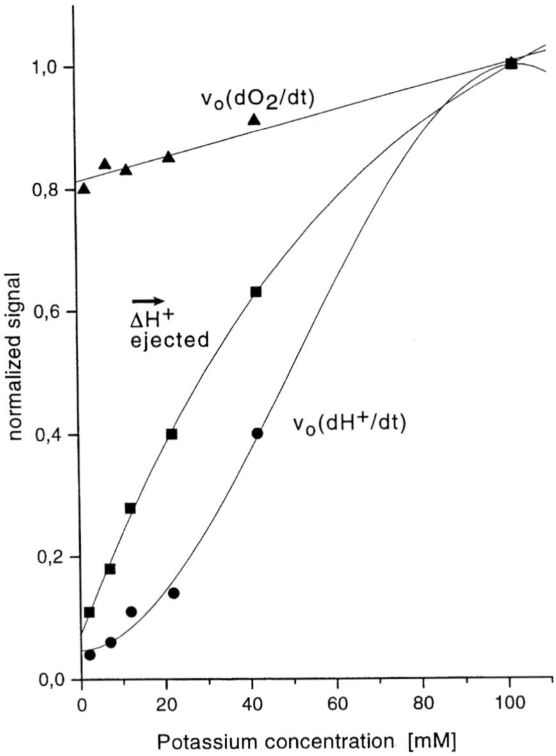

FIG. 1. The dependence of initial rates of respiration and proton ejection of *Sulfolobus acidocaldarius* cells on external potassium concentration. ΔH^+ indicates the amplitude of proton ejection. The data are normalized to the maximum values, i.e. 40 nmol O_2/min per mg, 260 nmol H^+/min per mg for the rates, and 330 nM/mg as the change of external proton concentration. The experiments began by the addition of a saturating oxygen pulse to an anaerobic cell suspension. Experimental details are as described in Moll & Schäfer (1988) and Moll (1987).

High K^+ concentrations exert almost negligible stimulation of respiratory rate in *Sulfolobus* (Fig. 1), but can stimulate the rate and amplitude of proton ejection by more than 10-fold. This phenomenon cannot be explained by any stoichiometric relation. It might involve both the counter-exchange of protons and potassium, as well as kinetic effects by K^+-induced conformational changes, or the blocking of symports and proton leaks by high [K^+]. In summary, it is likely that archaea maintain their cytosol at near neutral pH by the same mechanisms as eubacteria, although individual exchange carriers or electrogenic alkali ion pumps have not yet been characterized.

The persistence of the membrane potential in acidophiles has been attributed to a Donnan potential arising from non-diffusable charges, and to the possible contribution of a proton diffusion potential (Goulbourne et al 1986). The membrane-spanning cytochrome-aa_3 polypeptides of both *S. acidocaldarius* and *A. ambivalens* share an interesting property in relation to sequence, i.e. they have an asymmetrical statistical distribution of charged residues with a considerable excess of positive charges on the loops exposed to the inner membrane surface (Lübben et al 1992, Purschke et al 1997). This may create a Donnan potential of fixed charges that may be relevant in resting cells to maintain an inside-positive membrane potential. Given the intracellular and extracellular pH values, and taking the pK_a of protonable amino acid residues on the exposed surfaces into account, a 2.3–3.5-fold net excess of positive charges results on the intracellular side of subunit-I polypeptides, even under steady-state growth conditions.

However, respiring cells are far from thermodynamic equilibrium. More likely is the interconversion of $\Delta\Psi$ and ΔpH by mechanisms that involve electrogenic alkali cation transporters or electro-neutral cation H^+ exchangers, as originally described by Booth (1985). Although our own data on proton pumping in *S. acidocaldarius* suggest a crucial role of electrogenic K^+ uptake, the internal K^+ concentration (10–40 mM) is an order of magnitude below that of most neutrophiles ($K_i^+ > 100$ mM).

Aerobic redox-driven pumps

Primary H^+ pumps energized by respiratory redox systems play a central role in the extrusion of protons from the cytosol in acidophiles. Active H^+ pumping is suggested by H^+/O ratios measured in *S. acidocaldarius* following an anaerobic/aerobic transition (Moll & Schäfer 1988, Schäfer et al 1990, 1994). Optimally, an aerobic respiratory chain can provide three proton-translocating complexes: the NADH:coenzyme-Q reductase (complex I); the coenzyme-Q:cytochrome-c reductase (bc_1 complex); and the terminal cytochrome oxidase (complex IV). However, as reviewed in Schäfer (1996), neither a complex I nor a typical bc_1 complex have been identified in aerobic archaea. *Sulfolobales* do not contain c-type cytochromes, despite the presence of typical respiratory Rieske-FeS proteins (Schmidt et al 1996) that function in a different association than in classical bc_1 complexes. Irrespective of the primary donor of reducing equivalents—NADH dehydrogenase and succinate dehydrogenase, or sulfur-oxidizing systems as in *Acidianus*—electrons collect in a membrane-residing quinone pool (caldariella quinones). The reduced quinol(s) can be oxidized by various types of novel terminal oxidase super complexes, which then have to operate as the only respiratory proton pumps. Little is known about *Thermoplasmales*, but detailed protein chemical and genetic information is available for *Sulfolobus* (Lübben et al 1992, 1994, Iwasaki et al 1995) and *Acidianus* (Purschke et al 1997).

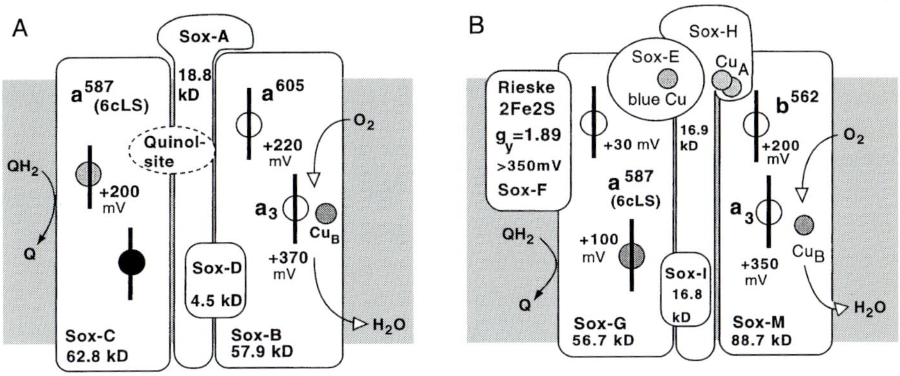

FIG. 2. Sketch of two respiratory complexes of *Sulfolobus acidocaldarius*, SoxABCD and SoxM. The shaded area represents the membrane. The boxes represent single polypeptide constituents, and their molecular masses, redox centres and respective redox potentials are shown. See text for details. (A) SoxABCD complex. (B) SoxM complex. The presence of the Cu centres in SoxH and SoxE of the SoxM complex is likely but has not yet been verified experimentally.

Figure 2 compares the composition of the two terminal oxidase complexes from *S. acidocaldarius*. The SoxABCD complex (panel A) is reminiscent of a functional fusion of an aa_3-type terminal oxidase with a b-type cytochrome like that operating in ubiquinol:cytochrome c reductases; this is cyt-a^{587} with haeme-A_S (Lübben 1995). The four subunits contain a total of five metal redox centres, and suggest the following electron pathway on the basis of their redox potentials:

$$QH_2 \rightarrow \text{Cytochrome-a}^{587}\ (\text{SoxC}) \rightarrow \text{haeme-a} \rightarrow \text{haeme-a}_3/\text{Cu}_B \rightarrow \text{oxygen.}$$

The quinol-binding site is assumed to be located at the interface of subunit II (SoxA) with both haeme-containing polypeptides, enabling the operation of a Q-cycle as sketched in Fig. 3. Indeed, with an artificially reconstituted vesicle system we have been able to demonstrate active H^+ pumping by the SoxABCD complex (Gleissner et al 1997). The determined H^+/O ratio for pumped protons ($\geqslant 1.2$) would be in accordance with a Q-cycle-like mechanism, if the aa_3 moiety itself does not contribute to pumping; otherwise it would be even higher. Although the isolated subunit-I (SoxB) of the complex was shown to catalyse the generation of a proton motive force when reconstituted into lipid vesicles (Gleissner et al 1994), this may simply occur by chemical charge separation, as can be concluded from structural considerations (see below).

Therefore, an alternative pumping mechanism is likely to operate in this complex. Multiple sequence alignments revealed that a class of terminal oxidases lacks essential amino acid residues that form one of the two channels for proton

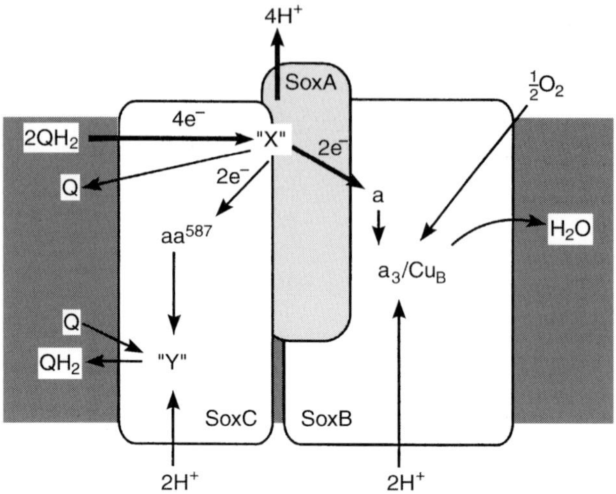

FIG. 3. Hypothetical scheme of a proton-translocating Q cycle in the terminal oxidase SoxABCD of *Sulfolobus acidocaldarius*. The X and Y centres represent high and low potential quinol-binding sites corresponding to the Q_o and Q_i sites of bc_1 complexes, respectively. The arrows indicate the fluxes of electrons and protons.

conduction to the binuclear haeme-a3/Cu_B redox centre in H^+-pumping cytochrome c oxidases (c.f. Fig. 4; Fetter et al 1995). This class includes *Sulfolobus* SoxABCD, *Acidianus* quinol oxidase, and oxidases from *Natronobacterium* and a *Thermus* species.

In contrast, the supercomplex SoxM of Fig. 2B combines all features of a classical bc_1 complex merged with a pumping terminal oxidase (Lübben et al 1994, Castresana et al 1995). It contains a total of eight redox metal centres and purifies as an integrated particle. Unfortunately, the isolated complex is devoid of sulfocyanin—a loosely associated blue Cu-protein likely to mimic the function of cytochrome c—and is therefore catalytically inactive. Sulfocyanin has recently been produced as recombinant protein (Komorowsky & Schäfer 1998), and it may help to reactivate the SoxM complex in reconstituted vesicle systems. The SoxM complex is encoded by a unique gene cluster (Castresana et al 1995). It has two functional parts that, based on structural and redox potential analyses, likely pump out two and four protons, respectively, per oxidized quinol. The Rieske FeS-protein with cytochrome a^{587} (SoxG) is likely to perform a Q-cycle using sulfocyanin as intermediate electron acceptor which then feeds electrons into the Cu_A-containing SoxH polypeptide that serves as donor to the SoxM terminal oxidase subunit. Conserved functional residues within this ba_3-type polypeptide suggest that it may act as a H^+ pump. This is based on structural models derived

helix nr.	Residue[a)]	*P.den.*	*H.sal.*	*S.ac.* SoxM	*S.ac.* SoxABCD	*A.amb.*	*N.pha.*	*T.th.*
h6	**E278**	+	+ (E296)	+ (E240)	- (V240)	- (V246)	- (V230)	- (I235)
loop 2 / 3	**D124**[b)]	+	+ (D193)	+ (D 91)	-	-	-	-
h6	Y280	+	+ (Y298)	+ (Y242)	+ (Y242)	+ (Y248)	+ (Y232)	+ (Y237)
h7	H326	+	+ (H344)	+ (H289)	+ (H288)	+ (H294)	+ (H278)	+ (H283)
h7	Y328	+	- (F346)	- (F291)	- (Q290)	+ (Y296)	+ (Y280)	- (F285)
h8	**K354**	+	+ (K384)	+ (K317)	+ (K329)	+ (K331)	+ (R320)	+ (R325)

FIG. 4. Amino acid residues involved in proton pumping by respiratory chain terminal oxidases. The sequences were translated from DNA sequences taken from the EMBL data bank. The numbering of residues includes the starting methionine in all cases. The folding pattern was adopted from the alignment to the 3D structure of the *Paracoccus dentrificans* oxidase serving as a scaffold. The boxed areas indicate residues essential in the D channel (upper box) and the K channel for protons (lower box). The + signs indicate that the respective residue is present with its position given in parentheses. The – signs indicate the absence of a residue and replacing residues (if present) are in parentheses. The dashed box encircles archaeal oxidases. [a)]Numbering according to *Paracoccus denitrificans*. [b)]Refers to whole motif ($-\mathrm{N}\,x_{10}\,\mathrm{D}\,x_6\mathrm{N}-$).

from the atomic coordinates of the two available master structures of terminal oxidases, i.e. of *Paracoccus denitrificans* (Iwata et al 1995) and beef heart mitochondria (Tsukihara et al 1996). The electron flow in the SoxM-complex is as follows, with the two putative pump sites enclosed by square brackets:

$$\mathrm{QH_2} \rightarrow \underbrace{[\mathrm{Cyt\text{-}a^{587}/(2Fe2S)}]}_{\text{pump 1 (?)}} \rightarrow \mathrm{SoxE{:}Cu} \rightarrow \underbrace{[\mathrm{SoxH{:}Cu_A} \rightarrow \text{hem-b}^{562} \rightarrow \text{haeme-a}_3/\mathrm{Cu_B}]}_{\text{pump 2 (?)}} \rightarrow \mathrm{O_2}$$

A primitive respiratory chain detected in aerobically grown *A. ambivalens* consists solely of an aa_3-type quinol oxidase (Purschke et al 1997). This complex contains six polypeptides but only subunit I (DoxB) identifies it as a member of the haeme/Cu oxidase superfamily; no additional redox centres are present in any of the other polypeptides that are totally dissimilar to known proteins on a sequence basis, and no other cytochromes are involved in respiration. The question of whether or not this oxidase is a proton pump has not been solved experimentally. Although all haeme- and Cu-ligating residues are invariantly conserved in their respective transmembrane helices, on a sequence basis the catalytic subunit has to be assigned to those aa_3 or ba_3 oxidases unable to pump protons. In Fig. 4 the latter are listed with respect to the replacement of essential residues of the proton-conducting D and K channels (Konstantinov et al 1997). In most cases specific polar and protonable residues are replaced by hydrophobic amino acids. The pathway of chemical protons for oxygen reduction (K channel) is conserved in all oxidases. However, exchanges affecting the D channel, especially the lack of Glu275 (*P. dentrificans* numbering), prevent active H^+ translocation (Verkovskaya et al 1997, Vygodina et al 1998). This glutamate residue is a key player in the pumping step of the proposed histidine cycle (Wikström et al 1994), and it may exert a gate function linked to the pumping channels via a water chain (Riistama et al 1997).

If the above generalization is incorrect, either alternative and hitherto unknown proton pathways and pumping mechanisms have to be postulated, or the net extrusion of protons may occur by chemical charge separation, i.e. by release of substrate (QH_2) protons outside and uptake of protons for oxygen reduction from inside the cell. The low capacity of this mechanism may be balanced by a high turnover. On the other hand, the system can operate at any conceivable proton gradient practically without thermodynamic back pressure. *A. ambivalens* may represent a typical example of the way such a mechanism operates. In spite of the enormous change in free energy of aerobic oxidation of sulfur to H_2SO_4 ($-\Delta G = 1014\,kJ/mol$), only a low growth rate is achieved (doubling time ($\geqslant 7\,h$), whereas high turnover rates of oxygen and sulfur consumption are observed.

Anaerobic thermoacidophilic archaea

Strongly acidophilic archaea tolerating a pH <2.5 have only been found amongst the order *Sulfolobales* and *Thermoplasmales* (Stetter 1996), the latter being less thermophilic. From the former order, *Stygioglobus azoricus* is an obligate and *Acidianus* a facultative anaerobe. Both can grow anaerobically, reducing S^0 by H_2 as the sole energy source. The free energy change of this reaction ($-29\,kJ/mol$), as derived from the difference of standard reduction potentials of 150 mV for the $H^+/1/2H_2$ and S^0/S^{2-} couples, is comparatively small. It may also decrease at low environmental hydrogen concentrations in natural habitats.

Much less is known about the electron transport systems of these anaerobes than those of aerobes. The same is also true for coupling between proton extrusion and sulfur respiration. In addition, many of the sulfur-metabolizing enzymes are soluble in the cytosolic space, and linkage to vectorial membrane reactions remains enigmatic. This holds equally for aerobic S^0 oxidation by *Acidianus*, and the pathways of reducing equivalents from the sulfur:oxygenase/reductase (Purschke et al 1997, Kletzin 1992) to the quinone pool are not known at all.

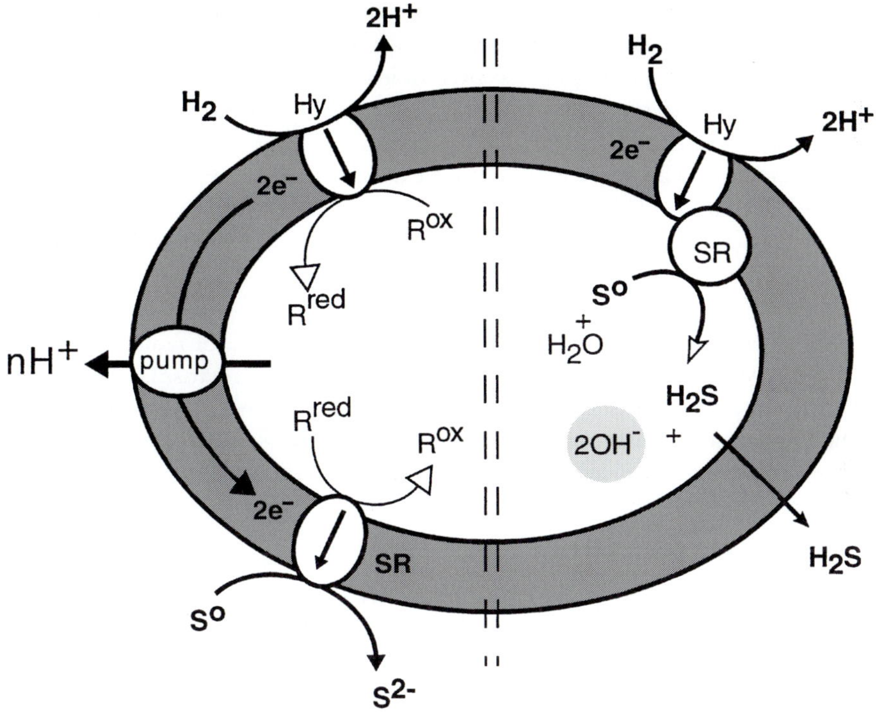

FIG. 5. Illustration of possible principles of proton extrusion in sulfur-reducing anaerobic acidophiles. Both schemes, left and right of the dashed line, imply the function of a hydrogenase (Hy) and a sulfur reductase (SR). Depending on the topology of these catalysts with respect to the membrane, either a pump is required (left) or the system might function by chemical charge separation (right). Arrows in the membrane indicate the direction of electron flux. R^{ox} and R^{red} symbolize the two forms of a putative (cytosolic) redox mediator.

Because of the weak acidity of the anaerobic sulfur respiration product, H_2S ($pK_a = 7.04$) cannot contribute to the acidification of the extracellular medium. Thus, the question of how sulfur reduction is linked to proton pumping becomes thermodynamically problematic. However, a membrane-bound hydrogenase has been purified from *A. ambivalens* (Kletzin 1994, Maier 1996), and sulfur reductases may be membrane linked, as has been found for other anaerobes (Maier 1996, Adams & Kletzin 1996). Although the orientation and topology of involved proteins is not clear in this specific case, one may hypothesize that the membrane-residing hydrogenase releases $2H^+$ outside and transfers reducing equivalents to an internally located sulfur reductase that immediately consumes intracellular $2H^+$ by forming H_2S which then leaves the cell (Fig. 5). This mechanism is analogous to chemical charge separation and formation of a proton motive force by non-pumping terminal oxidases, or by hydrogenases and heterodisulfide reductases of methanogens. Experimental proof has yet to be established. Recently an integral membrane protein complex from the eury-archaeon *Pyrodictium abyssi* has been described (Dirmeier et al 1998) which among its nine polypeptide subunits contains iron sulfur proteins, nickel, copper, b- and c-cytochromes. It exhibits both hydrogenase and sulfur reductase activities. Therefore, it may be a suitable candidate for energy conservation by generation of a proton motive force in this strictly anaerobic organism, though its orientation and topology within the plasma membrane remains to be investigated.

For both sulfur-reducing and sulfur-oxidizing acidophiles, the radically different assumption would have to be made that deacidification of the cytosol is achieved by a proton-pumping ATPase, if the energy derived from elemental sulfur utilization can be conserved exclusively by substrate-level phosphorylation (via adenosine phosphosulfate intermediates). This remains to be investigated.

Conclusions and perspectives

This short review does not present final solutions, rather it draws attention to open questions and future strategies. From this point of view, it appears that the principle mechanisms by which archaea cope with extreme acidity do not differ from those of other microbes. It has been shown that a number of novel membrane proteins and redox systems with unusual composition are involved in bioenergetics of aerobic archaea, whereas the details of primary energy transduction and proton extrusion mechanisms in anaerobic archaeal sulfur metabolism — such as the identification of P-type alkali cation pumps or H^+/cation exchangers — are still a matter of debate. In aerobes such as *S. acidocaldarius* the coexistence of parallel systems and the lack of specific inhibitors hamper the elucidation of their distinct physiological importance and properties.

Unfortunately, genetic systems suitable to solve these problems are not available to date. Therefore, the development of selection markers and stable shuttle vectors replicating in both *Escherichia coli*, for instance, and archaeal hosts, and withstanding extreme temperatures is a major challenge for further research. When transformation systems become accessible by pathway-directed deletions and/or by the construction of overproducers it should be achievable to distinguish single mechanisms and to produce archaeal membrane complexes in amounts suitable for detailed biophysical studies and eventually for crystallization.

References

Adams MWW, Kletzin A 1996 Oxidoreductase-type enzymes and redox proteins involved in fermentative metabolisms of hyperthermophilic Archaea. Adv Protein Chem 48:101–180

Bakker EP 1991 The role of alkali–cation transport in energy coupling of neutrophilic and acidophilic bacteria: an assessment of methods and concepts. FEMS (Fed Eur Microbiol Soc) Microbiol Rev 75: 319–334

Baumeister W, Lembcke G 1992 Structural features of archaebacterial cell envelopes. J Bioener Biomembr 24:567–575

Booth IR 1985 Regulation of cytoplasmic pH in bacteria. Microbiol Rev 49:359–378

Castresana J, Lübben M, Saraste M 1995 New archaebacterial genes coding for redox proteins: implications for the evolution of aerobic metabolism. J Mol Biol 250:202–210

Dirmeier R, Keller M, Frey G, Huber H, Stetter KO 1998 Purification and properties of an extremely thermostable membrane bound sulfur-reducing complex from the hyperthermophilic *Pyrodictium abyssi*. Eur J Biochem 252:486–491

Elferink MGL, De Wit JG, Driessen AJM, Konings WN 1993 Energy-transducing properties of primary proton pumps reconstituted into archaeal bipolar lipid vesicles. Eur J Biochem 214:917–925

Fetter JR, Qian J, Shapleigh J et al 1995 Possible proton relay pathways in cytochrome c oxidase. Proc Natl Acad Sci USA 92:1604–1608

Gleissner M, Elferink MGL, Driessen AJM, Konings WN, Anemüller S, Schäfer G 1994 Generation of proton-motive force by an archaeal terminal quinol oxidase from *Sulfolobus acidocaldarius*. Eur J Biochem 224:983–990

Gleissner M, Kaiser U, Antonopuolos E, Schäfer G 1997 The archaeal SoxABCD complex is a proton pump in *Sulfolobus acidocaldarius*. J Biol Chem 272:8417–8426

Goulbourne E Jr, Matin M, Zychlinsky E, Matin A 1986 Mechanism of ΔpH maintenance in active and inactive cells of an obligately acidophilic bacterium. J Bacteriol 166:59–65

Iwasaki T, Matsuura K, Oshima T 1995 Resolution of the aerobic respiratory system of the thermoacidophilic archaeon, *Sulfolobus* sp. strain 7. The archaeal, terminal oxidase supercomplex is a functional fusion of respiratory complexes III and IV with no c-type cytochromes. J Biol Chem 270:30881–30892

Iwata S, Ostermeier C, Ludwig B, Michel H 1995 Structure at 2.8 Å resolution of cytochrome c oxidase from *Paracoccus denitrificans*. Nature 376:660–669

Kletzin A 1992 Molecular characterization of the *sor* gene, which encodes the sulfur oxygenase/reductase of the thermo-acidophilic archaeon *Desulfurolobus* (*Acidianus*) *ambivalens*. J Bacteriol 174:5854–5859

Kletzin A 1994 Sulfur oxidation and reduction in archaea: sulfur oxygenase/reductase and hydrogenases from the extremely thermophilic and facultatively anaerobic archaeon *Desulfurolobus ambivalens*. Syst Appl Microbiol 16:534–543

Komatsu H, Chong PLG 1998 Low permeability of liposomal membranes composed of bipolar tetraether lipids from thermoacidophilic archaebacterium *Sulfolobus acidocaldarius*. Biochemistry 37:107–115

Komorowsky L, Schäfer G 1998 Heterologous expression of the *SoxE* gene of *Sulfolobus acidocaldarius* and characterization of this archael blue copper protein, in preparation

Konstantinov AA, Siletsky S, Mitchell D, Kaulen A, Gennis RB 1997 The roles of the two proton input channels in cytochrome c oxidase from *Rhodobacter sphaeroides* probed by the effects of site-directed mutations on time-resolved electrogenic intraprotein proton transfer. Proc Natl Acad Sci USA 94:9085–9090

Lübben M 1995 Cytochromes of archaeal electron transfer chains. Biochim Biophys Acta 1229:1–22

Lübben M, Schäfer G 1989 Chemiosmotic energy conversion of the thermoacidophile *Sulfolobus acidocaldarius*: oxidative phosphorylation and the presence of an F_0-related DCCD-binding proteolipid. J Bacteriol 171:6106–6116

Lübben M, Kolmerer B, Saraste M 1992 An archaebacterial terminal oxidase combines core structures of two mitochondrial respiratory complexes. EMBO J 11:805–812

Lübben M, Arnaud S, Castresana J, Warne A, Albracht SPJ, Saraste M 1994 A second terminal oxidase in *Sulfolobus acidocaldarius*. Eur J Biochem 224:151–159

Maier RJ 1996 Respiratory metabolism in hyperthermophilic organisms: hydrogenases, sulfur reductases, and electron transport factors that function at temperatures exceeding 100 °C. Adv Protein Chem 48:35–99

Matin A, Zychlinsky E, Keyhan M, Sachs G 1996 The capacity of *Heliobacter pylori* to generate ionic gradients at low pH is similar to that of bacteria which grow under strongly acidic conditions. Infect Immun 64:1434–1436

Mitchell P 1961 Coupling of phosphorylation to electron and hydrogen transfer by a chemiosmotic type of mechanism. Nature 191:144–148

Mitchell P 1963 Molecule, group and electron translocation through natural membranes. Biochem Soc Symp 22:142

Moll R 1987 Protonenbewegungen an der Membran des Archaeakteriums *Sulfolobus acidocaldarius*. PhD thesis, University of Hannover, Germany

Moll R, Schäfer G 1988 Chemiosmotic H^+ cycling across the plasma membrane of the thermoacidophilic archaebacterium *Sulfolobus acidocaldarius*. FEBS Lett 232:359–363

Purschke W, Schmidt CL, Petersen A, Anemüller S, Schäfer G 1997 The terminal quinol oxidase of the hyperthermophilic archaeon *Desulfurolobus ambivalens* exhibits unusual subunit structure and gene organization. J Bacteriol 179:1344–1353

Riistama S, Hummer G, Puustinen A, Dyer RB, Woodruff WH, Wikström M 1997 Bound water in the proton translocation mechanism of the haem–copper oxidases. FEBS Lett 414:275–280

Schäfer G 1996 Bioenergetics of the archaebacterium *Sulfolobus*. Biochim Biophys Acta 1277:163–200

Schäfer G, Anemüller S, Moll R, Meyer W, Lübben M 1990 Electron transport and energy conservation in the archaebacterium *Sulfolobus acidocaldarius*. FEMS (Fed Eur Microbiol Soc) Microbiol Rev 75:335–348

Schäfer G, Anemüller S, Moll R, Gleissner M, Schmidt CL 1994 Has *Sulfolobus* an archaic respiratory system? Structure function and genes of its components. Sys Appl Microbiol 16:544–555

Schmidt CL, Anemüller S, Schäfer G 1996 Two different respiratory Rieske proteins are expressed in the extreme thermoacidophilic crenarchaeon *Sulfolobus acidocaldarius*: cloning and sequencing of their genes. FEBS Lett 388:43–46

Stetter KO 1996 Hyperthermophilic procaryotes. FEMS (Fed Eur Microbiol Soc) Microbiol Rev 18:149–158

Tsukihara T, Aoyama H, Yamashita E et al 1996 The whole structure of the 13-subunit oxidized cytochrome c oxidase at 2.8 Angström. Science 272:1136–1144

Van de Vossenberg JCLM, Driessen AJM, Zillig W, Konings WN 1998 Bioenergetics and cytoplasmic membrane stability of the extreme acidophilic thermophilic archaeon *Picrophilus oshimae*. Extremophiles 2:67–74

Verkovskaya ML, García-Horsman A, Puustinen A et al 1997 Glutamic acid 286 in subunit I of cytochrome *bo* is involved in proton translocation. Proc Natl Acad Sci USA 94:10128–10131

Vygodina TW, Pecoraro C, Mitchell D, Gennis R, Konstantinov AA 1998 Mechanism of inhibition of electron transfer by amino acid replacement K362M in a proton channel of *Rhodobacter sphaeroides* cytochrome c oxidase. Biochemistry 37:3053–3061

Wikström M, Bogachev A, Finel M et al 1994 Mechanism of proton translocation by the respiratory oxidases. The histidine cycle. Biochim Biophys Acta 1187:106–111

Wittenberg R 1996 Charakterisierung der Elektronen Transportkette und Untersuchungen zur Bioenergetik von Natronobakterium pharaonis. PhD thesis, University of Dortmund, Germany

Woese CR 1993 The archaea: their history and significance. In: Kates M, Kushner DJ, Matheson AT (eds) The biochemistry of archaea. Elsevier Science, Amsterdam, p vvii–xxix

DISCUSSION

Krulwich: How strong are the data which suggest that the SoxABCD complex pumps protons, as opposed to mediating the expected generation of a proton gradient?

Schäfer: In collaboration with Will Konings' laboratory, we reconstituted lipid vesicles that contain only SoxB, i.e. the quinol-oxidizing aa_3 component of the complex. We showed that this part could generate a proton motive force when the system was activated with TMPD (N,N,N′,N′-tetramethyl-p-phenylene-diamine).

Krulwich: But that is just charge separation.

Schäfer: Yes. We were unable to tell if the protons are pumped or if they are chemical protons.

Poole: Can you explain, for the benefit of all, the meaning of 'chemical protons'?

Schäfer: Chemical protons are those which originate outside the vesicle (or cell) from the chemistry of water formation due to vectorial consumption of protons on the inside. The resultant excess of hydroxyl ions produces a gradient between the inside and the outside. In the presence of QH_2 the generation of chemical protons occurs by substrate proton release from coenzyme Q oxidation on the outside of the membrane. The terminal Q oxidase takes up protons for oxygen reduction inside the cell, producing the proton motive force. This is what we call the generation of chemical protons.

We subsequently isolated the SoxABCD complex as an integral protein, but we found that this was difficult to energize by a non-diffusable reductant. However, we solved this by co-reconstituting the complex with a Rieske iron sulfur protein. This complex can be energized when reconstituted in archaebacterial lipid vesicles by cytochrome c. An artificial cytochrome c oxidase is produced; so the vesicles could be energized with an impermeable reductant. Pulses of reduced cytochrome c do not generate chemical protons on the outside because all protons are pumped protons. The H^+/e^- ratio is about two, suggesting a Q cycle-like stoichiometric extrusion of H^+.

Padan: Is it possible to look at all these components in inverted membrane vesicles?

Schäfer: No. There is a problem with most archaebacteria in that they have an extremely rigid and firmly attached surface layer. This surface layer interdigitates with the plasma membrane; there is practically no periplasmic space. Therefore, it is impossible to obtain inverted membrane vesicles when you disrupt *Sulfolobus* cells, you can only obtain patches of membranes.

Konings: The talk on proton pumping by cytochrome oxidase reminds me of the old discussions of Peter Mitchell and Marten Wikström on the role of cytochrome oxidase in proton translocation (see Mitchell 1980). Your experiments on the effect of potassium on the proton/oxygen ratio suggest that there is a pathway that pumps protons and a pathway that does not. Therefore, there must be a mechanism that regulates the flow of electrons to different pathways.

Schäfer: That's an interesting idea but it's difficult to test because we don't have selective inhibitors for any of these electron transport pathways. Thermoacidophilic archaebacteria are insensitive to all known respiratory chain inhibitors with the exception of cyanide, which inhibits both terminal oxidases in *Sulfolobus acidocaldarius*. Also, at present these organisms are not accessible to genetic engineering; therefore, one of the projects in my laboratory is to try to establish a genetic system for *S. acidocaldarius* to get a handle on these pathways.

Skulachev: A simple explanation of your data is that the direction of the membrane potential should be the other way around, i.e. negative inside rather than positive. The terminal oxidase pumps protons, and a proton motive force is produced that is negative on the inside and positive on the outside. There is a large pH gradient, and so the system needs something to convert $\Delta\Psi$ to ΔpH. Potassium channels allow K^+ to equilibrate according to $\Delta\Psi$. Your K^+ potential measurement suggests that $\Delta\Psi$ is about -20 mV and this is the real potential that exists in growing cells. If there is a mixture of some healthy cells and some sick cells, the protons move into the sick cells, so that a large $\Delta\Psi$ (inside positive) is generated. This positive $\Delta\Psi$ is much larger than the negative $\Delta\Psi$ in the healthy cells. A former student of mine, M. Galperin, calculated the distribution of positive and negative charges in *Helicobacterium pylori* membrane proteins and found that they are arranged electrophoretically according to the so-called inside-positive rule. This rule is true for all membrane proteins for which X-ray data are available.

Konings: If this is the case then we should also see this membrane potential in intact membrane vesicles, and we don't.

Schäfer: In our experiments the membrane potential under steady-state conditions has an inside-positive direction. We measured this using probe distribution and correction factors, which is the method used by almost everyone else.

Skulachev: But decreasing the level of pumping is not sufficient to maintain the ΔpH. I am suggesting that the cells are not dead, rather they are sick. This explains your experiment that potassium ions increase the acidic pulse if you add oxygen.

Schäfer: But this implies that the sick cells are still pumping and respiring.

Skulachev: Yes, but the rates are reduced.

Fillingame: The potassium dependence may correspond to a low affinity potassium transport system. It is possible to test this by adding valinomycin. If it were true you would see this increase in proton pumping at much lower concentrations of potassium.

Schäfer: I agree. Another point I wanted to make is that when we added valinomycin, which destroys the membrane potential, we saw a higher rate of proton extrusion, but we did not see an increased amplitude.

Konings: The reason for this is that those ionophores do not solubilize easily in archaeal membranes.

Matin: If we assume that the cells are sick, Vladimir Skulachev may have a valid point when the external pH is 3.5, although I don't entirely agree with what he said. These organisms have an optimal pH of 1.5, and if you maintain the internal pH at 6.5 there is a large thermodynamic barrier against which to extrude H^+.

Skulachev: This is the case if there is a pump, but if chemical protons alone are involved the efficiency of the oxidase is twofold lower, which means that $\Delta\Psi$ must be twofold higher. My prediction is that the SoxABCD complex should operate at a more acidic pH. The second one should operate at a less acidic pH, when the efficiency can be higher.

Dimroth: If there is a reversed membrane potential as you suggest, there should be a mechanism that creates it. Do you know the nature of this mechanism?

Schäfer: The thought is that it is produced by a potassium pump, but we have been unable to find such a pump as yet.

Booth: The amount of potassium you need to create a reversed potential is about 10^4 ions, and you can't measure this by standard methods because even with 18 mM you have about 10^{16} ions per microlitre of cytoplasm.

Dimroth: But if a pump is present the potential should not be inverted. There has to be another mechanism, such as the one given by Vladimir Skulachev.

Konings: A potassium/proton symport would solve the problem, since this will be driven by the ΔpH (inside alkaline) and results in the translocation of positive charge from outside to inside.

Bogachev: This mechanism has to compensate for the negative potential generated by the respiratory chain, so in order to obtain a positive potential it is necessary to transport more cations inside the cell than the amount of protons pumped out by the respiratory chain. In the case of a potassium/proton symport, it would cause the cytoplasm to become acidified.

Booth: This is an argument that goes back a long time. When we reviewed this area in 1985 we looked at the data from Terry Krulwich on membrane vesicles and cells of *Bacillus acidocaldarius* (Guffanti et al 1984). In membrane vesicles, the system was normal in the conventional chemiosmotic sense, i.e. that you pump protons out and generate an alkaline intravesicular pH and membrane potential that is negative on the inside. However, there is a component missing because in cells there was a positive potential inside. When considering bulk proton movements and potassium movements, we have got to remember that most of the arguments about how much potassium has to be moved, and how many protons have to be moved, to generate the internal pH assume that it is necessary to regenerate ΔpH. There is no evidence for this. In fact, the evidence is the opposite. These membranes are proton tight, and the challenge to the pH gradient is small. Most of the protons pumped out of cells are recycled through ATPase and other proton-consuming pathways. Few protons move out and generate the pH gradient. If you could calculate how many by measuring the proton leak, you would have a measure of the actual work that has to be done per generation by potassium ion transport against proton movement. The problem of how many potassium ions have to be moved out is irrelevant. It could be as little as 10^4 cations per generation. If the system is tightly regulated by potential, then there is no problem; mammalian cells have solved these problems years ago.

Skulachev: This hypothesis can be solved using a flow cytometer that separates the cells.

Booth: A fluorimeter would do because you can measure the fluorescence of the cells that are sick by the uptake of the fluorescent dyes.

I have a comment on Günter Schäfer's experiments on potassium and respiration. We repeated these experiments in *Escherichia coli* and observed the same result. We depleted the cells of K^+ and incubated them at pH 5.3 (we didn't go down to pH 3.5), and we saw that the cells respire rapidly but we did not see a net proton extrusion. The membrane potential was about 170 mV, which is unusual because under steady-state growth the membrane potential of *E. coli* is about 100 mV. These cells have Kdp, so they have an efficient potassium transport mechanism, but the same results are also observed in Kdp^- cells. If Kdp is present, and it takes in potassium, the electrogenicity of Kdp inverts the membrane potential, such that there is a positive potential inside the cells. This is only a transient phenomenon in our case, but in your case it could be long term.

Schäfer: We have looked for Kdp- and Trk-like potassium transport systems in *S. acidocaldarius*, but we have not detected any.

Booth: If the K^+ concentration inside the cells is only 18–20 mM, there are about 360 mM negative charges on the DNA and RNA, and these charges must be compensated for with a cation. The question is, what is the cation that is accumulating in those cells? It must be sodium, ammonium or polyamine.

Schäfer: The DNA is bound to positively charged histone-like proteins and other small stabilizing proteins that have been characterized recently in *Methanogens*, *Thermococcus* and *Sulfolobus*.

Bogachev: How can you explain that the positive potential is resistant to gramicidin? Because this suggests that it is not a potential.

Schäfer: It is the diffusion potential of protons. Gramicidin is not ion specific only for alkali ions; it works with protons.

Bogachev: The diffusion potential has to be sensitive to gramicidin, so your data suggest that it is not a Donnan potential.

Schäfer: There may be a strong Donnan potential in the resting state, and it may be due to asymmetric charge distribution and the inside-positive rule. The helix-connecting loops of the terminal oxidase proteins have a 2.5–3.6-fold excess of positive charges inside, which may also create a Donnan potential in the equilibrated state, and not in the steady state.

Cook: I have a question regarding the expression of these two oxidases. In *E. coli* the proton-pumping oxidase is expressed at high oxygen concentration, but do you know anything about the expression of these two possible proton pumps?

Schäfer: We have studied this. Both pumps are expressed constitutively, and we could not change the ratio of expression of these two oxidases by changing any of the growth conditions. It is possible that their oxygen affinity is different, such that *Sulfolobus* can live under micro-aerobic conditions.

Cook: Did exponentially growing cells have the same expression as stationary phase cells?

Schäfer: Yes.

Dilworth: What does the proton-pumping exercise cost these organisms in terms of the percentage of the metabolic activity?

Schäfer: I don't know. If they can use a regular proton-pumping terminal oxidase, they may be able to extrude enough protons to survive under these conditions. Organisms such as *Acidianus ambivalens* only have one terminal oxidase, which belongs to the family of oxidases that cannot pump protons. This means that *A. ambivalens* can only generate a proton motive force using chemical protons. In this case, one would expect an extremely high respiratory turnover, which is what we observe. *A. ambivalens* is chemolitho-autotrophic and oxidizes sulfur to sulfuric acid, and its only carbon source is CO_2. It grows extremely slowly, although it uses up a lot of sulfur and oxygen.

Poole: The same is true for *Thiobacillus* sp.

Skulachev: Azotobacter vinelandii can grow at high oxygen concentration and low ammonium concentration. Under these conditions the respiratory chain contains non-coupled NADH-quinone reductase and cytochrome bd as the quinol oxidase. Bertsova et al (1997) have recently shown that the H^+/e^- ratio of cytochrome bd is one, and not two. This suggests that there is no pump, only chemical protons. The efficiency of the chain composed of non-coupled NADH-quinone reductase and cytochrome bd is fivefold lower than that of the normal respiratory chain, but the rate of respiration is much higher.

Cook: Is there any evidence that the oxidases you have described are essential for aerobic growth? Because in *E. coli* we can construct mutants that are defective in both cytochrome bd and bo, and these mutants still grow aerobically, suggesting that there are other yet undiscovered oxidases present in *E. coli*.

Schäfer: We have some indications that there may be another pathway in *S. acidocaldarius*, although at present we know little about this pathway. We have identified an iron sulfur protein that may be involved.

Fillingame: The *Bacillus stearothermophilus* lipid proton conductance versus temperature curve was similar to that of *E. coli*. How does *B. stearothermophilus* handle the proton conduction problem?

Konings: B. stearothermophilus has a respiratory chain that is 10-fold more active than *B. subtilis*. This high respiration rate causes protons to be pumped out against the leak of protons inward.

Fillingame: Does *B. stearothermophilus* have a decreased growth yield per unit of carbon source?

Konings: B. stearothermophilus has a much lower yield than *B. subtilis*.

Schäfer: This doesn't seem to be the case for *Sulfolobus*. It has a doubling time of about three hours, and we can grow it up to an OD of 1.9. We don't have a chemostat so we can't determine the yield per unit of carbon source.

Poole: Can you tell us more about haeme-A? Why did you describe it as being unusual?

Schäfer: A few years ago Lübben & Morand (1994) detected a type of haeme-A that has two additional prenyl residues in the hydrophobic side chains. They called this haeme-A_S, for *Sulfolobus*, although it is present in other archaea and has also been found in non-archaeal bacteria. Haeme-A_S is the predominant haeme used by this organism. Cytochrome a^{587} is structurally equivalent to cytochrome b, but it hosts haeme-A_S instead of haeme-B; the polarity around the haeme centres generates a spectrum similar to pyridin-haemochrome of isolated haeme-A.

Poole: Is there any structural information?

Schäfer: No. We know the redox potentials, and that there are two different cytochromes of the cytochrome a^{587} type. One is a product of the *soxC* gene of the SoxABCD complex, and the other is a product of a different gene (*soxG*) that is part of the SoxM complex. They have different haeme environments because they have different redox potentials: the haeme-a^{587} in the SoxABCD complex has a redox potential of about +220–250 mV; whereas that in the SoxM complex has two midpoint potentials, which are similar to b-type cytochromes of bc_1 complexes; the redox potentials are about +20 and +100 mV.

Poole: At what pH do you specify the redox potentials?

Schäfer: They are measured in a buffer at pH 6.5.

Matin: Do you have any ideas on how SoxABCD is threaded through the membrane? Is it completely on one side of the membrane?

Schäfer: No. SoxB, SoxC and SoxA are transmembrane proteins. SoxA has one predicted transmembrane helix and the others have 12.

Matin: Therefore, there are regions of this complex that are exposed to the external medium. The protein must be unusual if it can survive when the external pH is 1.0.

Schäfer: The exposed part may not be the functional part of the protein, it may just be a structural part.

Booth: Thiobacillus has a special protein called rusticyanin that is stable at pH 2.0 but not at pH 7.0.

Schäfer: The sulfocyanin protein of *S. acidocaldarius* has 27% similarity to rusticyanin. Another unusual protein in *S. acidocaldarius* that has an extracellular domain and is adapted to low pH is cytochrome b-558/566, which is a cytochrome that is not involved with electron transport. It is heavily glycosylated, indeed it may well be the only known glycosylated cytochrome. This glycosylation may protect it from unfavourable environments.

Poole: When you grow cells at pH values of around 1.0, how does one measure the pH? How does one calibrate the pH electrode?

Konings: That's a good question. It is a tricky thing to do.

Schäfer: We calibrate with commercial standards.

Poole: But they generally only go down to pH 4.

Epstein: You could use HCl, or HCl/KCl mixes.

Poole: I agree, but I wonder how many people actually do that because pH electrodes don't work well at pH 1.0.

Schäfer: For *S. acidocaldarius* we initially prepare the growth medium at pH 2, but we find that the pH drifts during growth.

References

Bertsova YV, Bogachev AV, Skulachev VP 1997 Generation of protonic potential by the bd-type quinol oxidase of *Azotobacter vinelandii*. FEBS Lett 414:369–372

Guffanti AA, Mann M, Sherman TL, Krulwich TA 1984 Patterns of electrochemical proton gradient formation by membrane vesicles from an obligatorily acidophilic bacterium. J Bacteriol 159:448–452

Lübben M, Morand K 1994 Novel prenylated hemes as cofactors of cytochrome oxidases. Archaea have modified hemes A and O. J Biol Chem 269:21473–21479

Mitchell P 1980 Proton motive cytochrome system of mitochondria. Ann NY Acad Sci 341: 564–584

pH homeostasis in acidophiles

A. Matin

Department of Microbiology and Immunology, Stanford University School of Medicine, Sherman Fairchild Science Building, Stanford, CA 94305, USA

Abstract. The acidophilic bacteria comprise an environmentally important group that includes pathogens. A fundamental requirement for existence in strongly acidic environments is generation of a large pH gradient (ΔpH) to maintain the cytoplasmic pH near neutrality and safeguard the acid-labile cell constituents. These organisms require the capacity to extrude H^+ across the thermodynamic barrier imposed by the ΔpH. They also need special transport proteins that can function in strongly acidic environments. The quintessential device by which the acidophiles fulfil these requirements is by generating a positive-inside membrane potential ($\Delta\Psi$) that mitigates both the force against which H^+ must be extruded, as well as the influxing force of protons into the cells. The $\Delta\Psi$ is generated through passive as well as active mechanisms. The former constitutes an H^+ diffusion as well as a Donnan potential, and the latter involves electrogenic circulation of Cl^-, and the operation of an electrogenic H^+/K^+ (Na^+) antiporter.

1999 Bacterial responses to pH. Wiley, Chichester (Novartis Foundation Symposium 221) p 152–166

A large number of bacteria and other organisms can resist, even grow, at acidic pH values. Enteric pathogens, for instance, must survive passage through the stomach where the pH over a 24 h period averages *c.* 1.5. Similarly, intracellular pathogens, namely *Salmonella*, *Brucella*, *Mycobacterium*, *Listeria*, and others, can replicate inside macrophages (Matin et al 1999a) where the pH is *c.* 5. What distinguishes the acidophilic organisms, as they are usually defined, is their requirement of pH values in the range of 2–4 for optimal growth. But this definition leaves the status of a bacterium like *Helicobacter pylori* ambiguous. *H. pylori* colonizes the human stomach and causes gastric ulcer disease. It grows optimally at neutral pH and the lowest pH at which it has been cultivated in the laboratory is 4–4.5 (Matin et al 1996). Yet, it can actively extrude ions at pH 3, and it also exhibits other bioenergetic attributes (Matin et al 1996) that uniquely define acidophilic bacteria (Cobley & Cox 1983, Matin 1990 a,b). Thus, although by conventional definition, it would not be considered an acidophilic bacterium, it resembles them at a deeper level. For this reason *H. pylori* is considered an acidophile in this chapter. It may indeed be that a better simulation of its environmental conditions would lead to its cultivation in the laboratory at pH values resembling its habitat.

Acidophiles are an important group of organisms, encompassing the three major biological lineages, i.e. archaea, bacteria and eukarya. Aside from the environmental niches associated with human disease, they occupy several other habitats: low-lying soils, aquatic environments, geothermal areas and mining regions. An understanding of the molecular and biophysical basis of the capacity to withstand a high H^+ concentration should be useful to control not only gastric ulcer disease, but also the pathogens mentioned above that recruit macrophages for intracellular growth. Moreover, many technologies based on the activities of acidophiles would also benefit from such. Examples include: the control of environmental pollution resulting from acid mine drainage and acid rain; biodesulfurization of fossil fuels; biomining (Brierly 1982); and low pH industrial fermentations. Indeed the range of the latter can be broadened by transfer of acidophilic character to industrially important organisms, thereby relaxing the sterility requirements during industrial fermentations.

Proton gradients

The two parameters connected with proton gradients, i.e. ΔpH and $\Delta\Psi$, have been measured in a large number of acidophilic bacteria, including their spheroplasts (Matin & Matin 1982) and membrane vesicles. A wide variety of techniques have been employed: accumulation of radiolabelled probes, as determined by flow dialysis or rapid centrifugation; cell lysis; nuclear magnetic resonance; and electron spin resonance (ESR) measurement (Hsung & Haug 1977, Oshima et al 1977, Cox et al 1979, Matin et al 1982, Matin 1990a). As in the work on other bacteria, the non-specific binding of probes has been a concern (Kashket 1985, Michels & Bakker 1985). However, different methods give similar results, including the ESR measurements (McLaggan et al 1989). The ESR technique exploits the phenomenon of exchange broadening (Melhorn et al 1982) to differentiate the signal of spin-labelled probes on two sides of the membrane. Impermeant paramagnetic molecules are used to quench the extracellular signal, thereby permitting a determination of the concentration of the probe inside the cells, vesicles, etc. Its usefulness in detecting non-specific probe binding is due to the fact that a membrane-associated probe produces a characteristic distortion of the ESR signal, which is not quenched when the cells or vesicles are sonified. In the measurements made with the acidophiles, no signal distortion was observed and the quenching occurred upon appropriate treatment (McLaggan et al 1989).

The proton motive force components for three actively respiring acidophilic bacteria are presented in Table 1; further information may be obtained from other reviews (Cobley & Cox 1983, Matin 1990a). In such cells, the cytoplasmic

TABLE 1 Proton gradients in active cells at pH_o 3

Organism	ΔpH^a	$\Delta\Psi$	Δp^b	*References*
Bacillus coagulans	−198 (6.6)	+50	−173	McLaggan et al 1989
Thermoplasma acidophilum	−150 (5.5)	+60	−90	Matin et al 1982, Matin & Matin 1982
Heliobacter pylori	−193 (6.2)	+81	−112	Matin et al 1996

[a]Numbers in parentheses indicate intracellular pH.
[b]All values are in mV.

pH is in the range of about 6, representing a pH gradient of up to 3.6 units, or some 190 mV. An unusual feature of these bacteria is that their $\Delta\Psi$ is positive inside. Thus, this $\Delta\Psi$ subtracts from the chemical H^+ gradient instead of augmenting it, as in other organisms. The magnitude of the proton motive force (Δp) is comparable to other organisms in many acidophiles, but not in all; the lower values could be a reflection of the fact that in many measurements dense cell suspensions were used that may have been oxygen limited (Zilberstein et al 1984).

The positive-inside $\Delta\Psi$ facilitates the existence of a large pH gradient that these bacteria must maintain to safeguard the integrity of acid-labile cytoplasmic components. By subtracting from the H^+ chemical gradient that the ΔpH imposes, the positive-inside $\Delta\Psi$ mitigates both the force against which protons must be extruded, as well as the force impelling them into the cells. The protective role of $\Delta\Psi$ becomes even more important in metabolically less active cells: as their capacity to extrude protons is compromised, the increasing $\Delta\Psi$ reduces the impelling force of H^+ influx, thus preventing rampant cytoplasmic protonation.

Nature of the membrane potential

The nature of the positive-inside membrane potential in the acidophilic mode of existence has attracted considerable interest. It is a complex phenomenon. In part, the $\Delta\Psi$ is due to two passive mechanisms, the H^+ diffusion potential and the Donnan potential. But the active circulation of ions other than H^+ also has a role.

H^+ diffusion potential

As the membrane capacitance is low, even a small influx of H^+ will generate a significant amount of positive-inside membrane potential. That this mechanism contributes to the $\Delta\Psi$ is suggested by several findings. As the diffusion potential

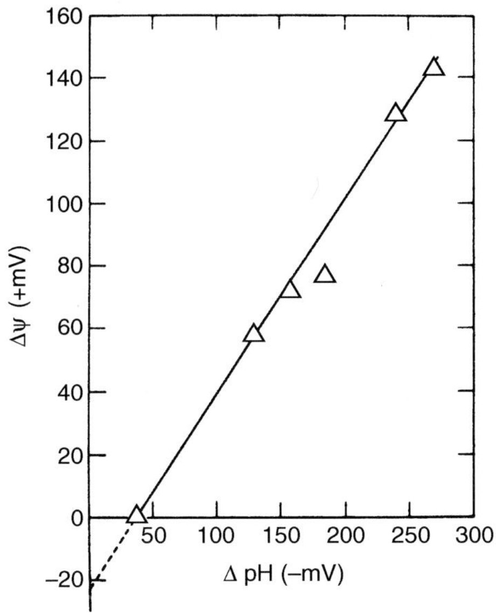

FIG. 1. $\Delta\Psi$ in an acidophile (*Thermoplasma acidophilus*) plotted as a function of ΔpH at different external pH values (Matin et al 1982).

is increased, for instance by decreasing the external pH (pH_o), the $\Delta\Psi$ increases proportionately (Fig. 1). Similarly, progressive inhibition of respiration, which promotes net H^+ influx, increases the $\Delta\Psi$ (Fig. 2). Evidence obtained with membrane vesicles of *Bacillus coagulans* has provided direct evidence that $\Delta\Psi$ generated by the H^+ diffusion potential can maintain a ΔpH in these bacteria. When everted membrane vesicles of this bacterium containing β-alanine (pH of 4.5) and 5 mM magnesium sulfate were suspended in pH 5.9 β-alanine buffer (thus imposing an initial ΔpH of 1.4), the intracellular pH (pH_i) decreased only slightly (from 78 to 70 mV) over the next 2 h. Meanwhile, a negative-inside $\Delta\Psi$ of over -50 mV was generated (Fig. 3), roughly balancing the outward-directed H^+ chemical gradient. These parameters were remarkably stable: at 45 h of incubation, a ΔpH of over 0.5 units was still retained, counterbalanced by the $\Delta\Psi$. Both the parameters were fully and rapidly collapsed by the addition of appropriate ionophores, e.g. K^+ plus valinomycin. Similar results were obtained with right-side-out (RSO) vesicles, in which the polarity of the pH gradient was reversed. In contrast, vesicles made from *Bacillus subtilis* failed to show any volume after brief incubation under the above conditions (Matin et al 1999b). Stable maintenance of ΔpH, balanced by a proton diffusion potential of opposite polarity, has been reported in artificial liposomes (Cafiso & Hubbell 1983), but to my knowledge this is the first report of this phenomenon in the membrane vesicles of a bacterium. Vesicles of *B. subtilis* failed to show this phenomenon.

Donnan potential

Figure 2 indicates that in whole cells of an acidophile, more H^+ influx per unit $\Delta\Psi$ generated occurred than is compatible with the role only of proton diffusion potential in the generation of the $\Delta\Psi$. Thus, protons are sequestered inside the cells. This sequestering arises from the pH_i buffering capacity. There is an abundance of titratable groups in the cytoplasm e.g. the amino acid side chains of proteins, phosphorylated groups of nucleic acids and those of metabolic intermediates. The amino groups from arginine, tyrosine and lysine residues outnumber the carboxyl groups from aspartate and glutamate, so that at neutral pH, where all of the amino groups are protonated, the protein molecules have a net positive charge. Decreasing the pH will increase this charge owing to protonation of the imidazole groups of histidine. Below pH 5, further increases in the charge will occur because of the protonation of carboxyl groups of glutamate and aspartate residues. Phosphorylated intermediates also have titratable groups in the region of pH 6. Thus, the non-diffusable net positive

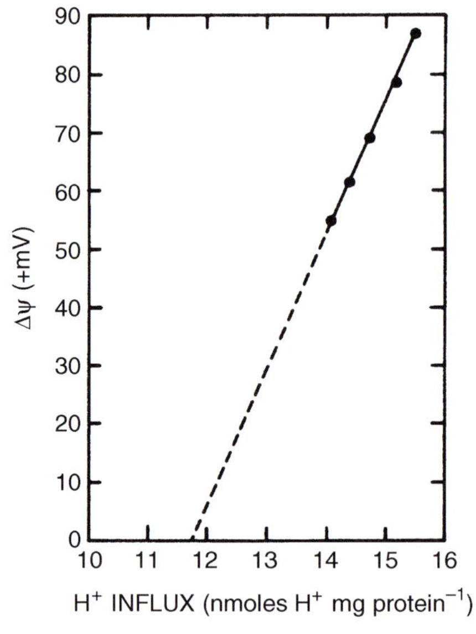

FIG. 2. $\Delta\Psi$ in cell suspensions of an acidophile (*Acidophilum facile*) treated with different azide concentrations plotted as a function of H^+ influx at the corresponding azide concentration (Goulbourne et al 1986).

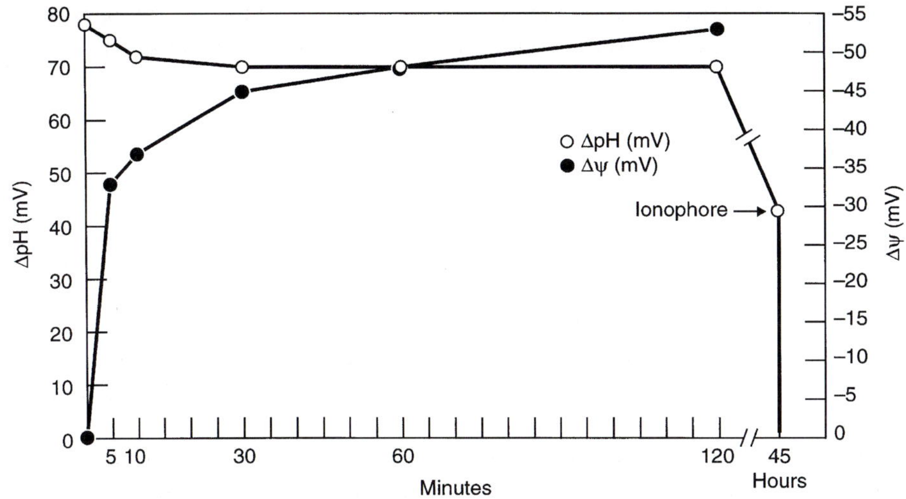

FIG. 3. Generation of $\Delta\Psi$ in everted membrane vesicles of *Bacillus coagulans*, as measured by the electron spin resonance technique. The initial internal pH was 4.5 and external pH, 5.9. See text for further details (Matin et al 1998b).

charge will increase as the cellular pH decreases due to H^+ influx and their sequestering by these molecules. The presence of an impermeable charge inside the cells creates a Donnan potential, and diffusable ions will reach a Donnan equilibrium to meet the demand for overall neutrality on both sides of the membrane (Donnan 1924).

Completely inactive, even dead, cells of acidophiles possess a large positive-inside membrane potential (Matin 1990a,b, Cobley & Cox 1983) that is refractory

TABLE 2 Proton gradients in inactive cells at pH_o 3

Organism	ΔpH[a]	$\Delta\Psi$	Δp[b]	*References*
Bacillus coagulans	−102 (4.7)	+95	−10	McLaggan et al 1989
Thermoplasma acidophilum	−126 (5.1)	+100	−26	Matin et al 1982, Matin & Matin 1982
Acidophilum facile	−132 (5.2)	+107	−25	Matin et al 1996

Inactivation due to starvation, or treatment with ionophores or metabolic poisons.
[a]Numbers in parentheses indicate intracellular pH.
[b]All values are in mV.

to collapse by treatment with any ionophore (Table 2), including a high concentration of SCN^- (Matin et al 1982). This $\Delta\Psi$ is due to the Donnan potential.

Chloride transport

Addition of Cl^- to a cell suspension of an acidophile caused a precipitous but transient decrease in the $\Delta\Psi$, indicating electrogenic Cl^- transport (Fig. 4). Addition of SO_4^{2-}, or PO_4^{3-} failed to produce this effect, indicating that the phenomenon was specific to Cl^-. The collapse of $\Delta\Psi$ from Cl^- addition was transient, but the new steady-state value was some 30% lower than the starting value. The regeneration of the membrane potential is probably due to a variety of causes, including active Cl^- extrusion as well as an electrogenic antiport, as discussed below.

Another indication that chloride movement is electrogenic was obtained by the use of membrane vesicles (of *B. coagulans*) that contained 0.5 M K_2SO_4 with or without other anions (pH_i 5). These vesicles were suspended in different concentrations of K_2SO_4 at pH 5 in an iso-osmolar environment (counter osmolyte Na_2SO_4), in order to impose potassium concentration gradients of different magnitudes. As measured by fluorescence dye quenching, the $\Delta\Psi$ generated by the addition of valinomycin was close to Nernstian with vesicles containing K_2SO_4 and KH_2PO_4 alone. In contrast, vesicles containing KCl exhibited markedly lower than Nernstian $\Delta\Psi$. Thus, while sulfate and phosphate were impermeant, electrogenic efflux of chloride provided a compensatory

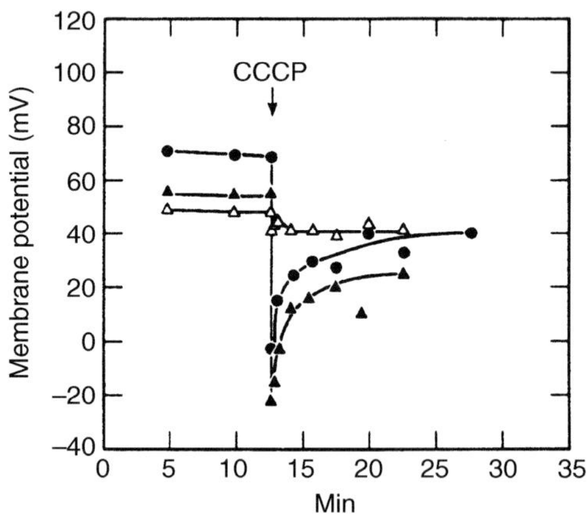

FIG. 4. Effect of addition of anions on the membrane potential of respiring *Bacillus coagulans*. Symbols: closed circles, 100 mM KCl; closed triangles, 100 mM NaCl; open triangles, 100 mM Na_2SO_4 (McLaggan et al 1990). CCCP, carbonyl cyanide m-chlorophenylhydrazone.

conductance to potassium exit from the vesicles, thus mitigating expression of the potassium diffusion potential (McLaggan et al 1990).

Since positive-inside membrane potential is a critical feature of the acidophilic mode of existence, their leakiness to Cl^- is paradoxical. What beneficial role this might have will be discussed below. However, at 300 mM concentration chloride becomes inhibitory to the growth of these bacteria; sulfate at a similar concentration has no inhibitory effect.

But Cl^- was inhibitory only at high concentrations; no effect on growth was seen up to a concentration of KCl or NaCl in the medium of 50 mM. In contrast, the lipophilic anion SCN^- drastically reduced the growth rate and yield at 1 mM, and was completely inhibitory at 10 mM concentration. This could be because Cl^- was less permeant than SCN^-. However, there is evidence for an active extrusion mechanism for Cl^-. In non-respiring cells of *B. coagulans*, Cl^- accumulated, but this accumulation was minimized in actively respiring cells. Thus, in the presence of 46 mM external Cl^-, respiring cells possessed 27 mM chloride, whereas a sample of the same culture treated with 10 mM azide (which abolished oxygen uptake by the cells) contained 112 mM chloride in a similar environment. This corresponds to a $\Delta\bar{\mu}_{Cl^-}$ of -63 and -37 mV for respiring and respiration-inhibited cells, respectively, and thus it appears that energized cells can actively exclude chloride from the cytoplasm. There was also an electrically silent component to chloride permeability mechanism with a PCl^- as high as 1.5×10^{-4}. This, and the activation energy calculations, suggested a specific mechanism for chloride flux (McLaggan et al 1990). Cl^- permeability of another acidophile has been reported (Alexander et al 1987).

Electrogenic cationic antiport

Suggestive evidence for the presence of an electrogenic H^+/K^+(or Na^+) antiporter comes from studies with membrane vesicles of *B. coagulans*. RSO membrane vesicles, loaded with 10 mM ATP at pH_i 5.0 (β-alanine buffer) were suspended in the same pH buffer. Addition of 20 mM K_2SO_4 to the external phase increased the internal pH to 5.5, and concurrently generated a negative-inside membrane potential of 30 mV. Thus, an electrogenic exchange of internal H^+ for more than one external K^+ probably occurred (Matin et al 1999b).

The roles of H^+ diffusion and Donnan potentials, and active ionic circulation

To provide our current view of the roles of these mechanisms in the acidophilic way of life, it is useful first to consider two extreme situations, that of actively respiring steady-state cells, and of completely inactive cells.

It is probable that most of the $\Delta\Psi$ exhibited by the fully active cells is H^+ diffusion potential. The $\Delta\Psi$ mitigates both the force against which protons must be extruded, as well as the force impelling them back into the cells. Consequently, H^+ circulation goes on essentially as it does in neutrophiles (Zychlinsky & Matin 1983a) with a similar magnitude of Δp, and coupling of Δp to the energy requiring processes of the cells such as ATP synthesis, nutrient uptake, etc. (Fig. 5).

In completely inactive cells (Table 2), one finds a ΔpH of up to 2 pH units, counter-balanced by a positive-inside membrane potential of about equal and opposite magnitude. The membrane potential under these conditions, as discussed above, is almost certainly a Donnan potential, and the fact that ΔpH exists in such cells does not have any physiological significance because this situation can be found in dead cells.

The interplay of H^+ diffusion potential, the Donnan potential and the active circulation of co- and counter-ions has most likely a major significance in cells fluctuating between active and inactive metabolic states. Clearly, as in other organisms, the ability of acidophiles to survive an inactive state is finite. Upon prolonged starvation, and/or exposure to other deleterious circumstances, they lose viability. In this state, of course, the equilibrium states of ΔpH and $\Delta\Psi$ and the propensity for co- and counter-ionic movements are of no help to the organism.

But of great importance for survival in nature is the ability to deal with semi-starvation conditions (Matin et al 1999a). As such conditions slow down the H^+

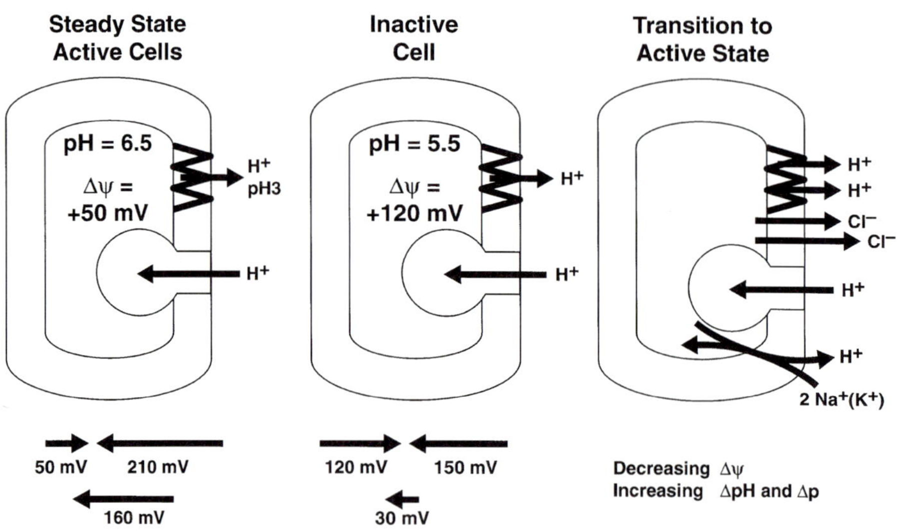

FIG. 5. Ionic circulation in active, inactive and rejuvenating cells of acidophiles (Matin et al 1999b). See text for details.

pumping capacity of the organism, there is a proportionate increase in the net H^+ influx into the cells. A mild influx perhaps initially is countered with an increase primarily of the H^+ diffusion potential. Soon, however, the influxing protons are neutralized by complexation with cellular macromolecules generating the Donnan potential, as discussed above. The protonation of the intracellular constituents is probably compatible with survival only up to a point, but starvation experiments and other evidence indicate that significant protection is afforded (Zychlinksky & Matin 1983b, Goulbourne et al 1986).

Upon return of nutrient proficiency, the cell must gear up for rapid growth and this would require extrusion of large amount of protons, the buffering of which protected it from imminent death during starvation. Further, this extrusion must be accomplished without generating a negative-inside $\Delta\Psi$, otherwise the energy barrier against which the H^+ must be extruded would become insurmountable. It is under these conditions that the other co- and counter-ionic movements assume primary significance. The electrically silent component of Cl^- transport probably permits maintenance of a low concentration of Cl^- or other permeant (McLaggan et al 1990) anions in the cells. In rejuvenating cells, co-movement of Cl^-, would make H^+ extrusion electrically silent. Moreover, the putative electrogenic antiporter would actively generate a positive-inside membrane potential, facilitating H^+ extrusion.

Special features/targets of acidophiles

The generation of ionic circulation at low pH appears to be a unique property of acidophiles (Matin et al 1996). Transport proteins traverse the membrane and are in part exposed to the external environment. Even the proteins required for cytochrome c assembly are transported into the bacterial periplasm (Page et al 1998), which is by and large in H^+ equilibrium with the external phase. For these and other exposed proteins to function at highly acidic pH values, they must have unique structural features. The sequence of many electron transport chain proteins in *H. pylori* does indeed appear to be different from their homologues found in neutrophilic organisms (Tomb et al 1997), and these differences may lie primarily in the exposed part of these proteins. Such unique structures can serve as epitopes for specifically targeting acidophilic pathogens like *H. pylori* (Matin et al 1996).

Acidophiles probably also possess unusual membrane features. In our own experiments, exposure of membrane vesicles of *B. subtilis* to a low pH led to their disintegration, as discussed above, whereas *B. coagulans* vesicles were stable and maintained H^+ gradients for prolonged periods. The occurrence of unusual ω-cyclo-hexyl fatty acids has been proposed (Sunamoto et al 1982) to play an important role in acidophilic bacteria by influencing the fluidity of their membranes, and this may have a role in this phenomenon. The acidophiles may also have unique protein components with isoelectric points facilitating generation of a Donnan potential to poise the cytoplasmic pH at near neutrality. These and other possibilities are now amenable to ready exploration given that the *H. pylori* genome has been sequenced and that tools for molecular dissection of bacteria are now in hand.

Acknowledgement

I thank Sara White for excellent secretarial help in the preparation of this article.

References

Alexander B, Leach S, Ingledew WJ 1987 The relationship between chemiosmotic parameters and sensitivity to anions and organic acids in the acidophile *Thiobacillus ferroxidans*. J Gen Microbiol 133:1171–1179

Brierly CL 1982 Microbiological mining. Sci Am 247:44

Cafiso DS, Hubbell WL 1983 Electrogenic H^+/OH^- movement across phospholipid vesicles measured by spin-labelled hydrophobic ions. Biophys J 44:49–57

Cobley JG, Cox JC 1983 Energy conservation in acidophilic bacteria. Microbiol Rev 47:579–595

Cox JC, Nicholls DG, Ingledew WJ 1979 Transmembrane electrical potential and transmembrane pH gradient in acidophile *Thiobacillus ferroxidans*. Biochem J 178:195–200

Donnan FG 1924 The theory of membrane equilibria. Chem Rev 1:73–90

Goulbourne E Jr, Matin M, Zychlinsky E, Matin A 1986 Mechanism of ΔpH maintenance in active and inactive cells of an obligately acidophilic bacterium. J Bacteriol 166:59–65

Hsung JC, Haug A 1977 Membrane potential of *Thermoplasma acidophila*. FEBS Lett 73:47–50

Kashket ER 1985 The proton motive force in bacteria: a critical assessment of methods. Ann Rev Microbiol 39:219–242

Matin A 1990a Keeping a neutral cytoplasm: the bioenergetics of obligate acidophiles. FEMS Microbiol Rev 75:307–318

Matin A 1990b Bioenergetics parameters and transport in obligate acidophiles. Biochim Biophys Acta 1018:267–270

Matin A, Matin MK 1982 The protonmotive force and the ΔpH in spheroplasts of an acidophilic bacterium (*Thiobacillus acidophilus*). J Gen Microbiol 128:3071–3075

Matin A, Wilson B, Zychlinsky E, Matin M 1982 Proton motive force and the physiological basis of ΔpH maintenance in *Thiobacillus acidophilus*. J Bacteriol 150:582–591

Matin A, Zychlinsky E, Keyhan M, Sachs G 1996 The capacity of *Helicbacter pylori* to generate ionic gradients at low pH is similar to that of bacteria which grow under strongly acidic conditions. Infect Immun 64:1434–1436

Matin A, Baetens M, Pandza S, Park C-H, Waggoner S 1999a Survival strategies in the stationary phase. In: Rosenberg E (ed) Microbial ecology and infectious disease. ASM Press, Washington, DC

Matin A, Neitchman W, Packer L, Keyhan M 1999b pH gradients and membrane potentials in membrane vesicles of *Bacillus coagulans*, in preparation

McLaggan D, Belkin S, Packer L, Matin A 1989 Electron spin resonance measurements of the effect of ionophores on the transmembrane pH gradient of an acidophilic bacterium. Arch Biochem Biophys 273:206–214

McLaggan D, Keyhan M, Matin A 1990 Chloride transport pathways and their bioenergetic implications in the obligate acidophile, *Bacillus coagulans*. J Bacteriol 172:1485–1490

Melhorn RJ, Candau P, Packer L 1982 Measurements of volumes and electrochemical gradients with spin probes in membrane vesicles. Methods Enzymol 88:751–762

Michels M, Bakker EP 1985 Generation of a large protonophore-sensitive proton motive force and pH difference in the acidophilic bacteria *Thermoplasma acidophilum* and *Bacilus acidocaldarius*. J Bacteriol 161:231–237

Oshima T, Arakawa H, Baba M 1977 Biochemical studies of acidophilic, thermophilic bacteria, *Bacillus acidocaldarius*: isolation of bacteria, intracellular pH, and stabilities of biopolymers. J Biochem 81:1107–1113

Page MD, Sambongi Y, Ferguson SJ 1998 Contrasting routes of *c*-type cytochrome assembly in mitochondria, chloroplasts and bacteria. Trends Biochem Sci 23:103–108

Sunamoto J, Wamoto K, Inoue K, Endo T, Nojima S 1982 Liposome membranes. XI. A suggestion of structural characteristic of acido-thermophilic bacterial membranes. Biochim Biophys Acta 685:283–288

Tomb JF, White O, Kerlavage AR et al 1997 The complete genome sequence of the gastric pathogen *Helicobacter pylori*. Nature 388:539–547

Zilberstein D, Agmon V, Schuldiner S, Padan E 1984 *Escherichia coli* intracellular pH, membrane potential, and cell growth. J Bacteriol 158:246–252

Zychlinsky E, Matin A 1983a Cytoplasmic pH homeostasis in an acidophilic bacterium, *Thiobacillus acidophilus*. J Bacteriol 156:1352–1355

Zychlinsky E, Matin A 1983b Effect of starvation on cytoplasmic pH, proton motive force, and viability of an acidophilic bacterium, *Thiobacillus acidophilus*. J Bacteriol 153:371–374

DISCUSSION

Konings: In the absence of energy, or in other conditions in which the cells do not metabolize effectively, protons move in until the proton motive force is essentially zero. This generates a stable situation because there is no reason why the protons should move, but can the cells survive forever under those conditions?

Matin: They don't survive forever. What presumably kills the cells is the excessive protonation of its essential proteins.

Konings: You said that the Donnan potential contributes to $\Delta\Psi$. This is true for every organism, as Peter Mitchell pointed out (Mitchell 1981). However, the contribution is small—in the order of 10–20 mV—and Peter Mitchell calculated that this contribution could not be significantly higher.

Matin: I am not sure that this necessarily applies to acidophiles. It's possible that acidophiles have special proteins that have a greater capacity to generate a positive potential.

Konings: But going beyond 20 mV requires an unreasonably high concentration of immobile negative charges. I suggest that you look at the effect on $\Delta\Psi$ and ΔpH in the presence of high concentrations of permeable ions because if you have a high gradient of permeable ions you should observe a lower $\Delta\Psi$.

Matin: If there is a high concentration of sodium chloride, for example, outside why would the Donnan potential decrease? Is it because the chloride ions would move in?

Konings: The chloride ions would move in, so the ratio of the charges would shift, and the Donnan potential is dependent on the localization of the negative charges on the inside and outside.

Epstein: What is the ionic extracellular salt concentration in your experiments?

Matin: The buffer we used is 0.1 M β-alanine.

Stock: How do you know that this is not a surface potential?

Matin: Because you would not see it decrease in the active cells. The large shifts in $\Delta\Psi$ in the transitions from an active to inactive state speak against a static mechanism.

Konings: The surface potential is not important in this case.

Stock: Why do you say that? When you measure the membrane potential across the envelope you include both the Donnan potential across outer portions of the envelope as well as the inner membrane.

Konings: But the surface potential has nothing to do with the membrane potential.

Padan: If a primary pump does not drive the observed positive potential, then at a certain electron transport rate proton leakage should occur. Effectively, this would result in electron transport uncoupling. You should look at the levels of ATP under these conditions, because ATP cannot be made when the electron transport rate is low.

Schäfer: We did this a long time ago in *Sulfolobus acidocaldarius* cells. We found that when we inhibited respiration the H^+/e^- ratio stayed constant, but following uncoupling, $\Delta\Psi$ became positive inside and ATP broke down, which coincides with the breakdown of the proton motive force (Lübben & Schäfer 1989).

Padan: The situation is different in *Escherichia coli* because in *E. coli* every proton that is extruded can be used to synthesize ATP. It is also the case for mitochondria. However, this is not true in your case because there is proton leakage.

Matin: In the inactive state the potential is due to the protons that are not extruded, and not the movement of protons back in. When the cells go from a completely inactive to an active state they displace a large number of protons. However, at a particular chemical gradient of protons, given the strength of the pump, it is not possible to remove all the protons, so a certain residual diffusion potential remains. One line of evidence that supports this is that if you raise the external pH to 4, then the ΔpH disappears because you decrease the proton gradient and the pump can now cope with the opposing force.

Booth: Helicobacterium pylori is an interesting organism. The paper by Meyer-Rosberg et al (1996) contains a bioenergetic analysis of *H. pylori*. At the range they studied it behaved just like *E. coli*. There are also some unusual observations with urease; it was the reverse of what everyone expected, i.e. urea was toxic at pH 7 in the absence of urease, whereas it was not toxic at low pH.

Matin: We found that in *H. pylori* at pH 3.0 the proton motive force is comparable to what is found in many acidophiles, even though we have not been able to cultivate the organism in the laboratory at that pH. When we increase the pH to 7.0 we observe the same bioenergetic characteristics that are found in neutrophilic bacteria. A current notion is that inside the gel which surrounds the stomach epithelium, the pH is much higher than in the lumen of the stomach (Engel et al 1984). I can't see how that would be the case because I cannot see how the gel can be a proton barrier. The pH of the gel has to be low, and in order for urease to neutralize that pH, the concentration of urea in the blood plasma has to be much higher than it is.

Booth: We talk about the stomach having a pH of 1.5, but that's only true for a fasted stomach. My gastroenterology colleagues tell me that within 5 min of eating a meal the stomach has a pH of 6.0, and it remains at pH 6.0 until the food passes out. You were talking about the fact that *H. pylori* has to have acid-resistant proteins in the periplasm, but it also has to be an organism that can cope with pH cycling. This is why I always liked the mucous idea, because this might protect *H. pylori* from rapid transitions in pH.

Matin: I have no doubt that this gel is protective, and that the urease has a role in enabling the bacteria to colonize. It is likely that it has to survive a large external pH range, not only within the stomach but also outside. However, over a 24 h period the average pH is 1.5. Of course it does depend on the host's eating habits: a lot of people eat regular meals and don't eat in between, and in these cases the pH most of the time is low.

Booth: It has been suggested that children acquire their buccal flora from their mother (Caufield et al 1988, Caufield & Walker 1989). They are now using this to trace the genealogy of the microbial flora. Is it the case for *H. pylori* that it comes in, and those people who are infected become infected at an early age but it only inflames later?

Matin: Fifty per cent of the world's population is infected with *H. pylori*, and from what I understand, it is acquired through ingestion. I'm not aware of any cases of inherited transfer or of any systematic evidence for later inflammation. It is possible that infection with *H. pylori* is a necessary but not sufficient condition for most ulcers.

Rowbury: The percentage is much higher in disadvantaged communities, but there are genetic resistance factors as well.

Bennett: You mentioned a chloride component in your scheme. Could you elaborate on the components involved in the control of chloride transport?

Matin: The chloride transport system appears to have three components, i.e. an active capacity to extrude Cl^-, and electrogenic and electrically silent mechanisms for chloride flux.

Bennett: How do you envisage that chloride ions are extruded specifically?

Matin: I imagine that there is a chloride pump, but I have no direct evidence for this.

Poole: Could you expand on your point on the breakout of proteins from the periplasm.

Matin: H. pylori thrives for considerable periods at low pH. If we assume that the pH in the periplasm is not substantially different from the extracellular pH, and I know that this is controversial, the constituents of the periplasm would have to be unusual because they are stable at low pH. Therefore, they may have unique epitopes that we could use to target *H. pylori* specifically. One problem with making vaccines against *H. pylori* is that antibodies directed against it cross-react with gastric epithelial cells (Sipponen et al 1993). Thus, epitopes unique to *H. pylori* are needed.

Park: One of the differences between *Campylobacter jejuni* and *H. pylori*, given that they are closely related, is that only *H. pylori* has a sheath on the flagellum. Is it possible that the sheath on the flagellum is an adaptation that enables the flagellum to be stable at low pH?

Matin: It is possible, but at present this entire area remains unexplored.

References

Caufield PW, Walker TM 1989 Genetic diversity within *Streptococcus mutans* evident from chromosomal DNA restriction fragment polymorphisms. J Clin Microbiol 27:274–278 (erratum 1989 J Clin Microbiol 27:1918)

Caufield PW, Ratanapridakul K, Allen DN, Cutter GR 1988 Plasmid-containing strains of *Streptococcus mutans* cluster within family and racial cohorts: implications for natural transmission. Infect Immun 56:3216–3220

Engel E, Peskoff A, Kauffman GL Jr, Grossman MI 1984 Analysis of hydrogen ion concentration in the gastric gel mucus layer. Am J Physiol 247:G321–G338

Lübben M, Schäfer G 1989 Chemiosmotic energy conversion of the archaebacterial thermoacidophile *Sulfolobus acidocaldarius*: oxidative phosphorylation and the presence of an F_0-related N,N′-DCCD-binding proteolipid. J Bacteriol 171:6106–6116

Meyer-Rosberg K, Scott DR, Rex D, Melchers K, Sachs G 1996 The effect of environmental pH on the proton motive force of *Helicobacter pylori*. Gastroenterol 111:886–900

Mitchell P 1981 Bioenergetic aspects of unity in biochemistry: evolution of the concept of ligand conduction in chemical, osmotic, and chemiosmotic reaction mechanisms. In: Semenza G (ed) Of oxygen, fuels, and living matter. Part 1. Wiley, Chichester, p 1–160

Sipponen P, Siurala N, Goodwin CS 1993 Histology and ultrastructure of *H. pylori* infections: gastritis, duodenitis and peptic ulceration, and their relevance as pre-cancerous conditions. In: Goodwin CS, Worsley BW (eds) *Helicobacter pylori*. CRC Press, Boca Raton, FL, p 37–62

pH tolerance in *Bacillus*: alkaliphiles versus non-alkaliphiles

Terry Ann Krulwich, Arthur A. Guffanti and Masahiro Ito*

*Department of Biochemistry, Mount Sinai School of Medicine, New York, NY 10029, USA, and *Department of Life Sciences, Toyo University, Oura-gun, Gunma, Japan*

Abstract. Monovalent cation/proton antiporters that catalyse electrogenic uptake of H^+ in exchange for cytoplasmic K^+ and/or Na^+ are centrally involved in bacterial pH homeostasis under alkaline challenge. Systematic attempts have identified some, but not yet all, of the genes encoding such antiporters that participate in pH homeostasis in the neutrophilic *Bacillus subtilis* and the extremely alkaliphilic *Bacillus firmus* OF4. In each organism there are at least three distinct antiporters involved in pH homeostasis. They differ in cation requirement, with pH homeostasis specifically utilizing Na^+/H^+ antiport in the alkaliphile and using either Na^+ or K^+/H^+ antiport in *B. subtilis*. Some of the antiporters involved in pH homeostasis are constitutive and are in place to respond to sudden pH shifts, but there is also an inducible component. At least two sets of homologous antiporters (NhaC and Mrp/Pha) function in both alkaliphiles and neutrophiles. An additional antiporter of a different transport protein family, the Gram-positive tetracycline-metal/H^+ antiporter, is important in pH homeostasis in *B. subtilis* but has not yet been shown to be present in any alkaliphile. There are also differences outside of the antiporters themselves that contribute to the greater capacity of the alkaliphiles for pH homeostasis, including cation re-entry capacity and possible surface properties.

1999 Bacterial responses to pH. Wiley, Chichester (Novartis Foundation Symposium 221) p 167–182

The completion of sequencing projects for several prokaryotic genomes leads to the expectation that the study of complex physiological processes will be facilitated because the gene discovery aspect will be made easier by the enormous database. As yet, this gene discovery phase — the identification of the transport systems, and the enzymes and molecules that support bacterial pH homeostasis — is far from complete in any single prokaryote. This chapter will focus on the status and comparison of the findings to date in two species of *Bacillus* that are capable of thriving in rather different ranges of pH on non-fermentable carbon sources, i.e. approximately pH 5.5–8.5 for *Bacillus subtilis* and approximately pH 7.5–11 for alkaliphilic *Bacillus firmus* OF4 (Krulwich et al 1994, Sturr et al 1994). On fermentable carbon sources, the difference in the upper pH limit for growth is retained. However, the optimal pH range for growth shifts downward somewhat, and the relative importance of different active transport, metabolic and passive mechanisms for pH homeostasis may change (Gilmour & Krulwich 1997, Krulwich et al 1997). For pH homeostasis in the alkaline pH range, most of the focus of investigation and identification has been on the monovalent cation/proton antiporters that are central to the process. These antiporters exchange cytoplasmic Na^+ and/or K^+ for external H^+ with a H^+/Na^+ or K^+ ratio greater than unity, i.e. they are electrogenic antiporters that can catalyse net proton uptake during primary extrusion of protons by a proton-translocating respiratory chain (Booth 1985, Krulwich et al 1994, Padan & Schuldiner 1996, Krulwich et al 1997). Mutations in specific antiporter-encoding genes have resulted in loss or partial loss of the normal capacity for pH homeostasis in several bacterial species (Hamamoto et al 1994, Krulwich et al 1997) but in no bacterial strain have all the genes encoding participating antiporters yet been identified and characterized. Since there are multiple antiporters, there are opportunities for second-site mutations changing the expression of one antiporter in order to compensate for mutational loss of another (Cheng et al 1996a). In the well-studied *Escherichia coli*, even a mutant that has deletions or disruptions in three antiporters exhibits no major high pH-related phenotype in the absence of added Na^+, whose cytotoxicity is greatly elevated in such strains (Ohyama et al 1994, Padan & Schuldiner 1996). Probably as yet unidentified K^+/H^+ antiporters have taken over, as such transporters have been described in *E. coli* (Plack & Rosen 1980).

Although there is much more to be achieved in the antiporter gene identification and characterization realm, the broader ion cycle that supports the work of those antiporters must ultimately include other elements. The primary force generator(s) required to establish the transmembrane ion gradients that energize the secondary antiporters may differ in efficacy at high pH in different organisms. Perhaps there will yet turn out to be some antiporters that are themselves directly coupled to energy. There is an ABC-type Na^+ extrusion system in *B. subtilis* but it is not apparently involved in pH homeostasis (Cheng et al 1997). There is an indication that *E. coli* may have a primary sodium resistance mechanism but, again, not clearly with an involvement in pH homeostasis *per se* (Harel-Bronstein et al 1995). Because of sequence similarities to several primary ion-translocating complexes, a primary energy relationship has been considered for one of the emerging antiporter systems, i.e. Pha/Mrp, but the evidence to date supports a secondary nature for that antiport (Hamamoto et al 1994, Putnoky et al 1998, Ito et al 1998). In addition to primary force generators, there must be replenishment of the cytoplasmic cation substrate for antiport to support ongoing antiport activity. The hypotheses concerning these elements in the active cycle for pH homeostasis, as well as hypothesized elements in pH homeostasis outside of that active cycle, will be briefly considered in light of emerging data.

Overall capacity for adjusting to a sudden alkaline pH shift

As shown in Table 1, malate-grown and energized *B. subtilis* and alkaliphilic *B. firmus* OF4 can both regulate and maintain a cytoplasmic pH close to 7.5. When cells equilibrated at pH 7.5 are suddenly shifted to medium at pH 8.5, the alkaliphile only exhibits this pH homeostasis in the presence of Na^+ whereas in *B. subtilis* K^+ is also efficacious. A mutant strain of *B. subtilis* in which the multifunctional *tetA*(L) antiporter gene is disrupted exhibits neither Na^+- nor K^+-dependent pH homeostasis in such shift experiments. The less rigorous challenge of growth indicates that the mutant strain retains some capacity for growth at pH values > 8. This apparently depends upon a second-site mutation that elevates the expression of a partially compensatory antiporter (Cheng et al 1996a, W. Wang & T.A. Krulwich, unpublished results 1998). *B. subtilis* cells that are subjected to a pH $8.5 \rightarrow 10.5$ shift exhibit no capacity for pH homeostasis, whereas *B. firmus* OF4 exhibits the capacity to retain its pre-shift cytoplasmic pH as long as Na^+ is present. In malate-grown and energized cells, the entire ΔpH of over 2 units is dependent upon Na^+, whereas there is a significant Na^+-independent component in glucose-grown and energized cells (Krulwich et al 1994, Gilmour & Krulwich 1997). Among other possible factors is a much higher cytoplasmic buffering capacity in alkaliphilic cells growing fermentatively vs. aerobically (Ruis & Loren 1998).

The experiments shown in Table 1 were conducted at 100 mM Na^+ or K^+. The dominant antiporters in pH homeostasis are likely to possess a high V_{max} and might, as is the case for TetA(L), have a relatively high K_m (Guffanti & Krulwich 1995). In both *B. subtilis* and *B. firmus* OF4 there are data indicating that different antiporters function optimally over different concentration ranges of the cytoplasmic cation, as most clearly shown for Na^+. Figure 1 provides a schematic summary of the antiporters that are currently shown or hypothesized by analogy to play roles in the capacities documented in Table 1.

Antiporters involved in pH homeostasis in *Bacillus subtilis*

In *B. subtilis*, roles for two antiporters in pH homeostasis, i.e. TetA(L) and Mrp, have been at least partially defined. A third antiporter, NhaC, is a homologue of an alkaliphile antiporter of known function (Ito et al 1997) and is also hypothesized to have a role. The TetA(L) antiporter is a multifunctional antiporter that is the product of the long-recognized, chromosomal *tetA*(L) gene that is one of the efflux type of tetracycline (Tc)-resistance genes. As with other Tet proteins, from

TABLE 1 Dependence of pH homeostasis on Na^+ and K^+ upon an alkaline shift in external pH

		pH_i 10 min after pH shift				
		pH 7.5→8.5[a]			*pH 8.5→10.5*[b]	
Strain	*Growth and energization substrate*	*Choline*	*NaCl*	*KCl*	*Na_2CO_3*	*K_2CO_3*
Bacillus firmus OF4	Glucose	–	–	–	8.4 ± 0.2	9.2 ± 0.1
Bacillus firmus OF4	Malate	8.5 ± 0.1	7.5 ± 0.1	8.4 ± 0.1	8.4 ± 0.1	10.5 ± 0.1
Bacillus subtilis BD99	Malate	8.5 ± 0.2	7.5 ± 0.1	7.6 ± 0.1	10.5 ± 0.1	10.5 ± 0.1
Bacillus subtilis JC112 (Δ*tetA* [L])	Malate	8.5 ± 0.2	8.5 ± 0.2	8.5 ± 0.2	–	–

[a]Cells were equilibrated at pH 7.5 in the presence of 100 mM choline, KCl or NaCl. The external pH was quickly raised to pH 8.5 and, after 10 min, the cytoplasmic pH was determined as described in Cheng et al (1996a).
[b]Cells were equilibrated in 50 mM $NaCO_3$ or K_2CO_3 buffer at pH 8.5. The external pH was rapidly raised to 10.5 and, after 10 min, the internal pH was determined as described in Cheng et al (1996a).
Values given are mean ($n = 6$) ± S.D. –, not determined.

both Gram-negative and Gram-positive bacteria (McMurry et al 1980, Yamaguchi et al 1990), TetA(L) catalyses Tc-divalent metal/H^+ exchange (Guffanti & Krulwich 1995). Disruption of the *B. subtilis tetA* (L) gene leads to an increased sensitivity to both low Tc levels (Cheng et al 1996a) and cobalt ions (A. A. Guffanti, unpublished results 1998). However, the more unanticipated finding was that such mutants were alkali and Na^+ sensitive and had a major impairment in both Na^+- and K^+-dependent pH homeostasis and Na^+ resistance (Table 1; Cheng et al 1996a). It is also likely that this protein catalyses metal-coupled efflux of some endogenous substrate at neutral pH, accounting for the reduced growth rate and greater cobalt sensitivity of mutants with non-functional TetA(L) at pH 7 in the absence of added Tc or Na^+ (Cheng et al 1996a, A. A. Guffanti, unpublished results 1998). Studies of the purified and reconstituted TetA(L) transporter confirmed that this protein catalyses both Tc-cobalt/H^+ and $Na^+(K^+)/H^+$ antiport (Cheng et al 1996b). The data were consistent, as would be expected for a participant in pH homeostasis, with an electrogenic antiport. This is in contrast to the electroneutral antiport proposed for Gram-negative Tet proteins (Kaneko et al 1985). The choice of a Tet protein to 'partner' as a monovalent cation/proton

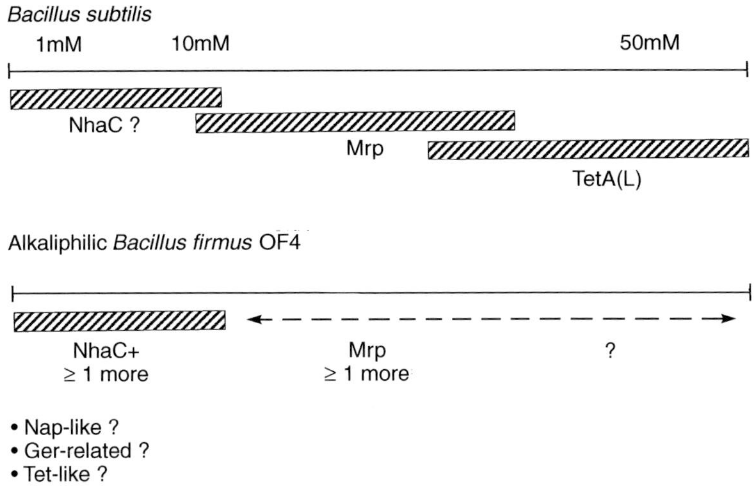

FIG. 1. Diagrammatic representation of status of identification of Na^+/H^+ antiporters that participate in pH homeostasis in different ranges of Na^+ concentration in *Bacillus subtilis* and in *Bacillus firmus* OF4. The genes indicated at the bottom left of the figure and followed by question marks are candidates for additional antiporters that may have roles in one or both of these *Bacillus* species.

antiporter is a satisfying one a priori. Tc is most toxic at low external pH values at which the inwardly directed proton gradient drives non-carrier-mediated accumulation of the protonated antibiotic (Yamaguchi et al 1991). Up-regulation of a Na^+/H^+ antiporter that would reduce or abolish that pH gradient should increase Tc resistance modestly and create a linkage even before the development of multifunctional capacity (Cheng et al 1996c). Na^+ is most toxic at alkaline pH, where the pH homeostasis role of the antiporter would also be important, and thus the multiple functions could have different pH optima and compete little.

When the sequence of the *B. subtilis* genome was completed, several candidates for additional antiporter-encoding genes were evident. So far data have been obtained that support roles for two of these. First, a seven cistron homologue to the *pha* region of *Rhizobium meliloti* (Putnoky et al 1998) was found. In *R. meliloti*, this complex has been proposed to encode a K^+/H^+ antiporter that is required for adjustment of the root nodule bacteria to conditions within the nodule and hence for nitrogen fixation. Homology is also notable to several other proton-translocating complexes, including NADH dehydrogenase (Weidner et al 1993) and the Hyf locus of *E. coli* (Andrews et al 1997). The Pha genes are strong homologues of an incomplete operon that was implicated as a major Na^+/H^+ antiporter-encoding region in alkaliphilic *Bacillus lentus* C-125; a point mutant in the first ORF had a non-alkaliphilic phenotype and apparently reduced secondary Na^+/H^+ antiport activity (Hamamoto et al 1994). We endeavoured to characterize this locus in *B. subtilis*, as well as in the alkaliphilic *B. firmus* OF4. The full sequence was available in *B. subtilis*, having been sequenced by Oudega et al (1997); the locus from *B. firmus* OF4 was cloned using PCR primers based on sequence from the regions that were known for two *Bacillus* loci. As shown in Table 2, all the ORFs of the four partially or completed sequences showed significant sequence similarity to one another, with the two alkaliphilic species showing the greatest similarity, although *B. lentus* C-125 is probably more closely related overall to *B. subtilis* than to *B. firmus* OF4. All seven predicted products of the operons for which complete sequence is known are expected to be hydrophobic.

Disruptions of two different genes of the *pha* homologue from *B. subtilis* were made. Data obtained from pH shift experiments, Na^+ flux experiments and a series of growth experiments indicate that the *B. subtilis* locus has a major role in resistance to Na^+ cytotoxicity and a role in pH homeostasis in a range of Na^+ concentrations below that where TetA(L) functions optimally (Fig. 1). It is also involved in resistance to bile salts, and thus our laboratory has suggested the name Mrp for multiple resistance and pH locus (Ito et al 1998). Initial data indicate that the seven-cistron locus constitutes an operon which must be intact for the antiporter to be expressed and/or active. The full range of functions for the locus and roles of each gene product as catalysts, regulators or assembly factors are yet to be clarified.

Preliminary data indicate that the compensatory mutation in mutants of *B. subtilis* that are disrupted in *tetA*(L) does not result in up-regulation of the *mrp* region but does result in the up-regulation of one (of two) homologues of the *nhaC*-encoded antiporter of *B. firmus* OF4 (W. Wang, unpublished results 1998). As shown in Fig. 2, among the two *B. subtilis* paralogues of the alkaliphile NhaC, YheL exhibits a hydropathy profile somewhat more closely related to that of the alkaliphile, whereas YqkI is slightly more closely related to a putative antiporter from *Haemophilus influenzae*. The closer relationship of YheL to NhaC is reflected in sequence similarity as well. Indeed, the *B. subtilis* mutant that lacks a functional TetA(L) expresses more YheL, but not more Mrp or YqkI, than wild-type *B. subtilis* (W. Wang, unpublished results 1998).

Antiporters involved in pH homeostasis in *Bacillus firmus*

Attempts were made several years ago to assay the aggregate Na^+/H^+ antiport activity of membranes from *B. subtilis* and *B. firmus* OF4 under roughly comparable conditions (Krulwich et al 1994). The alkaliphile was assessed to have about 10-fold the activity of *B. subtilis*. Perhaps the Na^+ specificity of pH homeostasis is necessitated by this high activity, which might otherwise lead to

TABLE 2 Per cent sequence identity and similarity among alkaliphilic and non-alkaliphilic Mrp homologues

Comparison		*Per cent identity*	*Per cent similarity*	*Overlap*
OF4-MrpA	vs. C125-MrpA	79.0	84.3	804 aa
	vs. BS-MrpA	58.0	67.6	800 aa
	vs. RM-PhaA	40.3	49.6	697 aa
	vs. RM-PhaB	54.9	64.7	51 aa
C125-MrpA	vs. BS-MrpA	57.5	67.5	800 aa
	vs. RM-PhaA	41.6	50.4	681 aa
	vs. RM-PhaB	36.5	47.8	115 aa
BS-MrpA	vs. RM-PhaA	39.0	50.0	704 aa
	vs. RM-PhaB	56.9	72.5	51 aa
OF4-MrpB	vs. C125-MrpB	80.9	87.9	141 aa
	vs. BS-MrpB	51.7	66.4	143 aa
	vs. RM-PhaB	31.1	47.9	119 aa
C125-MrpB	vs. BS-MrpB	57.0	66.9	142 aa
	vs. RM-PhaB	36.5	47.8	115 aa
BS-MrpB	vs. RM-PhaB	39.5	51.3	119 aa
OF4-MrpC	vs. C125-MrpC	87.5	92.0	112 aa
	vs. BS-MrpC	63.3	76.1	109 aa
	vs. RM-PhaC	40.4	50.0	108 aa
C125-MrpC	vs. BS-MrpC	63.0	76.9	108 aa
	vs. RM-PhaC	38.5	50.0	108 aa
BS-MrpC	vs. RM-PhaC	43.8	52.8	93 aa
OF4-MrpD	vs. BS-MrpD	51.2	62.3	475 aa
	vs. RM-PhaD	37.1	46.6	533 aa
BS-MrpD	vs. RM-PhaD	35.9	48.5	355 aa
OF4-MrpE	vs. BS-MrpE	49.1	59.8	112 aa
	vs. RM-PhaE	24.6	38.1	135 aa
BS-MrpE	vs. RM-PhaE	17.0	32.1	107 aa
OF4-MrpF	vs. BS-MrpF	46.1	61.8	89 aa
	vs. RM-PhaF	29.8	42.9	84 aa
BS-MrpF	vs. RM-PhaF	28.0	43.9	82 aa
OF4-MrpG	vs. BS-MrpG	45.1	63.7	102 aa
	vs. RM-PhaG	36.5	49.4	85 aa
BS-MrpG	vs. RM-PhaG	30.4	38.0	79 aa

Sequence similarities and identities are shown for alkaliphilic *Bacillus firmus* OF4 Mrp (OF4) and partial operon of *Bacillus lentus* C-125 Mrp (C125), and for non-alkaliphilic *Bacillus subtilis* Mrp (BS) and *Rhizobium melitoti* (RM).

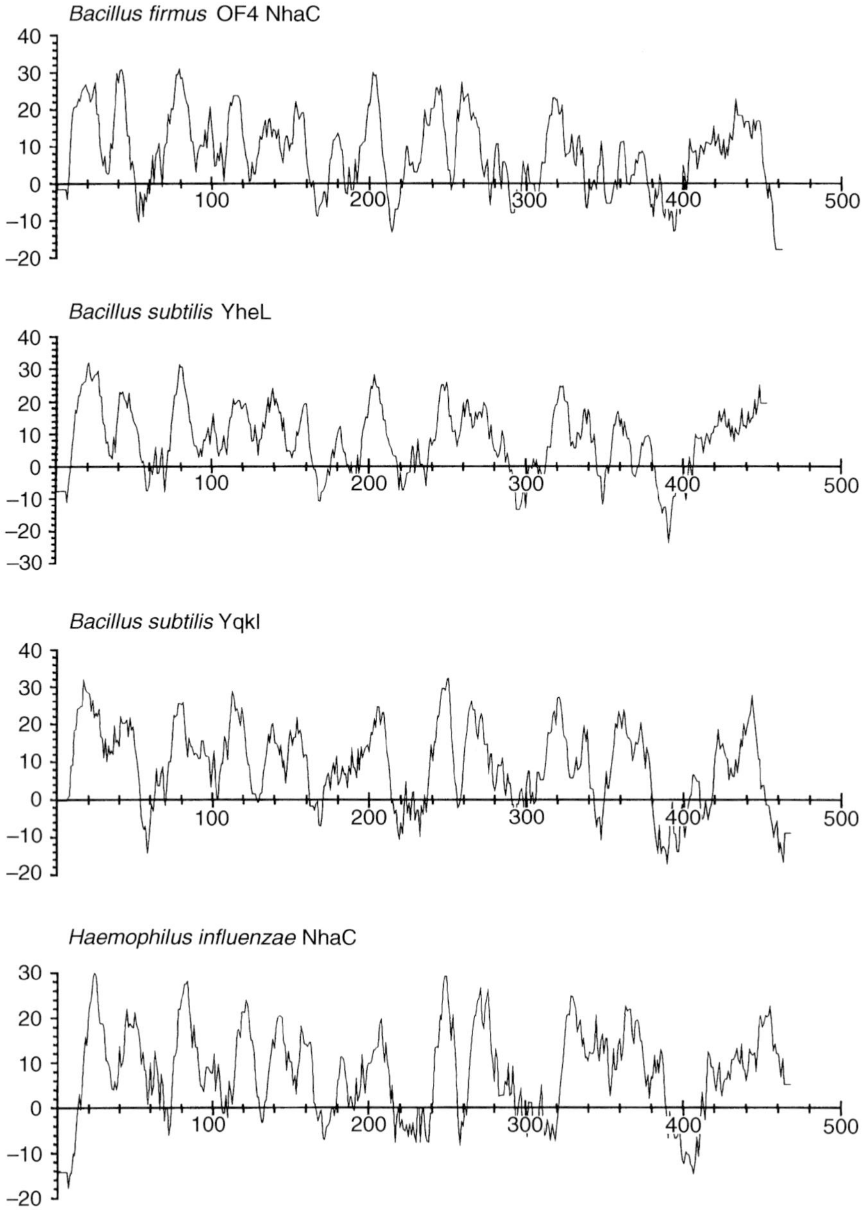

FIG. 2. Hydropathic profiles of *Bacillus firmus* OF4 NhaC, and of *Bacillus subtilis* and *Haemophilus influenzae* homologues.

reduction of the cytoplasmic K^+ concentration. The dominant antiporter(s) are not yet clear. The *nhaC* gene of alkaliphilic *B. firmus* OF4 was the first antiporter-encoding gene to be identified and cloned from an alkaliphile (Ivey et al 1991). Currently modelled on the basis of hydropathy plots as a protein with 12 transmembrane segments (Fig. 3), it was first cloned in a truncated, active form that complemented antiporter-deficient, Na^+-sensitive *E. coli* strains. Cloned *nhaC*, however, did not complement pH homeostasis-minus mutants of either *B. subtilis* or *B. firmus* OF4 for their inability to regulate their cytoplasmic pH during a pH shift. Consistently, disruption of the *nhaC* gene did not yield a non-alkaliphilic phenotype. Rather, the phenotype of the *nhaC*-disrupted mutant, N13, was consistent with this high affinity antiporter having a role in adaptation to high pH at low Na^+ concentrations and also having a major role in Na^+ extrusion at near neutral pH (Ito et al 1997). Interestingly, the alkaliphile actually requires higher Na^+ for growth at near-neutral than at alkaline pH. Perhaps the higher cytoplasmic proton concentration at near-neutral pH inhibits antiporter and/or solute symporters. Studies of pH adaptation in pH 7.5- vs. pH 10.5-grown cells

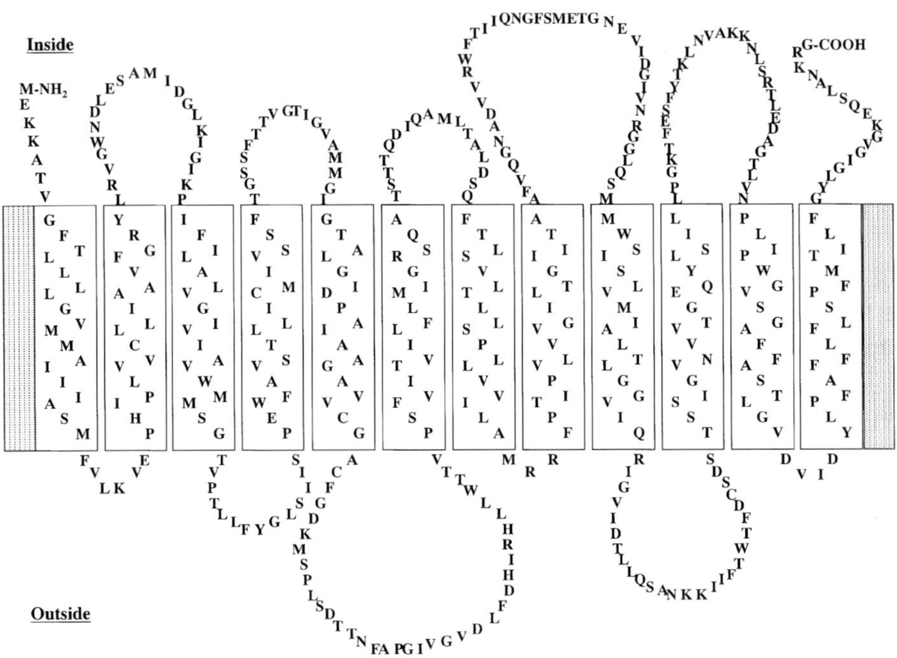

FIG. 3. A schematic representation of the topology of *Bacillus firmus* OF4 NhaC in the membrane based on hydropathy predictions, such as those shown in Fig. 2, and alignments with other transporters that contain 12 putative membrane-spanning regions.

of wild-type and N13 indicated that NhaC is expressed at both pH values and that there are at least two additional antiporters: one is a high affinity antiporter that is induced at high pH; and another is a largely constitutive, lower affinity antiporter that might be the dominant one. Mrp would be a strong candidate for the latter antiporter based on analogy with the findings in *B. lentus* C-125 (Hamamoto et al 1994); notably, this alkaliphile also has multiple antiporters of importance even if Mrp is dominant (Kitada et al 1994).

Other antiporters, transporters and macromolecules

There are candidates for additional antiporter-encoding genes that may be among the complement of antiporters that covers a broad range of Na^+. These include *napA*-like genes (Waser et al 1992, Tani et al 1996), one of which as been found in a spore germination locus in *Bacillus megaterium*. These antiporters are structurally related to Kef systems for K^+ efflux that are apparent channels (Booth et al 1996). There is the interesting possibility that some of the other putative transporters in *ger* loci (Moir & Smith 1990) will also be part of such a complement.

For the alkaliphile in particular, with its reliance upon Na^+-dependent pH homeostasis, re-entry routes for Na^+ are crucial. They are also abundant in the form of the large number of Na^+-coupled solute uptake systems in the alkaliphile (Krulwich et al 1985,1997). However, it is likely that for situations in which such solutes may not abound, pH-regulated Na^+ channels may be an important mode of Na^+ re-entry, e.g. the Na^+ channel associated with the flagella of alkaliphiles (Sugiyama 1995).

Finally, there may be devices that increase the cation concentration near the coupling membrane. Such a function has been hypothesized for a small, putative Na^+-binding protein encoded in the alkaliphile *nhaC* operon (Ito et al 1997). Acidic cell wall polymers have also been suggested to play such a role, which would explain the observation that the presence of such polymers is essential for growth of *B. lentus* C-125 at highly alkaline pH (Aono et al 1995). Recently, *B. firmus* OF4 has been shown to induce a particular S-layer protein at elevated pH (R. Gilmour, unpublished results 1998). For all of these polymers, however, there is no evidence for direct involvement in pH homeostasis as opposed to some common function such as Mg^{2+} sequestration that becomes otherwise growth limiting at high pH.

Conclusions

The central participants in the active cycle that enable prokaryotes to achieve pH homeostasis at alkaline pH are a group of structurally diverse, electrogenic cation/proton antiporters. In a given bacterium, there are several distinct antiporters that function in different ranges of cation concentration. Although antiporters with relatively high V_{max} values may tend to be the dominant antiporter, overexpression of alternate antiporters can at least partially compensate for its mutational loss. Extremely alkaliphilic bacteria specifically rely upon Na^+/H^+ antiport, e.g. *B. firmus* OF4, whereas non-alkaliphilic bacteria such as *B. subtilis* can utilize either $Na^+(K^+)/H^+$ antiporters, some of which may be homologues of alkaliphilic antiporters with broader cation specificity. Cationic polymers are apparently important in cellular adaptation to alkaline growth conditions but there is no evidence yet for the direct involvement of these polymers in pH homeostasis.

Acknowledgements

This work was supported by research grants GM28454 and GM52837 from the National Institutes of General Medical Sciences to T.A.K. and a grant from The Inoue Enryo Memorial Foundation for Promoting Science to M.I.

References

Andrews SC, Berks BC, McClay J et al 1997 A 12-cistron *Escherichia coli* (*hyf*) encoding a putative proton-translocating formate hydrogenlyase system. Microbiology 143:3633–3647

Aono R, Ito M, Jolin KN, Horikoshi K 1995 A high cell wall negative charge is necessary for the growth of the alkaliphilic *Bacillus lentus* C-125 at elevated pH. Microbiology 141:2955–2964

Booth IR 1985 Regulation of cytoplasmic pH in bacteria. Microbiol Rev 49:359–378

Booth IR, Jones MA, Mclaggan D et al 1996 Bacterial ion channels In: Konings WN, Kaback HR, Lolkema JS (eds) Handbook of biological physics II. Transport in membranes. Elsevier Science, Amsterdam, p 693–729

Cheng J, Guffanti AA, Wang W, Krulwich TA, Bechhofer DH 1996a Chromosomal *tetA*(L) gene of *Bacillus subtilis*: regulation of expression and physiology of a *tetA*(L) deletion strain. J Bacteriol 178:2853–2860

Cheng J, Hicks DB, Krulwich TA 1996b The purified *Bacillus subtilis* tetracycline efflux protein TetA(L) reconstitutes both tetracycline-cobalt/H^+ and $Na^+(K^+)/H^+$ exchange. Proc Natl Acad Sci USA 93:14446–14451

Cheng J, Baldwin K, Guffanti AA, Krulwich TA 1996c Na^+/H^+ antiport activity conferred by *Bacillus subtilis tetA*(L), a 5′ truncation product of *tetA*(L), and related plasmid genes upon *Escherichia coli*. Antimicrob Agents Chemother 40:852–857

Cheng J, Guffanti AA, Krulwich TA 1997 A two-gene ABC-type transport system that extrudes Na^+ in *Bacillus subtilis* is induced by ethanol or protonophore. Molec Microbiol 23:1107–1120

Gilmour R, Krulwich TA 1997 Construction and characterization of a mutant of alkaliphilic *Bacillus firmus* OF4 with a disrupted *cta* operon and purification of a novel cytochrome *bd*. J Bacteriol 179:863–870

Guffanti AA, Krulwich TA 1995 Tetracycline/H^+ antiport and Na^+/H^+ antiport catalyzed by the *Bacillus subtilis* TetA(L) transporter expressed in *Escherichia coli*. J Bacteriol 177:4557–4561

Hamamoto T, Hashimoto M, Hino M et al 1994 Characterization of a gene responsible for the Na^+/H^+ antiporter system of alkalophilic *Bacillus* species strain C-125. Mol Microbiol 14: 939–946

Harel-Bronstein M, Dibrov P, Olami Y, Pinner E, Schuldiner S, Padan E 1995 MH1, a second-site revertant of an *Escherichia coli* mutant lacking Na^+/H^+ antiporters ($\Delta nhaA\, \Delta nhaB$), regains Na^+-resistance and a capacity to excrete Na^+ in a$\Delta\bar{\mu}_H$+-independent fashion. J Bacteriol 270:3816–3822

Ito M, Guffanti AA, Zemsky J, Ivey DM, Krulwich TA 1997 Role of the *nhaC*-encoded Na^+/H^+ antiporter of alkaliphilic *Bacillus firmus* OF4. J Bacteriol 179:3851–3857

Ito M, Guffanti AA, Krulwich TA 1998 *mrp*: a multigene, multiple resistance-pH adaptation locus in *Bacillus subtilis*. Abstr General Meeting American Society for Microbiology, p 344

Ivey DM, Guffanti AA, Bossewitch JS, Padan E, Krulwich TA 1991 Molecular cloning and sequencing of a gene from alkaliphilic *Bacillus firmus* OF4 that functionally complements an *Escherichia coli* strain carrying a deletion in the *nhaA* Na^+/H^+ antiporter gene. J Biol Chem 266:23483–23489

Kaneko M, Yamaguchi A, Sawai T 1985 Energetics of tetracycline efflux system encoded by *Tn*10 in *Escherichia coli*. FEBS Lett 193:194–198

Kitada M, Hashimoto M, Kudo T, Horikoshi K 1994 Properties of two different Na^+/H^+ antiport systems in alkaliphilic *Bacillus* sp. Strain C-125. J Bacteriol 162:768–772

Krulwich TA, Federbush JG, Guffanti AA 1985 Presence of a non-metabolizable solute that is translocated with Na^+ enhances Na^+-dependent pH homeostasis in alkalophilic *Bacillus*. J Biol Chem 260:4055–4058

Krulwich TA, Cheng J, Guffanti AA 1994 The role of monovalent cation/proton antiporters in Na^+-resistance and pH homeostasis in *Bacillus*: an alkaliphile *vs* a neutralophile. J Exp Biol 196:457–470

Krulwich TA, Ito M, Gilmour R, Guffanti AA 1997 Mechanisms of cytoplasmic pH regulation in alkaliphilic strains of *Bacillus*. Extremophiles 1:163–169

McMurry L, Petrucci RE Jr, Levy SB 1980 Active efflux of tetracycline encoded by four genetically different tetracycline resistance determinants in *Escherichia coli*. Proc Natl Acad Sci USA 77:3974–3977

Moir A, Smith DA 1990 The genetics of bacterial spore germination. Ann Rev Microbiol 44:531–553

Ohyama T, Igarashi K, Kobayashi H 1994 Physiological role of the *chaA* gene in sodium and calcium circulations at a high pH in *Escherichia coli*. J Bacteriol 176:4311–4315

Oudega B, Koningstein G, Rodriguez L et al 1997 Analysis of the *Bacillus subtilis* genome: cloning and nucleotide sequence of a 62 kb region between 275-(*rrnB*) and 284-(*pai*). Microbiology 143:2769–2774

Padan E, Schuldiner S 1996 Bacterial Na^+/H^+ antiporters — molecular biology, biochemistry and physiology. In: Konings WN, Kaback HR, Lolkema JS (eds) Handbook of biological physics II. Transport in membranes. Elsevier Science, Amsterdam, p 501–531

Plack RH Jr, Rosen BP 1980 Cation/proton antiport systems in *Escherichia coli*: absence of potassium/proton antiporter activity in a pH-sensitive mutant. J Biol Chem 255:3824–3825

Putnoky P, Kereszt A, Nakamura T et al 1998 The *pha* gene cluster of *Rhizobium meliloti* involved in pH adaptation and symbiosis encodes a novel type of K^+-efflux system. Mol Microbiol 28:1091–1101

Ruis N, Loren JG 1998 Buffering capacity and membrane H^+ conductance of neutrophilic and alkalophilic Gram-positive bacteria. Appl Environ Microbiol 64:1344–1349

Sturr MG, Guffanti AA, Krulwich TA 1994 Growth and bioenergetics of alkaliphilic *Bacillus firmus* OF4 in continuous culture at high pH. J Bacteriol 176:3111–3116

Sugiyama S 1995 Na^+-driven flagellar motors as a likely Na^+ re-entry pathway in alkaliphilic bacteria. Mol Microbiol 15:592

Tani K, Watanabe T, Matsuda H, Nasu M, Kondo M 1996 Cloning and sequencing of the spore germination gene of *Bacillus megaterium* ATCC 12872: similarities to the NaH-antiporter gene of *Enterococcus hirae*. Microbiol Immunol 40:99–105

Waser M, Hess-Bienz D, Davies K, Solioz M 1992 Cloning and disruption of a putative Na/H-antiporter gene of *Enterococcus hirae*. J Biol Chem 267:5396–5400

Weidner U, Geier, S, Ptock A, Friedrich T, Leif H, Weiss H 1993 The gene locus of the proton-translocating NADH: ubiquinone oxidoreductase in *Escherichia coli*. J Mol Biol 233:109–122

Yamaguchi A, Udegawa T, Sawai T 1990 Transport of divalent cations with tetracycline as mediated by the transposon *Tn*10-encoded tetracyline resistance protein. J Biol Chem 265:4809–4813

Yamaguchi A, Ohmori H, Kaneko-Ohdera M, Nomura T, Sawai T 1991 ΔpH-dependent accumulation of tetracycline in *Escherichia coli*. Antimicrob Agents Chemother 35:53–56

DISCUSSION

Quivey: We are interested in regulation of ATPases. Do you know anything about the regulation of some of these antiporter redundancies? Is the up-regulation of transcription involved or a shift into the optimal pH range?

Krulwich: We have done many studies on the regulation of Na^+/H^+ antiporter activity and on some of the individual antiporter species. We first tried to sort out the effects of transcription versus translation versus direct effects on activity. With respect to the bulk antiport activity, that of alkaliphilic cells is 10-fold higher than *Bacillus subtilis* under equivalent conditions, and is higher still in some of the variants that can be selected from the chemostat at pH 11.4. If you lower the pH, the activity is greatly decreased, so the antiporters are strongly regulated at the level of activity. We have done some preliminary northern analyses and shown that NhaC in *B. subtilis* may be regulated at the transcriptional level. We don't know anything about the details of this regulation, except that preliminary experiments suggest NhaC is up-regulated in the Tet mutant. The locus we know most about, however, is TetA(L) in *B. subtilis*. There is a major translational component that is regulated by tetracycline. Tet regulation involves a novel stem loop mechanism. We don't know if this multifunctional antiporter is also regulated by cobalt, but the efflux of cobalt and heavy metals by the tetracycline antiport is an important physiological feature of that antiport. In addition, there is a transcriptional component of Tet regulation that responds, but more modestly than the translational effects, to sodium, tetracycline, potassium and high pH. The translational component of regulation causes a higher induction of Tet than transcriptional regulation, i.e. to levels that are toxic to the cell. My assumption is that the translational regulation of Tet is 'do or die'. When confronted by significant antibiotic concentrations, cells make enough Tet such that they are inhibited and the cells become leaky, but it's worth it because otherwise they will be killed by tetracycline. When Tet is not present, the mechanism is not there. Regulation then occurs at the level of transcription of the kind typical for most membrane proteins. This gives the required results in terms of the boost in $Na^+(K^+)/H^+$ antiport-related activities you want without making the cells leak. This is just the beginning of the story. We don't know yet how this co-ordinates with changes in pH.

Quivey: In the chemostat have you ever let Na^+/H^+ antiporter activity come back down and looked at lifetimes?

Krulwich: No.

Skulachev: What is the ion specificity of the tetracycline antiporter?

Krulwich: There are two substrate modes: (1) the tetracycline cobalt complex; and (2) the monovalent cation set of substrates. It is not possible to transport cobalt, manganese or magnesium alone. Yamaguchi et al (1990) showed for the Tn10 transporter, and we've shown the same thing for the TetA(L) transporter (Guffanti & Krulwich 1995, Cheng et al 1996), that radioactive cobalt is not transported in the absence of tetracycline, just as antibiotics are not transported without an appropriate cation, and there is a 1:1 ratio in the complex. In that site you can use cobalt, magnesium or manganese. Calcium didn't do much. The monovalent cation site can use either potassium or sodium, and it is inhibited by lithium, although I don't know if there's a lithium/proton antiport.

Matin: You mentioned that the Tet system was subject to both transcription and translation regulation. What are the factors that induce this system?

Krulwich: Tetracycline induces the translation of Tet by about 26-fold. The transcriptional component is more modest, of the order of two- to fourfold, and it is induced by sodium, potassium, high pH and Tet alone.

Matin: You say that if too much is synthesized it would be harmful to the bacteria. However, in most situations bacteria have a self-limiting response. There is usually a repressor, the concentration of which increases at the same time as the induction of the antibiotic-transporting pump.

Krulwich: In the case of some of the antibiotic resistance mechanisms, there is instead a translational stop mechanism, so that high levels of expression are only induced when the antibiotic is present. The paradigm for this is the erythromycin resistance system that was studied years ago by Horinouchi & Weisblum (1980) and since by Dubnau (1984) and others.

Stock: Why doesn't potassium have a protective effect in alkaliphiles?

Krulwich: Our hypothesis is that alkaliphiles have to maintain such a high antiport activity all the time that they risk reducing the levels of cytoplasmic potassium, and so they have learned to be sodium specific in their antiport that supports pH homeostasis.

Stock: But in the experiments where cells were protected by sodium you add extracellular potassium so the potassium gradient is probably going in the opposite direction, directed inward rather than outward.

Krulwich: That doesn't matter. The antiporters involved in pH homeostasis are electrogenic, and this is the 'efflux limb' of a cycle. The cation effluxes in the antiport part of the cycle, but the potassium or sodium has to come from somewhere. There is an 'entry limb' also to replenish that cation in the cytoplasm.

Stock: If the cells are in low potassium there is an electrochemical gradient of potassium directed outward. When potassium is added to the medium the gradient is in the opposite direction, so if the potassium/proton antiport system were active under these conditions, it would be working in the opposite direction and this would not give protection.

Krulwich: Not if it is electrogenic. The electrogenicity is crucial. We have shown this for both the sodium/proton antiport and the potassium/proton antiport. There is no question that these ions are driven out against their chemical gradient but with their total electrochemical gradient. The sum of these two gradients is the driving force, and there is a large transmembrane potential. These antiports are catalysing charge movement. They are not just exchanging one potassium ion for one proton, they're exchanging, hypothetically, one potassium ion for two protons. Therefore, the result of one net turnover is movement of positive charge inside, so the transmembrane potential is a major part of the driving force. Our interpretation of the alkaliphiles is that they don't recycle because they can't afford to keep losing potassium. They may have not have as many potassium uptake systems as *Escherichia coli*.

Konings: I would like to elaborate on Terry Krulwich's talk. The problem is that *Bacillus alkalophilus* likes to couple everything to protons, whereas for thermophiles the situation is easier. *Thermoalkalkibacter borgoriae* grows at pH 9.5–10.0 and has an internal pH of 7.5–8.0, so the pH gradient is about +120 mV. It pumps out sodium via an ATPase, and in this way generates a sodium gradient in which $\Delta\bar{\mu}_{Na^+}$ is -60 mV. It also generates a $\Delta\Psi$ of -120 mV, so the total sodium motive force is -180 mV. These membranes are permeable to protons, and the protons move inwards as a result of the $\Delta\Psi$ and generate the reversed pH gradient of +120 mV. Therefore, the $\Delta\Psi$ is 120 mV, the ΔpH is +120 mV and the proton motive force is zero. The organism can therefore live happily at high pH because it maintains its internal pH in this way at pH 7.5. The situation for this thermophilic alkaliphile is much easier, therefore, by using sodium ions instead of protons as the coupling ion.

Glenn: Can you elaborate on the amplification of genes in the alkaliphiles, particularly in relation to whether it is constant or whether the extent of amplification is determined by any environmental stresses?

Krulwich: Again, the system we know most about is the Tet system. The original discovery of the *B. subtilis* Tet gene was subsequent to its amplification, so I thought amplification was going to be a normal mechanism of regulation, in addition perhaps to the translational mechanism (Wilson & Morgan 1985, Ives & Bott 1990). However, if you look carefully at the conditions in which the amplifications first occurred, you find that a second gene copy is always involved. If you introduce a second gene copy, you've really caused the amplification. We wondered whether it would happen otherwise, and Bechhofer & Stasinopoulos (1998) have subsequently observed a modest amplification of Tet. Like many other amplifications it involves being near the origin, which the *tetA*(L) gene is, and again it involves two copies, but which are generated naturally and internally. We don't yet know whether amplification is involved for NhaC. We have speculated that it is because we know there are six copies of the direct repeats, some of which are clustered in that region rather than being distributed randomly.

Quivey: I'm interested in the S-layer that you see at high pH. When we grow *Streptococcus mutans* in the chemostat at low pH, we observe high levels of external carbohydrate and increased wall growth, and I would like to know if you have observed this also.

Krulwich: We don't know yet whether there are two, and that what we are seeing is one of them turned on at high pH. We looked for candidates on our gels for the other hypothetical protein that would be increased at low pH, and we have identified candidate proteins.

References

Bechhofer DB, Stasinopoulos SJ 1998 *tetA*(L) mutants of a tetracycline-sensitive strain of *Bacillus subtilis* with the polynucleotide phosphorylase gene deleted. J Bacteriol 180: 3470–3473

Cheng J, Hicks DB, Krulwich TA 1996 The purified *Bacillus subtilis* tetracycline efflux protein TetA(L) reconstitutes both tetracycline-cobalt/H^+ and $Na^+(K^+)/H^+$ exchange. Proc Natl Acad Sci USA 93:14446–14451

Dubnau D 1984 Translational attenuation: the regulation of bacterial resistance to the macrolide-lincosamide-streptogramin B antibiotics. CRC Crit Rev Biochem 16:103–132

Guffanti AA, Krulwich TA 1995 Tetracycline/H^+ antiport and Na^+/H^+ antiport catalyzed by the *Bacillus subtilis* TetA(L) transporter expressed in *Escherichia coli*. J Bacteriol 177:4557–4561

Horinouchi S, Weisblum B 1980 Posttranscriptional modification of mRNA conformation: a mechanism that regulates erthyromycin-induced resistance. Proc Natl Acad Sci USA 77:7079–7083

Ives CL, Bott KF 1990 Characterization of chromosomal DNA amplifications with associated tetracycline resistance in *Bacillus subtilis*. J Bacteriol 172:4936–4944

Wilson CR, Morgan AE 1985 Chromosomal-DNA amplification in *Bacillus subtilis*. J Bacteriol 163:445–453

Yamaguchi A, Udagawa T, Sawai T 1990 Transport of divalent cations with tetracycline as mediated by the transposon *Tn*10-encoded tetracyline resistance protein. J Biol Chem 265:4809–4813

The molecular mechanism of regulation of the NhaA Na^+/H^+ antiporter of *Escherichia coli*, a key transporter in the adaptation to Na^+ and H^+

Etana Padan, Yoram Gerchman, Abraham Rimon, Andrea Rothman, Nir Dover and Orna Carmel-Harel

The Hebrew University of Jerusalem, Department of Microbial and Molecular Ecology, Institute of Life Sciences, 91904 Jerusalem, Israel

Abstract. The NhaA Na^+/H^+ antiporter is the main system responsible for adaptation to Na^+ and alkaline pH (in the presence of Na^+) in *Escherichia coli* and many other enteric bacteria. It is under intricate control. At the protein level it is regulated directly by pH, one of its regulatory signals. A pH shift from 7 to 8.5 activates the antiporter and, in a fashion correlated with the activity change, confers a conformation change that, in isolated membrane vesicles, is reflected in the exposure of trypsin-cleavable sites. H225 and G338 are essential for the pH response of NhaA. *nhaA* transcription is dependent on NhaR, a positive regulator of the LysR family, and is regulated by Na^+, the other environmental signal. Na^+ affects the NhaR/*nhaA* interaction directly by changing the footprint of NhaR on *nhaA* in a pH-dependent fashion. The expression of *nhaA* is also under global regulation of H-NS. We suggest that the pattern of regulation of *nhaA* found in *E. coli* is a paradigm for the response of proteins and genes to H^+ and Na^+, the most common ions that challenge every cell.

1999 Bacterial responses to pH. Wiley, Chichester (Novartis Foundation Symposium 221) p 183–199

In the early 1970s, Peter Mitchell and colleagues (Mitchell 1961, Mitchell & Moyle 1967, West & Mitchell 1974) discovered an activity in biological membranes that couples the fluxes of Na^+ and H^+. He coined the term Na^+/H^+ antiporters for the proteins responsible for this activity and suggested that they are involved in the homeostasis of both Na^+ and H^+ in cells. Since then, Na^+/H^+ antiporter activity has been found in cytoplasmic membranes of many cells and in organellar membranes (review in Padan & Schuldiner 1994). Up to now only one species of

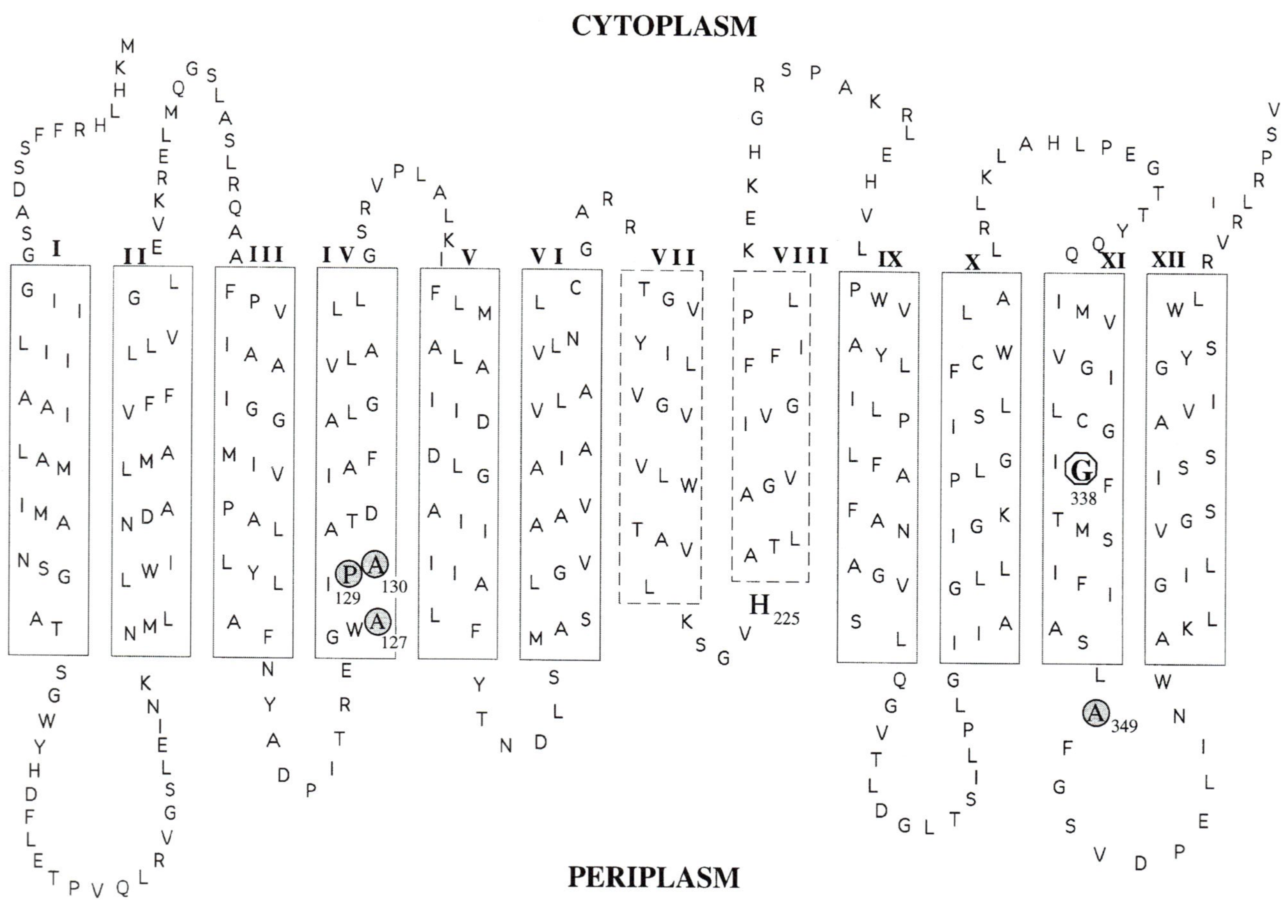
CYTOPLASM
I
II
III
IV
V
VI
VII
VIII
IX
X
XI
XII
H225
338
349
129
130
127
PERIPLASM

bacterium has been found to lack a Na^+/H^+ antiporter, and this is accordingly restricted in its pH range of growth (Speelmans et al 1993).

nhaA and *nhaB* genes

A clue to the molecular nature of the antiporters was obtained when we identified, mapped and cloned two genes, *nhaA* (Goldberg et al 1987, Karpel et al 1988) and *nhaB* (Pinner et al 1992), and made single and double deletion mutants (Padan et al 1989, Pinner et al 1993). The single deletions showed that *nhaA* is the major indispensable gene for the adaptation to Na^+ and Li^+, and for withstanding alkaline pH in the presence of Na^+, whereas *nhaB* is the housekeeping gene. There was no Na^+/H^+ antiporter activity in the double deletion mutant, and the cells were highly sensitive to H^+ and Na^+ (Pinner et al 1993). This sensitivity is cured upon transformation with antiporter genes. The double deletion mutant has since become important for the cloning and expression of both homologous and heterologous antiporter genes (Padan & Schuldiner 1996).

The DNA sequences of *nhaA* and *nhaB* are predicted to encode proteins that have little homology, but they have similar putative secondary structures, composed of 12 transmembrane segments connected with hydrophilic loops (Padan & Schuldiner 1994, 1996, Rothman et al 1996). This model is typical of secondary transporters that utilize chemiosmotic energy as a driving force.

The NhaA model has recently been substantiated by determining the topology of NhaA across the membrane utilizing *phoA* fusions (Manoil & Beckwith 1985) combined with epitope mapping and exposure to proteolysis from either side of the membrane (Rothman et al 1996). Most interestingly, even at this low resolution of structure we could identify two short transmembrane segments (VII and VIII) in NhaA (Fig. 1).

NhaA and NhaB proteins

Both the NhaA and NhaB proteins have been identified, overexpressed, purified and reconstituted in a functional form in proteoliposomes. This was first achieved by classical biochemistry (Taglicht et al 1991, Pinner et al 1994), and then by fusing the 5′ end of NhaA to the strong *tac* promoter and the 3′ end to sequences encoding

---

FIG. 1. Two-dimensional model of NhaA (based on that of Rothman et al 1996) with the location of the mutants affecting its response to pH. The one-letter amino acid code is used. Putative transmembrane segments are shown in boxes, and are connected by hydrophilic segments. I–XII indicate the numbers of the transmembrane segments. The residues involved in the pH response of NhaA are shown; H225 enlarged and in bold, G338 enlarged and in white octagonal; the locations of the mutations A127V, A130T, P129L and A349V suppressing G338S are encircled and highlighted in grey.

polyhistidine, a His-tag (Olami et al 1997). These allow strong expression and efficient affinity purification of the protein on a Ni^{2+} NTA column (Qiagen, Hilden, Germany; Olami et al 1997). With this procedure 3–4 mg of pure protein is obtained per litre of cells.

The purified reconstituted antiporters are as active as the native proteins in the membrane, in terms of kinetics, turnover numbers and bioenergetics, and can therefore be used in studies of structural biology, for example. (It is important to emphasize here that the structures of the transporters have not yet been resolved.)

Direct biochemical studies of the protein revealed that the stoichiometry of the exchange reaction is $2H^+/Na^+$ for NhaA (Taglicht et al 1993) and $3H^+/2\ Na^+$ for NhaB (Pinner et al 1994), and that NhaA, but not NhaB, is highly sensitive to pH (Taglicht et al 1991, Pinner et al 1994). Such high pH sensitivity is characteristic of antiporters, as well as other transporters and proteins involved in pH regulation (Aronson 1985, Grinstein et al 1992, Wakabayashi et al 1992, 1994, Dell et al 1994, Nakamura et al 1995, Hall et al 1995, Slonczewski & Foster 1996, Sekler et al 1996, Bearson et al 1997).

The pH response of NhaA

The V_{max} of NhaA increases by three orders of magnitude over the pH range 7.0–8.5 (Taglicht et al 1991, Padan & Schuldiner 1996). Hence, this molecule must be equipped with a 'sensor' that senses external pH directly. In addition, it must undergo a conformational change such that the signal is transduced into a change in V_{max}. This conformational change has been demonstrated by probing intact membrane vesicles at various pH values with trypsin (Rothman et al 1997). There are many trypsin-cleavable sites in the loops connecting the transmembrane segments of NhaA. However, the protein remains relatively uncleaved at low pH, in contrast to high pH when the trypsin-cleavable sites become exposed (Fig. 2). Identifying the residues involved in this dramatic response to pH has become a major focus of our work.

There are two strategies for identifying these residues. One involves site-directed mutagenesis of likely residues encoded by plasmid NhaA, followed by introduction of the mutagenized plasmid into the Δ*nhaA*-Δ*nhaB* host and observing the effect of the mutation on the pH response. The second approach involves the selection of cells that have a defective pH response from a bacterial population that contain randomly mutagenized plasmid-borne *nhaA*.

Histidine 225

Histidine residues have a p*K* of 6 in solution, and are therefore likely candidates. We tested the pH response of the eight histidine residues in NhaA, and showed that

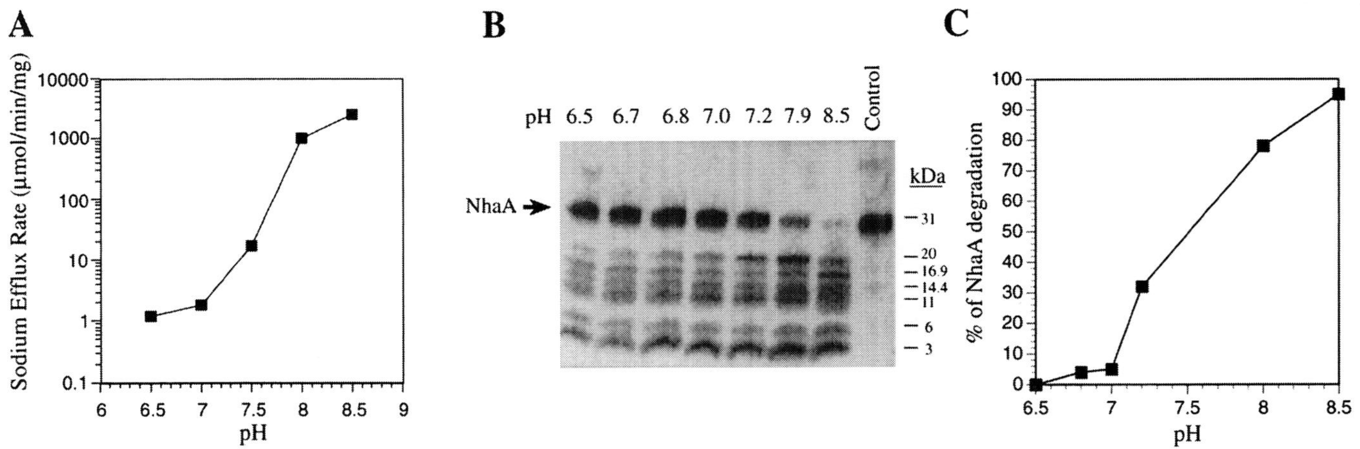

FIG. 2. The digestion of NhaA by trypsin and the activity of the protein show similar dependences on pH. (A) The pH profile of the activity of NhaA, purified and reconstituted in proteoliposomes (Taglicht et al 1991). (B & C) Solubilized membranes containing NhaA labelled specifically with [^{35}S]methionine were suspended to a final concentration of 0.147 mg of protein/ml and incubated with 2.6 μg/ml trypsin at the pH values indicated. Reactions were terminated by addition of trypsin inhibitor and were analysed on a 16.5% tricine gel. In the control lane, the inhibitor was added prior to the trypsin (pH 8.5; for details see Rothman et al 1997). The radioactive gel was exposed to a phosphorimager (shown in B) and expressed graphically in C (the degree of digestion observed at pH 6.5 was defined as zero).

only H225 was important for pH regulation (Fig. 1; Gerchman et al 1993, Rimon et al 1995). Both Ser and Cys can replace H225, so we first concluded that polarity rather than charge is essential at this site. However, we subsequently found that charge at position 225 does affect pH response. Substitution of this residue with arginine (H225R) resulted in an acidic shift, whereas substitution with aspartate (H225D) caused an alkaline shift. Substitution with alanine (H22SA), which is not hydrophilic or charged, resulted in a protein that did not respond to pH (Rimon et al 1995). These conclusions have recently been confirmed by Noumi et al (1997).

H225C is functional and responds to pH (Olami et al 1997), so we therefore used this mutant, together with site-directed modification (Altenbach et al 1990, Akabas et al 1992, 1994) using SH-reagents, to look at the topology and reactivity of H225. We first removed all the native cysteine residues from the NhaA antiporter (because they may have complicated the interpretation of the results), and found that this protein was active and pH sensitive (Olami et al 1997), suggesting that the native cysteines of NhaA are dispensable. We then demonstrated that a combined cysteine-less/H225C mutant was also active and pH sensitive, and we therefore probed this mutant with SH reagents. In addition, we constructed His-tagged derivatives, so that we could probe directly the purified proteins reconstituted in proteoliposomes (Olami et al 1997).

The classical SH reagent N-ethyl maleimide (NEM) had no effect on the wild-type or the cysteine-less NhaA, but inhibited the combined cysteine-less/H225C mutant (Olami et al 1997). To quantitate the degree of NEM alkylation in the membrane we used [^{14}C]-NEM and His-tagged derivatives of NhaA (wild-type, and cysteine-less and cysteine-less/H225C mutants). We exposed the membranes to [^{14}C]-NEM, then affinity-purified the proteins on a Ni^{2+} column, separated them with SDS-PAGE and autoradiographed the gels. As expected, the cysteine-less control was not labelled, and in contrast the cysteine-less/H225C mutant was labelled at least 10-fold more strongly than all three native cysteines of the wild-type protein together. The cysteine-less/H225C mutant reached saturation at 0.5 mM NEM at an alkyl/NhaA stoichiometry of 1, whereas the native cysteines were not saturated even at 2.5 mM (Olami et al 1997).

This difference cannot be ascribed to the impermeability of NEM through the membrane, because right-side-out (RSO) and inside-out (ISO) membrane vesicles produced identical results. Given that deprotonation of cysteine is required for alkylation by NEM, we have suggested that the native cysteines reside within the protein in a hydrophobic environment, which prevents deprotonation. In contrast, H225C is exposed in an aqueous environment and is therefore readily alkylated. According to our model, the native cysteines reside inside the transmembrane segments, whereas H225C is located at the edge of the shortest transmembrane segment (VIII) facing the bulk of the protein (Fig. 1).

The quantitative alkylation by NEM allowed us to determine the topology of H225 using impermeant SH reagents (as described in Altenbach et al 1990, Karlin 1993, Akabas et al 1994). The thinking behind this approach is that residues exposed to the periplasm will react with an impermeant probe in RSO vesicles but not in ISO vesicles. We exposed RSO and ISO membrane vesicles of His-tagged cysteine-less/H225C mutants to the impermeant reagents methanethiosulfonate ethyl-trimethyl ammonium bromide (MTSET; Akabas et al 1992, Stauffer & Karlin 1994) and p-chloromercuribenzosulfonate (PCMBS). The H225C residues left free were then titrated by alkylation with [^{14}C]-NEM as described above and in Olami et al (1997). The results showed that H225C is periplasmic (Fig. 1).

To determine the length of the transmembrane segment VIII we performed the same protocol with E241, predicted to be located on the internal side of the membrane at the edge of transmembrane segments VIII opposing H225. E241 faces the cytoplasm, and is exposed to impermeant probes only in ISO membrane vesicles. Taken together these results support the model of NhaA (Rothman et al 1996) and the existence of short transmembrane segments in this transporter (Fig. 1).

The short transmembrane segments, as well as the critical H225 residue, are conserved among the NhaA family members; both were found in the recently cloned *nhaA* from *Vibrio alginolyticus* (Nakamura et al 1995).

Glycine 338

NhaA is a prerequisite for growth at alkaline pH in the presence of Na^+. Δ*nhaA*-Δ*nhaB* cells do not grow in the presence of Na^+ at neutral or alkaline pH. Transformation with plasmid *nhaB*, which does not have a pH response and has a lower H^+/Na^+ stoichiometry, cure the cells at neutral pH, whereas plasmid *nhaA*, which has a pH response and a high H^+/Na^+ stoichiometry cures the cells both at neutral and alkaline pH. On this basis, we made an educated guess as to the phenotype that would be produced by pH response-impaired *nhaA*, i.e. that cells would grow normally at neutral pH in the presence of Na^+ but not at alkaline pH in the presence of Na^+.

We randomly mutagenized plasmid *nhaA*, transformed it to Δ*nhaA*-Δ*nhaB* and, by replica plating, isolated cells that grow in the presence of Na^+ at neutral but not at alkaline pH. We found that these cells do grow at alkaline pH in the presence of K^+, suggesting that the combination of high pH and Na^+ is deleterious to the cells rather than the osmolarity. We cloned and sequenced one of these mutants, and found that it had the point mutation G338S (Fig. 1).

The G338S mutant is expressed and is stable in the membrane at alkaline pH. It is as active as the wild-type at alkaline pH both in isolated membrane vesicles, as well

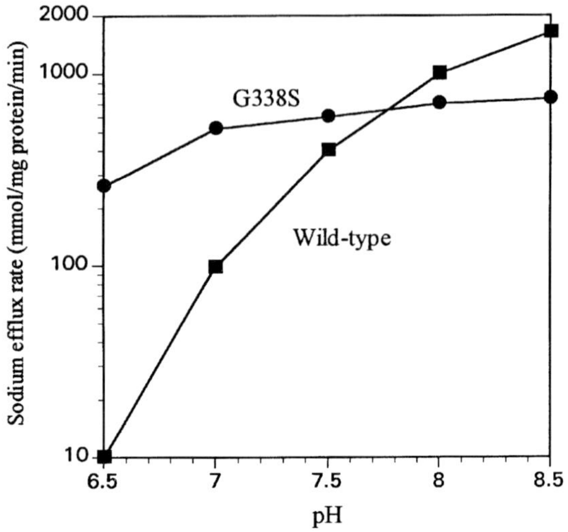

FIG. 3. The pH dependence of His-tagged G338S antiporter. His-tagged derivatives of the wild-type NhaA (■) and its mutant G338S (●) were overexpressed and purified. Proteoliposomes for Na^+ efflux and the assay were as described in Olami et al (1997).

as purified and reconstituted in proteoliposomes. It is also electrogenic. Strikingly, it loses pH control and remains open throughout the pH range 6.0–9.0 (Fig. 3).

Hence, the pH control of NhaA is required for growth at alkaline pH in the presence of Na^+. These conclusions have been substantiated by the isolation of mutations suppressing the G338S phenotype, which allow growth at alkaline pH in the presence of Na^+. In all suppressants pH control is restored. We suggest that the set point of NhaA pH control functions as an energy valve. The mutations suppressing G338S (in transmembrane segment XI) cluster in transmembrane segment IV in the vicinity of D133, which has been claimed to be involved in the Na^+-coupling site of NhaA (Inoue et al 1995).

We suggest that transmembrane segments IV and XI interact. The pH response occurs with a huge conformational change that can be probed by trypsinolysis. As expected, at acidic pH the wild-type is completely resistant to trypsinolysis, but is digested at alkaline pH, whereas the pH-relaxed antiporter G338S is digested throughout the entire pH range. Thus, NhaA reacts directly in response to pH, i.e. the environmental signal, and represents the most direct, efficient and rapid pattern of regulation.

nhaA is induced specifically by Na^+ and Li^+

The next level of regulation of NhaA is expression at the gene level. Experiments with *nha*A′-′*lac*Z protein fusions (Karpel et al 1991) and northern analysis (Dover et al 1996) showed that Na^+ and Li^+ are the environmental signals that turn on *nhaA*, and that alkaline pH potentiates the effect of these ions. This demonstrated for the first time that there is a unique regulatory network responding specifically to Na^+, rather than to osmolarity or ionic strength. Interestingly, a similar role has recently been assigned to Na^+ in the regulation of expression of the Na^+/ATPase of *Enterococcus hirae* (Murata et al 1996).

The NhaR regulator

The regulatory gene of *nhaA* is *nha*R, which maps downstream to *nhaA*. It encodes a positive regulator that works in trans. Δ*nha*R confers Na^+ and Li^+ sensitivity in spite of the presence of *nhaA* (Rahav-Manor et al 1992). When applied as a multicopy plasmid, it increases dramatically the expression of the *nhaA'-'lacZ* protein fusion (Rahav-Manor et al 1992). NhaR is homologous to the LysR family of positive regulators involved in the response of bacteria to various environmental stresses (Henikoff et al 1988, Christman et al 1989, Rahav-Manor et al 1992). The immediate signal for the *nha*R-dependent induction of *nhaA* is intracellular rather than extracellular Na^+. This was demonstrated by changing the intracellular concentration of Na^+ without changing its extracellular concentration. Multicopy *nhaA* decreased the intracellular concentration of Na^+ and inhibited expression of *nhaA'-'lacZ* by producing cells with high levels of antiporter (Rahav-Manor et al 1992). Deleting both *nhaA* and *nhaB* increased the intracellular concentration of Na^+ and enhanced the expression of the *nhaA'-'lacZ* fusion protein (Dover et al 1996).

The molecular mechanism of the interaction between Na^+, *nhaA*-DNA and NhaR

We demonstrated that NhaR may have a Na^+ 'sensor' by studying an E134G point mutation in *nha*R (Carmel et al 1994). This mutant increased the affinity of the regulatory system to Na^+ and increased resistance to Li^+. These results suggest that the interaction between Na^+, NhaR and the target regulatory sequences of *nhaA* is direct, so our next step was to study the direct interaction between these components. We affinity purified NhaR to homogeneity by constructing a His-tagged NhaR that was as active as the wild-type protein, and observed one homogenous peak at 60 kDa following HPLC separation (Carmel et al 1997). Both wild-type NhaR (Rahav-Manor et al 1992) and its His-tagged derivative (Carmel et al 1997) bind specifically to DNA sequences overlapping the promoter region of *nhaA*. To identify the regulatory sequences of *nhaA* recognized by His-tagged NhaR, we tested the mobility of PCR-amplified fragments of *nhaA* DNA of various lengths in a gel retardation assay in the presence or absence of His-tagged NhaR (Carmel et al 1997). Fragments that bound the regulator contained at least one LysR consensus motive (Goethals et al 1992, Schell 1993). The smallest fragment overlapping all binding sequences was located between −120 to +14 bp, and it contained three LysR consensus motives. Na^+ had no effect on the gel retardation pattern, suggesting that a change in the footprint rather than the affinity is affected by Na^+ . This behaviour is characteristic of the LysR family of regulators (Schell 1993, Toledano et al 1994) .

The footprint of NhaR on *nhaA* is Na^+ sensitive

The DNase I footprint of His-tagged NhaR on *nhaA* revealed a long segment spanning about 90 bases that aligned with the segment binding His-tagged NhaR and the three consensus motives of LysR family members (Goethals et al 1992, Schell 1993). Na^+, at concentrations of up to 100 mM, had no effect on the DNAase I footprint. This negative result, however, is due to the DNAase protection assay itself, which is not sensitive enough and is limited to the minor groove of the DNA (Sasse-Dwight & Gralla 1991).

A dimethyl sulfate (DMS) methylation protection footprint, both *in vitro* and *in vivo*, revealed a specific effect of Na^+ (Carmel et al 1997). DMS methylates G and to a lesser extent A, mainly in the major groove of the DNA (Sasse-Dwight & Gralla 1991), and we used it to identify the footprint of His-tagged NhaR. We expected to observe that DNA bases not protected by the regulator would be methylated by DMS, whereas the protected bases would not. After *in vivo* and *in vitro* treatment with DMS, we isolated and treated the DNA with piperidine, which splits methylated bases only. Analysis of the products obtained *in vitro*, by primer extension, showed that bases −24, −29, −60 and −92 were protected by His-tagged NhaR. NhaR-mediated protection *in vitro* of bases −24, −29 and −92 was not affected by either 100 mM K^+ or 100 mM Na^+, whereas base −60 was exposed to methylation and subsequently cleaved by 100 mM Na^+ (but not 100 mM K^+). Na^+ titration analysis revealed that concentration range of Na^+ affecting the footprint was within the range expected from the intracellular concentration (10–20 mM). In line with the finding that the expression of *nhaA′-′lacZ* is markedly potentiated *in vivo* by pH (Karpel et al 1991), the interaction of Na^+ with base −60 but not with either bases −24 and −29 was dramatically affected by pH. Na^+ affected the NhaR/*nhaA* interaction at base −60 at alkaline pH (7.5–8.5) but not at acidic pH (6.5).

We also carried out this DMS methylation protection assay *in vivo*. As expected from the *in vitro* results, bases −24 and −29 were protected and not affected by either Na^+ or K^+; whereas base −60 was protected in the presence of K^+ but not in the presence of Na^+. However, base −92, which *in vitro* was protected by the regulator but unaffected by the presence of Na^+, was exposed *in vivo* by Na^+ but not by K^+. The Na^+-specific effects on the footprint of NhaR on *nhaA* required NhaR, since they were not observed in a Δ*nha*R strain (Carmel et al 1997). We therefore suggest that Na^+ has a direct effect on the interaction of NhaR with G^{-60} of *nhaA*, and an indirect effect on the interaction of NhaR with G^{-92}. The latter most probably requires either a certain topology of the DNA or another factor existing only *in vivo*. Taken together this work suggests that NhaR is both the sensor and the transducer of the Na^+ signal. It regulates the expression of *nhaA*, and it undergoes a conformational change upon Na^+ binding. This change results in a pH-dependent decreased binding of NhaR to G^{-60} both *in vivo* and *in vitro*, and a decreased binding of NhaR to G^{-92} *in vivo*. In addition, the expression of *nhaA* is under global regulation of H-NS (Dover et al 1996), which may involve a change in the topology of the DNA via an unknown mechanism (Tupper et al 1994).

Summary

We have established that the NhaA Na^+/H^+ antiporter, the main system responsible for adaptation to Na^+ and alkaline pH (in the presence of Na^+) in *E. coli* and many other enteric bacteria, is under intricate control (Padan & Schuldiner 1996). At the protein level it is regulated directly by pH, one of its regulatory signals. At the gene level its transcription is dependent on NhaR, a positive regulator of the LysR family, and is regulated by Na^+, the other environmental signal. Na^+ affects the NhaR/*nhaA* interaction directly by changing the footprint of NhaR on *nhaA* in a pH dependent fashion. Since H^+ and Na^+ are the most common ions that challenge every cell, we suggest that the pattern of regulation of *nhaA* found in *E. coli* is a paradigm for the mechanism and response of proteins and genes to these ions.

Acknowledgements

The research in the laboratory of E.P. is supported by grants from the Israel Science Foundation, administered by the Israel Academy of Sciences and Humanities, and the BMBF and the BMBF's International Bureau at the DLR (German–Israeli Project Cooperation on Future-Oriented Topics [DIP]). We also thank the Moshe Shilo Center for Marine Biogeochemistry.

References

Akabas MH, Stauffer DA, Xu M, Karlin A 1992 Acetylcholine receptor channel structure probed in cysteine-substitution mutants. Science 258:307–310

Akabas MH, Kaufmann C, Archdeacon P, Karlin A 1994 Identification of acetylcholine receptor channel-lining residues in the entire M2 segment of the α-subunit. Neuron 13:919–927

Altenbach C, Marti T, Khorana H, Hubbell W 1990 Transmembrane protein structure: spin labelling of bacteriorhodopsin mutants. Science 248:1088–1092

Aronson PS 1985 Kinetic properties of the plasma membrane Na^+-H^+ exchanger. Annu Rev Physiol 47:545–560

Bearson S, Bearson B, Foster JW 1997 Acid stress responses in enterobacteria. FEMS Microbiol Lett 147:173–180

Carmel O, Dover N, Rahav-Manor O et al 1994 A single amino acid substitution (G134-Ala) in NhaR1 increases the inducibility by Na^+ of the product of *nhaA*, a Na^+/H^+ antiporter gene in *Escherichia coli*. EMBO J 13:1981–1989

Carmel O, Rahav-Manor O, Dover N, Shaanan B, Padan E 1997 The Na^+ specific interaction between the LysR-type regulator, NhaR and the *nhaA* gene, encoding the Na^+/H^+ antiporter of *Escherichia coli*. EMBO J 16:5922–5929

Christman M, Storz G, Ames B 1989 OxyR, a positive regulator of hydrogen peroxide-inducible genes in *Escherichia coli* and *Salmonella typhimurium* is homologous to a family of bacterial regulatory proteins. Proc Natl Acad Sci USA 86:3484–3488

Dell CL, Neely MN, Olson ER 1994 Altered pH and lysine signalling mutants of *cad*C, a gene encoding a membrane-bound transcriptional activator of the *Escherichia coli cad*BA operon. Mol Microbiol 14:7–16

Dover N, Higgins C, Carmel O, Rimon A, Pinner E, Padan E 1996 Na^+-induced transcription of *nhaA*, which encodes an Na^+/H^+ antiporter in *Escherichia coli*, is positively regulated by *nha*R and affected by *hns*. J Bacteriol 178:6508–6517

Gerchman Y, Olami Y, Rimon A, Taglicht D, Schuldiner S, Padan E 1993 Histidine 226 is part of the pH sensor of NhaA, a Na^+/H^+ antiporter in *Escherichia coli*. Proc Natl Acad Sci USA 90:1212–1216

Goethals K, Van Montagu M, Holsters M 1992 Conserved motifs in a divergent *nod* box of *Azorhizobium cauhnodans* ORS571 reveal a common structure in promoters regulated by LysR-type proteins. Proc Natl Acad Sci USA 89:1646–1650

Goldberg BG, Arbel T, Chen J et al 1987 Characterization of Na^+/H^+ antiporter gene of *E. coli*. Proc Natl Acad Sci USA 84:2615–2619

Grinstein S, Woodside M, Sardet C, Pouyssegur J, Rotin D 1992 Activation of the Na^+/H^+ antiporter during cell volume regulation. Evidence for a phosphorylation-independent mechanism. J Biol Chem 267:23823–23828

Hall HK, Karem KL, Foster JW 1995 Molecular responses of microbes to environmental pH stress. Adv Microb Physiol 37:229–272

Henikoff S, Haughn GW, Calvo JM, Wallace JC 1988 A large family of bacterial activator proteins. Proc Natl Acad Sci USA 85:6602–6606

Inoue H, Noumi T, Tsuchiya T, Kanazawa H 1995 Essential aspartic acid residues, Asp-133, Asp-163 and Asp-164, in the transmembrane helices of a Na^+/H^+ antiporter (NhaA) from *Escherichia coli*. FEBS Lett 363:264–268

Karlin A 1993 Structure of nicotinic acetylcholine receptors. Curr Opin Neurobiol 3:299–309

Karpel R, Olami Y, Taglicht D, Schuldiner S, Padan E 1988 Sequencing of the gene *ant* which effects the Na^+/H^+ antiporter activity in *Escherichia coli*. J Biol Chem 263:10408–10414

Karpel R, Alon T, Glaser G, Schuldiner S, Padan E 1991 Expression of a sodium proton antiporter (NhaA) in *Escherichia coli* is inducted by Na^+ and Li^+ ions. J Biol Chem 266:21753–21759

Manoil C, Beckwith J 1985 Tn*phoA*: a probe for protein export signals. Proc Natl Acad Sci USA 82:8129–8133

Mitchell P 1961 Coupling of phosphorylation to electron and hydrogen transfer by a chemiosmotic type of mechanism. Nature 191:144–146

Mitchell P, Moyle J 1967 Respiration-driven proton translocation in rat liver mitochondria. Biochem J 105:1147–1162

Murata T, Yamato I, Igarashi K, Kakinuma Y 1996 Intracellular Na^+ regulates transcription of the *ntp* operon encoding a vacuolar-type Na^+ translocating ATPase in *Enterococcus hirae*. J Biol Chem 271:23661–23666

Nakamura T, Komano Y, Unemoto T 1995 Three aspartic residues in membrane-spanning regions of Na^+/H^+ antiporter from *Vibrio alginolyticus* play a role in the activity of the carrier. Biochim Biophys Acta 1230:170–176

Noumi T, Inoue H, Sakurai T, Tsuchiya T, Kanazawa H 1997 Identification and characterization of functional residues in a Na^+/H^+ antiporter (NhaA) from *Escherichia coli* by random mutagenesis. J Biochem (Tokyo) 121:661–670

Olami Y, Rimon A, Gerchman Y, Rothman A, Padan E 1997 Histidine 225, a residue of the NhaA-Na^+/H^+ antiporter of *Escherichia coli* is exposed and faces the cell exterior. J Biol Chem 272:1761–1768

Padan E, Schuldiner S 1994 Molecular physiology of Na^+/H^+ antiporters, key transporters in circulation of Na^+ and H^+ in cells. Biochim Biophys Acta 1185:129–151

Padan E, Schuldiner S 1996 Bacterial Na^+/H^+ antiporters: molecular biology, biochemistry and physiology. In: Konings WN, Kaback HR, Lolkema JS (eds) The handbook of biological physics II. Transport in membranes. Elsevier Science, Amsterdam, p 501–531

Padan E, Maisler N, Taglicht D, Karpel R, Schuldiner S 1989 Deletion of *ant* in *Escherichia coli* reveals its function in adaptation to high salinity and an alternative Na^+/H^+ antiporter system(s). J Biol Chem 264:20297–20302

Pinner E, Padan E, Schuldiner S 1992 Cloning, sequencing and expression of *nhaB* gene encoding a Na^+/H^+ antiporter in *Escherichia coli*. J Biol Chem 267:11064–11068

Pinner E, Kotler Y, Padan E, Schuldiner S 1993 Physiological role of NhaB, a specific Na^+/H^+ antiporter in *Escherichia coli*. J Biol Chem 268:1729–1734

Pinner E, Padan E, Schuldiner S 1994 Kinetic properties of NhaB, a Na^+/H^+ antiporter from *Escherichia coli*. J Biol Chem 269:26274–26479

Rahav-Manor O, Carmel O, Karpel R et al 1992 NhaR, a protein homologous to a family of bacterial regulatory proteins (LysR), regulates *nhaA*, the sodium proton antiporter gene in *Escherichia coli*. J Biol Chem 267:10433–10438

Rimon A, Gerchman Y, Olami Y, Schuldiner S, Padan E 1995 Replacements of histidine 226 of NhaA-Na^+/H^+ antiporter of *Escherichia coli*. Cysteine (H226C) or serine (H226S) retain both normal activity and pH sensitivity, aspartate (H226D) shifts the pH profile toward basic pH, and alanine (H226A) inactivates the carrier at all pH values. J Biol Chem 270:26813–26817

Rothman A, Padan E, Schuldiner S 1996 Topological analysis of NhaA, a Na^+/H^+ antiporter from *Escherichia coli*. J Biol Chem 271:32288–32292

Rothman A, Gerchman Y, Padan E, Schuldiner S 1997 Probing the conformation of NhaA, a Na^+/H^+ antiporter from *Escherichia coli*, with trypsin. Biochemistry 36:14572–14576

Sasse-Dwight S, Gralla J 1991 Footprinting protein–DNA complexes *in vivo*. Methods Enzymol 208:146–168

Schell M 1993 Molecular biology of the LysR family of transcriptional regulators. Annu Rev Microbiol 47:597–626

Sekler I, Kobayashi S, Kopito RR 1996 A cluster of cytoplasmic histidine residues specifies pH dependence of the AE2 plasma membrane anion exchanger. Cell 86:929–935

Slonczewski JL, Foster JW 1996 pH regulation and survival in extreme acid or base. In: Neidhardt FC, Curtiss III R, Ingraham JL et al (eds) *Escherichia coli* and *Salmonella*: cellular and molecular biology, 2nd edn. ASM Press, Washington, DC, p 1539–1549

Speelmans G, Poolman B, Abee T, Konings WN 1993 Energy transduction in the thermophilic anaerobic bacterium *Clostridium fervidus* is exclusively coupled to sodium ions. Proc Natl Acad Sci USA 90:7975–7979

Stauffer DA, Karlin A 1994 The electrostatic potential of the acetylcholine binding sites in the nicotinic receptor probed by reactions of binding-site cysteines with charged methanethiosulfonates. Biochemistry 33:6840–6849

Taglicht D, Padan E, Schuldiner S 1991 Overproduction and purification of a functional Na^+/H^+ antiporter coded by *nhaA* (*ant*) from *Escherichia coli*. J Biol Chem 266:11289–11294

Taglicht D, Padan E, Schuldiner S 1993 Proton-sodium stoichiometry of NhaA, an electrogenic antiporter from *Escherichia coli*. J Biol Chem 268:5382–5387

Toledano M, Kullik I, Trinh F, Baird P, Schneider T, Storz G 1994 Redox dependent shift of OxyR-DNA contacts along an extended DNA-binding site: a mechanism for differential promoter selection. Cell 78:897–909

Tupper AE, Owen-Hughes TA, Ussery DW et al 1994 The chromatin-associated protein H-NS alters DNA topology *in vitro*. EMBO J 13:258–268

Wakabayashi S, Sardet C, Fafournoux P, Pouysségur J 1992 The Na^+/H^+ antiporter cytoplasmic domain mediates growth factor signals and controls 'H^+-sensing'. Proc Natl Acad Sci USA 89:2424–2428

Wakabayashi S, Bertrand B, Shigekawa M, Fafournoux P, Pouysségur J 1994 Growth factor activation and 'H^+-sensing' of the Na^+/H^+ exchanger isoform 1 (NHE 1). Evidence for an additional mechanism not requiring direct phosphorylation. J Biol Chem 269:5583–5588

West IC, Mitchell P 1974 Proton/sodium antiport in *Escherichia coli*. Biochem J 144:87–90

DISCUSSION

Foster: You said that pH affected the ability of NhaR to regulate the expression of the genes? How does this work?

Padan: We can reconstitute the situation *in vitro* with both native NhaR and recombinant protein (His-tagged), and DNA (Carmel et al 1997). It seems that the interplay between protons and sodium is important because they can compete for similar sites, and this competition is pH dependent. There is evidence for such competition in the Na^+/K^+ ATPase (Polvani et al 1989). It just depends on the pH.

Foster: Do you see the same effect *in vivo*?

Padan: Yes. Using *nhaA'-'lacZ* fusions we showed that Na^+ is the inducer, but alkaline pH potentiates expression (Karpel et al 1991).

Foster: Does the internal pH change that much?

Padan: We know that the internal pH changes before the pH homeostasis is established (Zilberstein et al 1982).

Stock: Does sodium act as an allosteric effector to shift the pH profile, or at low pH do you saturate the carrier so that protons are shuttled back and forth without effect. In other words, is the effect at the active site for transport or at some other locus?

Padan: That's a good question. We thought it wasn't at the active site because we didn't see any effects of sodium on the pH response. However, the suppressors of the new mutant all cluster in helix 4, which has a special aspartate residue (D133) that may be involved in sodium binding (Noumi et al 1997).

Dimroth: You showed that in the sodium/proton antiporter mutant, sodium ions are toxic at high pH. I understand that the antiporter is required to regulate the pH, but it appears that it is also required for sodium homeostasis. Why is sodium toxic to the cells?

Padan: There is no good answer to this (reviewed in Padan & Schuldiner 1993). It is known that sodium is toxic to protein synthesis requiring potassium. Potassium ions also compete with sodium ions, so if the cells are loaded with potassium the cells are protected.

Booth: The binding sites of the enzymes modulated by potassium are about 10 mM, and when sodium becomes toxic the potassium levels don't fall much below 300 mM. I agree that sodium is toxic, but we don't know why. We shouldn't be looking at these enzymes that require potassium because sodium doesn't inhibit those enzymes, and there is so much potassium there anyway.

Fillingame: Is that also true for protein synthesis?

Booth: Yes. Only 10 mM is required for protein synthesis.

Fillingame: I have a question about the short helix, transmembrane 8. What is the evidence that it traverses the membrane?

Padan: We did cysteine scanning on both sides. In a cysteine-less NhaA mutant containing H225C it is possible to titrate each cysteine by [4C]N-ethyl maleimide (NEM). Hence, using non-permeant alkylating agents such as p-chloromercuri-benzosulfonate (PCMBS) and oriented membrane vesicles it is possible to compete with the NEM alkylation only from the side to which the cysteine is exposed and deduce the topology of H225C (Olami et al 1997). It was found to be exposed to the periplasm. With this approach the short transmembrane segments were confirmed.

Fillingame: I am sceptical of using NEM, particularly if you are only looking at two residues and not scanning the entire helix.

Padan: But how can you explain the competition between NEM and PCMBS? You could say that there is a channel that goes into the transmembrane domain, but it's still short.

Fillingame: If you show that protection with both right-side-out and inside-out vesicles induce the inverse effect then I would be convinced. I am not convinced that an impermeable sulfhydryl reagent exists.

Padan: This is exactly what we did, and we found an inverse effect of PCMBS in the two types of oriented membranes.

Poole: Is there any new structural information that sheds light on these short helices?

Padan: We have 2D crystals, but the resolution is not good. We are now trying to generate monoclonal antibodies for obtaining 3D crystals, so that we can do structural analysis by X-ray diffraction (E. Padan, M. Venturi, C. Hunte & H. Michel, unpublished observations 1998). At present we have four monoclonals, three of which recognize the native conformation. It may be of interest to the people in this audience to know how we generated these monoclonal antibodies so rapidly. The usual process for soluble proteins is to do an ELISA. However, ELISAs don't work for membrane proteins because you can't use detergents in ELISAs and because ELISAs are performed at pH 9.5, which will probably cause the proteins to denature. We then discovered Ni^{+}-NTA-coated plates (Quiagen, Hilden, Germany). The protein can hook onto the plates via the His-tag in the presence of detergent and at pH 7.5, and be presented to the antibody in a native conformation.

Schäfer: You mentioned that the resolution of the 2D crystals was poor, but is it possible to tell whether NhaA is a dimer or monomer in this crystalline matrix?

Padan: It is not yet possible to tell.

Schäfer: How sure are you about the structure of these two transmembrane helices? Did you just use structure prediction programs?

Padan: No. We did cysteine scanning.

Schäfer: But this doesn't necessarily tell you that you have a helix. It may be a barrel arrangement of β strands, like in porins for example.

Padan: I agree.

Booth: I would like to ask a question about G338S. How does that mutation exert its inhibitory effect on growth?

Padan: For the suppressor's phenotype it is clear that down-regulation of NhaA is critical for growth. So the question then becomes, why does unregulated NhaA kill the cells? I originally thought that it was due to over-acidification, but there are two reasons why this is not the case: it is not possible to rescue the cells by weak bases that alkalinize the cytoplasm; and alkaline pH should improve growth, but it doesn't. What I now think is that the cells are already energetically stressed under these conditions, so if you put in an extra load by an uncontrolled antiport the residual energy is drained.

Booth: It seemed to me that the mutant had the same activity at pH 8.5 as the wild-type, so I couldn't see why you had introduced an extra load. If they both had the same activity under those conditions, how do you construe that there is an extra load?

Padan: The initial rate is indeed similar, but after the acidification of the cytoplasm has occurred the wild-type shuts down immediately, whereas the mutant does not.

Bennett: In your NhaR-binding experiments where you observed that binding responded to sodium and high pH, did you also look at whether it responded to cations such as lithium or potassium?

Padan: It does not respond to potassium, so we used potassium as the control. We also tested lithium, and we found that it responds in a fashion similar to Na^{+} (Karpel et al 1991).

References

Carmel O, Rahav-Manor O, Dover N, Shaanan B, Padan E 1997 The Na^+-specific interaction between the LysR-type regulator, NhaR and the *nhaA* gene, encoding the Na^+/H^+ antiporter of *Escherichia coli*. EMBO J 16: 5922–5929

Karpel R, Alon T, Glaser G, Schuldiner S, Padan E 1991 Expression of a sodium proton antiporter (NhaA) in *Escherichia coli* is inducted by Na^+ and Li^+ ions. J Biol Chem 266:21753–21759

Noumi T, Inoue H, Sakurai T, Tsuchiya T, Kanazawa H 1997 Identification and characterization of functional residues in a Na^+/H^+ antiporter (NhaA) from *Escherichia coli* by random mutagenesis. J Biochem (Tokyo) 121:661–670

Olami Y, Rimon A, Gerchman Y, Rothman A, Padan E 1997 Histidine 225, a residue of the NhaA-Na^+/H^+ antiporter of *Escherichia coli* is exposed and faces the cell exterior. J Biol Chem 272:1761–1768

Padan E, Schuldiner S 1993 Na^+ transport systems in prokaryotes. In: Bakker EP (ed) Alkali cation transport systems in prokaryotes. CRC Press, Boca Raton, FL, p 3–27

Polvani C, Sachs G, Blostein R 1989 Sodium ions as substitutes for protons in the gastric H,K-ATPase. J Biol Chem 264:17854–17859

Zilberstein D, Agmon V, Schuldiner S, Padan E 1982 The sodium/proton antiporter is part of the pH homeostasis mechanism in *Escherichia coli*. J Biol Chem 257:3687–3691

Bacterial energetics at high pH: what happens to the H^+ cycle when the extracellular H^+ concentration decreases?

Vladimir P. Skulachev

Department of Bioenergetics, A. N. Belozersky Institute of Physico-Chemical Biology, Moscow State University, Moscow 119899, Russia

Abstract. A decrease in the extracellular H^+ concentration creates difficulties for membrane-linked energetics in bacteria employing H^+ as the coupling ion. At high extracellular pH (pH_o), H^+ ions pumped from the cell by, say, the respiratory chain, are immediately neutralized by the alkaline extracellular medium. Under such conditions, the only driving force that might compel outer H^+ ions to return to cytosol and perform their function is the electric potential difference across the cytoplasmic membrane ($\Delta\Psi$). However, when ΔpH in the opposite direction is equal to, e.g., 2 pH units (intracellular pH = 7.5 at pH_o = 9.5), $\Delta\Psi$ would be so high that the risk of membrane electric breakdown would increase. This is why some bacteria deal with high pH by, for example, replacing H^+ by Na^+ as the coupling ion rather than by increasing $\Delta\Psi$. It has been shown in several species of bacteria that the alkalinization of the growth medium induces primary Na^+ pumps (e.g. Na^+-motive respiratory chain enzymes and Na^+ ATPase). Electrogenic Na^+ efflux via these pumps produces an electrochemical Na^+ potential difference ($\Delta\bar{\mu}_{Na^+}$) composed of $\Delta\Psi$ and ΔpNa^+. $\Delta\bar{\mu}_{Na^+}$ can be used to perform various types of membrane-linked work. The $\Delta\Psi$ constituent of $\Delta\bar{\mu}_{Na^+}$ may maintain electrophoretic influx of H^+ such that the alkalinization of cytoplasm is prevented. The latter function may be supported by a mechanism based on the uphill influx of Cl^- instead of Na^+. This seems to be the case for alkaliphilic and halophilic *Natronobacter pharaonis*. There is an indication that not only Na^+ but also Ca^{2+} may substitute for H^+ in *Gleobacter violaceus* growing at high pH.

1999 Bacterial responses to pH. Wiley, Chichester (Novartis Foundation Symposium 221) p 200–217

Alkaliphilic and alkalitolerant bacteria: a bioenergetic paradox

According to the chemiosmotic H^+ cycle concept, bacteria carry out: (i) an uphill export of H^+ ions at the expense of respiration or light energy, thus producing a transmembrane electrochemical H^+ potential difference ($\Delta\bar{\mu}_{H^+}$); and (ii) a downhill import of H^+ ions coupled to $\Delta\bar{\mu}_{H^+}$ utilization for ATP synthesis, transport of solutes or rotation of flagella (Mitchell 1966, Skulachev 1977). This scheme presents difficulties for bacteria living at pH values higher than the intracellular pH (pH_i). Here the efflux of H^+ ions, if it exists, occurs from a compartment with a high H^+ concentration to one that has a lower H^+ concentration. This is the downhill process, if one takes into account only the ΔpH constituent of $\Delta\bar{\mu}_{H^+}$. Thus, the H^+ influx is no longer favourable from a thermodynamic point of view and, hence, it cannot support the function of $\Delta\bar{\mu}_{H^+}$ consumers. The problem may be solved by an increase in another $\Delta\bar{\mu}_{H^+}$ constituent, namely the transmembrane electric potential difference (ΔΨ) but, as will be discussed below, increased ΔΨ cannot be the only mechanism by which bacteria adapt energetically to high pH.

Possible ways to solve the problem

Theoretically, there are several ways to overcome difficulties arising from high extracellular pH (pH_o).

(1) The production of a transmembrane electric potential (ΔΨ) that is large enough to counterbalance unfavourable ΔpH

This strategy is used by bacteria when the pH increases from acidic to neutral or slightly alkaline values (Padan et al 1981). It does not work in more alkaline media because ΔΨ would have to be so high that the probability of the electric breakdown of the bacterial membrane would dramatically increase (Skulachev 1988).

(2) Organization of a local H^+ cycle

It has been proposed that H^+ ions expelled from cytosol are not diluted in the extracellular medium but are immediately taken up and used by the cell (Guffanti et al 1978, 1981, Hicks & Krulwich 1995). However, it is difficult to explain how the neutralization of expelled H^+ ions by the alkaline extracellular medium is prevented. Small increases in the H^+ conductance of a coupling membrane result in the uncoupling of the $\Delta\bar{\mu}_{H^+}$-producing and $\Delta\bar{\mu}_{H^+}$-consuming processes. Therefore, the prevention of $\Delta\bar{\mu}_{H^+}$ dissipation and the organization of a local H^+ cycle requires the existence of a low H^+-conducting membrane outside the cytoplasmic membrane. An alternative possibility is that when pH_o is alkaline the generators and consumers of $\Delta\bar{\mu}_{H^+}$ in the cytoplasmic membrane (which at neutral pH catalyse H^+ transfer across the membrane) catalyse H^+ transfer along the membrane. Different enzymes should be involved in these two types of energy coupling in alkaliphiles vs. neutrophiles, although there is no evidence to date that this is the case. There is also no evidence that respiratory or photosynthetic phosphorylation occurs in open membrane sheets or within enzyme complexes, which should be possible if H^+ ions were directly translocated from, for example, a respiratory enzyme to H^+ ATP synthase.

(3) Change in direction of the transmembrane H^+

This solution involves the import, rather than the export, of H^+ ions by generators of $\Delta\bar{\mu}_{H^+}$. In this scenario the directions of $\Delta\Psi$ and ΔpH would be the same, so the problem, formally speaking, would be solved. However, there are no examples of such an unusual $\Delta\Psi$ direction (i.e. inside positive); in all living cells studied the cytoplasm is negative with respect to the extracellular medium. An inside-positive $\Delta\Psi$ inevitably entails the opposite orientation of all the membrane proteins electrophoretically arranged in the membrane (Skulachev 1988), which is why such a solution is not used by alkaliphilic or alkalitolerant bacteria.

(4) Formation of mesosomes, chromatophores or thylakoids

The formation of intracellular vesicles where energy transduction processes are localized would solve the problem only partially, i.e. it would enable the coupling of respiration and photosynthetic electron transport to ATP synthesis. It would not solve the problems of uphill transport of solutes and mechanical work (motility) because the corresponding processes are localized in the cytoplasmic membrane and not in intracellular vesicles.

(5) Employment of substrate-level phosphorylations to form ATP

This mechanism has the same disadvantage as that mentioned in solution (4). It is not a universal solution because respiring and phototrophic bacteria do live in conditions of high pH.

(6) Decreased efficiency of $\Delta\bar{\mu}_{H^+}$ consumers

One might imagine an H^+ ATP synthase that at high pH_o uses the downhill transport of, for example, 12 H^+ ions (instead of 3 H^+ ions at neutral pH) to form one ATP molecule. If this were the case, a fourfold lower $\Delta\bar{\mu}_{H^+}$ (45 mV instead of 180 mV) would be sufficient to support oxidative phosphorylation. Mechanistically, this means that at high pH the H^+ ATP synthase rotor should operate in a 12-step (instead of three-step) fashion. The disadvantage of such a system is that it operates with low efficiency, and there have been no reports of decreased efficiencies of solute porters or flagellar motors in alkaline media. Moreover, at high pH, $\Delta\bar{\mu}_{H^+}$ decreases to below 45 mV (Padan et al 1981).

Further increases in the H^+ /ATP stoichiometry are unlikely for a typical H^+ ATP synthase containing 12 *c*-subunits and, hence, 12 aspartate or glutamate residues that accept protons half-way from the outer to inner surface of the cytoplasmic membrane.

(7) Replacement of H^+ by another ion

This radical and efficient answer to the problem was suggested by Guffanti et al (1978), and it has been shown that in some alkalitolerant bacteria Na^+ substitutes for H^+ as the coupling ion.

The Na^+ cycle in some alkalitolerant bacteria

The marine bacterium *Vibrio alginolyticus* coexists with alga that during daylight hours cause an alkaline pH shift because of their photosynthetic conversion of CO_2 to carbohydrates. The opposite pH shift (from high to neutral pH) occurs during the night. *V. alginolyticus* is well prepared for such oscillations in H^+ concentration. Its respiratory chain contains the *bo*-type H^+-motive terminal $CoQH_2$ oxidase (Smirnova et al 1990, Kim et al 1991, Miyoshi-Akiyama et al 1993) and the Na^+-motive NADH-CoQ reductase (Tokuda & Unemoto 1981, 1982, 1984, Tokuda 1984, Dibrov et al 1986a), the latter being induced by high pH_o.

The Na^+-motive NADH-CoQ reductase differs from its functional analogue, the H^+-motive NADH-CoQ reductase, since it is composed of no more than six subunits, in contrast to the latter enzyme which is composed of between 16 and $\geqslant 40$ subunits (Rich et al 1995, Pfenninger-Li et al 1996, Dimroth 1997). Amino acid differences are quite different. Moreover, the H^+-motive enzyme contains FMN and six FeS clusters, whereas the Na^+-motive enzyme contains FAD and one 2Fe-2S cluster (Rich et al 1995, Pfenninger-Li & Dimroth 1995). The Na^+-motive enzyme is resistant to typical inhibitors of the H^+-motive NADH-CoQ reductase, but is extremely sensitive to 2-heptyl-4-hydroxy-quinoline N-oxide (Tokuda & Unemoto 1982, Tokuda 1984, Dibrov et al 1986a). Low concentrations of Ag^+ convert the Na^+-motive enzyme from a Na^+ pump to a Na^+ channel (Semeykina & Skulachev 1990), and a similar effect is observed in animal Na^+/K^+ ATPases (Hussain et al 1994, B. M. Anner, personal communication 1995).

Bogachev et al (1997) measured the Na^+/e^- stoichiometry of the Na^+-NADH-CoQ reductase span in the *V. alginolyticus* respiratory chain, and found it to be close to 1 (Fig. 1). Similar measurements for the *E. coli* H^+-NADH-CoA reductase gave a value of about 1.5 (Bogachev et al 1996), suggesting that the Na^+-motive enzyme has a lower thermodynamic efficiency. This may be due to the simple structure of the enzyme.

Na^+-motive NADH-CoQ reductase activity has also been identified in *Vibrio costicola* (Udagawa et al 1986), *Vibrio parahaemolyticus*, halotolerant bacterium Ba1 (Ken-Dror et al 1984), marine or halotolerant bacteria of the genera *Alkaligenes*, *Alteromonas*, *Flavobacterium*, *Deleya*, *Halovibrio* and *Pseudomonas* (see Skulachev 1988, 1994 for references), and also in *Haemophilus influenzae* (Hayashi et al 1996) and *Klebsiella pneumoniae* (see Dimroth 1997 for references).

Under some conditions, the terminal step of the respiratory chain of some bacteria can also involve a Na^+-motive enzyme, e.g. when *Bacillus halodurans* FTU

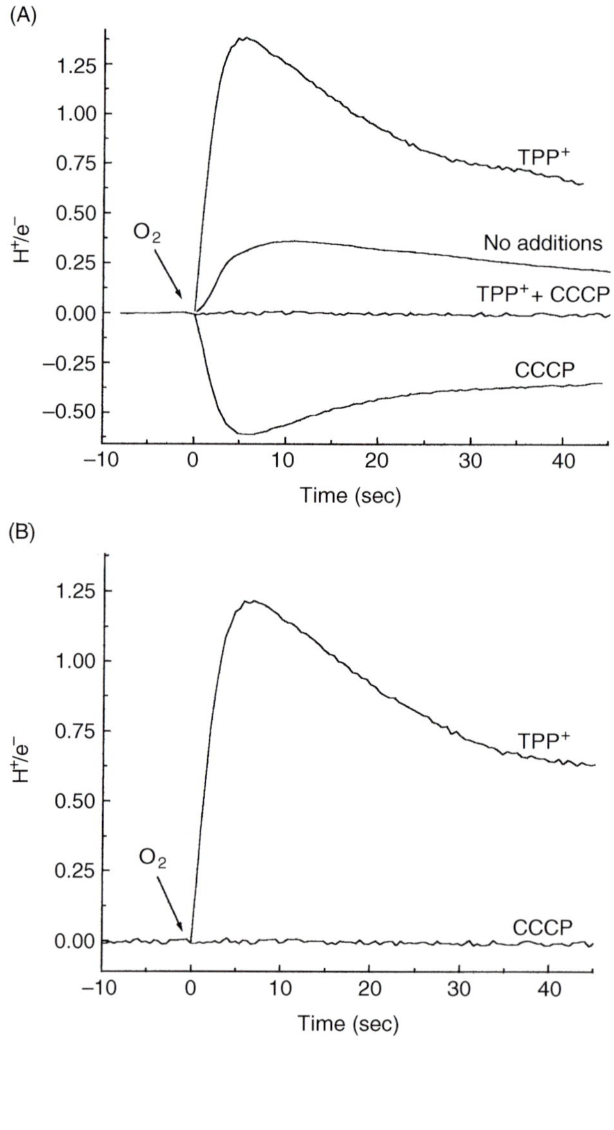

or *E. coli* are grown in conditions of low $\Delta\bar{\mu}_{H^+}$ (i.e. at high pH, in the presence of uncoupler or of a low cyanide concentrations; Semeykina et al 1989, Kostyrko et al 1991, Avetisyan et al 1991, 1992, Azarkina 1995). The existence of Na^+-motive respiration in *E. coli* was later confirmed by Padan's and Wikström's groups (Harel-Bronstein et al 1995, Verkhovskaya et al 1996). Webster and co-workers (Efiok & Webster 1990, Park et al 1996) have also suggested that a Na^+-motive terminal oxidase is present in *Vitreoscilla*.

In *V. alginolyticus* the uphill transport of solutes (Tokuda et al 1982), the rotation of the flagellum (Chernyak et al 1983, Dibrov et al 1986b) and respiratory phosphorylation (Dibrov et al 1986a) can be driven by $\Delta\bar{\mu}_{Na^+}$. The osmotic and mechanical work of many other marine or halotolerant bacteria also involve Na^+ (reviewed in Skulachev 1988, 1994, Dimroth 1997). In *V. alginolyticus* a single polar $\Delta\bar{\mu}_{Na^+}$-consuming flagellum is replaced by many lateral $\Delta\bar{\mu}_{H^+}$-consuming flagella when the bacterium is grown on solid agar instead of seawater (Atsumi et al 1992). Apparently, this switch is due to both the disappearance of high $[Na^+]_o$ and alga-linked alkalinization.

Some metanogenes produce CH_4 from CH_3COOH and H_2CO_3, and in these bacteria an alkaline pH shift takes place when acidic substances are consumed. Some energy-conserving steps of metanogenesis-linked electron transport are also Na^+ motive, the resultant $\Delta\bar{\mu}_{Na^+}$ being utilized either by Na^+ ATP synthase or by an energy-consuming reaction during metanogenesis (for reviews, see Müller et al 1993, Skulachev 1994, Dimroth 1997). Na^+ ATP synthase is the sole ATP-producing enzyme when H^+ ATP synthase, also present in metanogenes, is specifically inhibited by rhodamine (Smigan et al 1994) or pH_o is alkaline (Smigan et al 1995).

FIG. 1. The H^+/e^- stoichiometry in *Vibrio alginolyticus* cells. (A) Wild-type. The acidification of the sample without carbonyl cyanide m-chlorophenylhydrazone (CCCP) and tetraphenyl phosphonium (TPP^+) is due to the operation of the H^+-motive terminal oxidase. This effect is stimulated by the penetrating cation TPP^+, which is taken up electrophoretically by the cells. The TPP^+ uptake is accompanied by the conversion of the $\Delta\Psi$ constituent of $\Delta\bar{\mu}_{H^+}$ to ΔpH. Addition of CCCP abolishes contribution of the H^+-motive oxidase and causes conversion of $\Delta\Psi$, produced by the Na^+-NADH-CoQ reductase, to ΔpH due to electrophoretic H^+ influx. This ΔpH is abolished by TPP^+, which competes with H^+ for $\Delta\Psi$. In the presence of CCCP, the Na^+/e^- stoichiometry is equal to H^+/e^- because each Na^+ pumped out is electrophoretically counterbalanced by H^+ taken up by the cell. The Na^+/e^- value of about 0.7 measured in this experiment increases up to about 1.0, provided that we take into account the contribution of the non-coupled NADH-CoQ reductase that is also present in *V. alginolyticus*. (B) Mutant Nap1 lacking the Na^+-motive NADH-CoQ reductase. There is no CCCP-induced alkalinization in the mutant deficient in the Na^+-NADH-CoQ reductase. Additions: 15 μM CCCP; 5 mM TPP^+. A small amount of O_2 dissolved in aerated water was added to anaerobic bacterial suspension (arrow; Bogachev et al 1997).

In fermenting *Enterococcus hirae*, both an F-type H^+ ATPase and a V-type Na^+ ATPase have been described. The H^+-to-Na^+ switch is induced by alkaline pH, the addition of an uncoupler or a mutation in the H^+ ATPase (Kakinuma & Iragashi 1990, 1995). In addition, Prowe et al (1996) discovered a Na^+ ATPase in the thermoalkaliphilic anaerobic bacterium LBS3.

It should be stressed that at high pH the substitution of H^+ by Na^+ not only solves the problem of energy coupling but may also be an effective mechanism for the prevention of cytoplasm alkalinization when pH_o rises. Figure 2 summarizes possible strategies that bacteria may use to maintain cytosolic pH

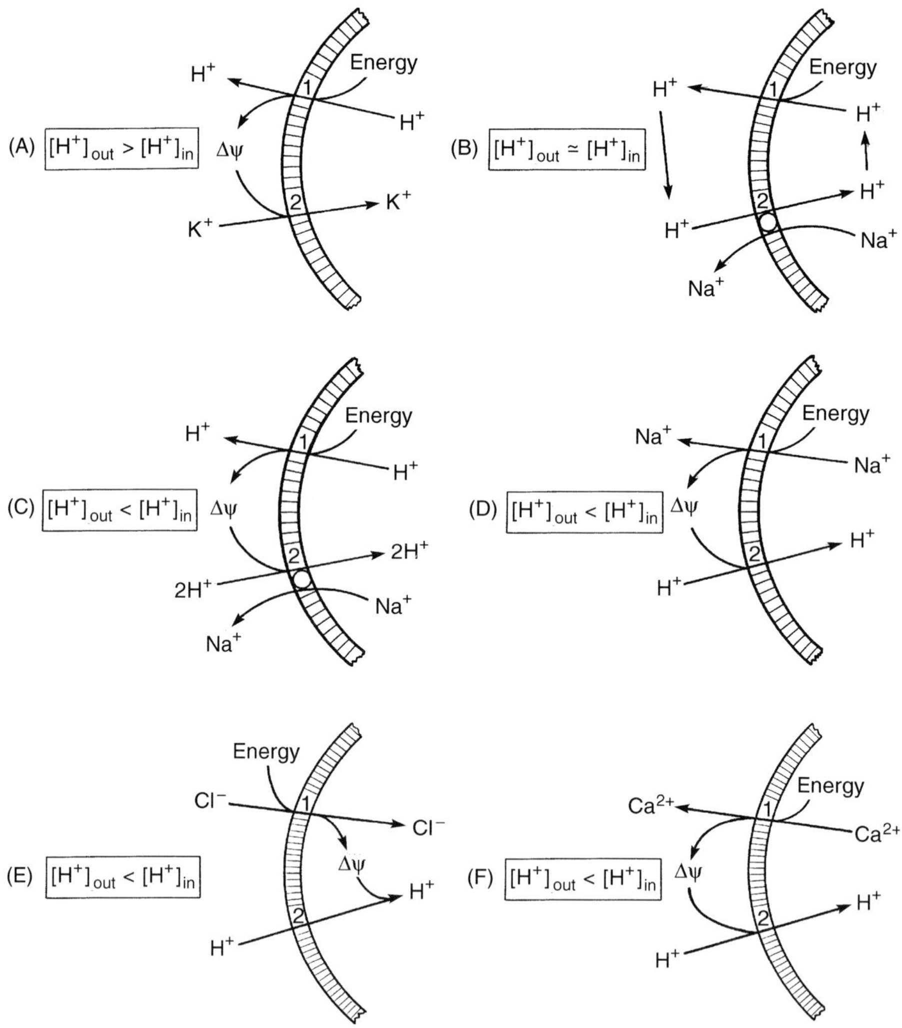

homeostasis at various pH_o values. At acidic pH (Fig. 2A) H^+ ions are expelled by the respiratory chain, resulting in the generation of $\Delta\Psi$. Electrophoretic K^+ influx then converts $\Delta\Psi$ to ΔpH, and as a result the pH_i becomes higher than pH_o.

At neutral pH there is no pH_i/pH_o imbalance. The circulation of H^+ ions supports various types of membrane-linked work and, in particular, Na^+ extrusion by electroneutral Na^+/H^+ antiporters (Fig. 2B).

When the extracellular pH increases, the antiports become electrogenic, such that $\Delta\Psi$ is used to return more H^+ into the cell than was pumped out by, for example, the respiratory chain (Fig. 2C). However, the operation of such a scheme is limited by the increased risk of the electric breakdown of the membrane at high $\Delta\Psi$ (see above).

There is no such risk if Na^+, instead of H^+, is pumped out from the cell by the respiratory chain. In this case, the acidification of the cytoplasm may be achieved simply by increased cytoplasmic membrane H^+ conductance (electrophoretic H^+ uniport, Fig. 2D). This situation may be especially useful at high pH when the electrogenic Na^+/nH^+ antiporter fails to help. For example, at extracellular pH 11.5 *Bacillus alkalophilus* maintains the internal pH at around 9 and a $\Delta\Psi$ of about 160 mV (Guffanti et al 1978). Thus, $\Delta\Psi$ is almost equal to ΔpH, as is predicted in Fig. 2D.

Role of Cl^- in an alkaliphilic halobacterium

The optimal conditions for growth of the archaeal species *Natronobacter pharaonis*, which is from the soda lakes of the Wadi Natrum of upper Egypt, are 4 M salt and pH 9.2. Engelgard and co-workers found that at $pH_o = 9.2$ and in aerobic conditions *N. pharaonis* cells maintain $\Delta\Psi = 280$ mV and $pH_i = 7.7$ (R. Wittenberg, B. Bechinger, M. Lübben & M. Engelgard, personal communication 1997). They concluded that the resulting $\Delta\bar{\mu}_{H^+}$ of about 180 mV is sufficient for the ATP synthesis catalysed by H^+-motive respiration and H^+-driven ATP synthase. In

FIG. 2. Possible mechanisms of pH homeostasis in bacteria. (A) Acidophiles. Energy supplied by H^+-motive respiration, photosynthetic electron transfer or ATP hydrolysis, i.e. $\Delta\bar{\mu}_{H^+}$ generators (1), is utilized to form $\Delta\Psi$, which is converted to ΔpH (inside alkaline) by an H^+ uniporter (2). (B) Neutrophiles. Energy from the $\Delta\bar{\mu}_{H^+}$ generators (1) is utilized by the $\Delta\bar{\mu}_{H^+}$ consumers, and in particular by a Na^+ efflux mechanism, e.g. an electroneutral Na^+/H^+ antiporter (2). (C) Alkaliphiles. $\Delta\bar{\mu}_{H^+}$ generators (1) are as in (B) but Na^+ is expelled by an electrogenic $Na^+/2H^+$ antiporter utilizing high $\Delta\Psi$ to import H^+ in spite of a ΔpH in an unfavourable direction (2; inside acidic). (D) Alkaliphiles. Na^+ is pumped put by a Na^+ pump, i.e. Na^+-motive respiration or Na^+ ATPase (1), and $\Delta\Psi$ is used to import H^+ by a H^+ uniporter (2). (E) Alkaliphilic *Natronobacter pharaonis*. Cl^- is pumped into the cell by halorhodopsin (1). $\Delta\Psi$ is utilized by import of H^+ via an H^+ uniporter (2). (F) Alkalitolerant *Gleobacter violaceus*. The Ca^{2+} efflux by a Ca^{2+} pump (1) forms $\Delta\Psi$, which is used to take up H^+ by a H^+ uniporter (2).

addition, they showed that decreased $[O_2]$ results in a decreased $\Delta\Psi$ and [ATP], and that this can be prevented or reversed by illumination with light at a wavelength of 500–600 nm. This photophosphorylation is assumed to be supported by $\Delta\Psi$ generated by halorhodopsin, a light-driven Cl^- pump (for review, see Skulachev 1988) present in *N. pharaonis* (Bivin & Stoeckenius 1986), and it may be carried out by co-operation of halorhodopsin electrogenically pumping Cl^- to the cytosol and H^+ ATP synthase electrophoretically transporting H^+ in the same direction. Besides ATP formation, the latter process might be used as a mechanism to maintain neutral pH_i when pH_o is high (Fig. 2E). Alternatively, ATP synthesis may be catalysed by a putative Cl^--driven ATP synthase that organizes, in co-operation with halorhodopsin, a Cl^- cycle. This idea was put forward in 1996 by M. Engelgard (personal communication).

ATP-supported Cl^- transport has already been described in eukaryotes. For example, in the salt gland of the plant *Limonium vulgaris*, Cl^- ATPase pumps Cl^- from cytosol to the outer space, and in the mollusc *Aplysia califormica*, Cl^- ATPase pumps Cl^- in the opposite direction, i.e. from the medium to the cytosol. This 110 kDa enzyme belongs to the P-type family of ATPases. In addition, a F-type Cl^- ATPase has been found in *Acetabularia acetabulum* that pumps Cl^- from the outside to the inside (see Skulachev 1988, Avetisyan et al 1998 for references).

In our attempts to verify the Cl^- cycle hypothesis (Avetisyan et al 1998) we found that the photophosphorylation in cyanide-treated *N. pharaonis* cells (Fig. 3A) is inhibited by low concentrations of the Cl^-/OH^- antiporter, triphenyltin (Fig. 3B), which supports the existence of a Cl^- cycle in this species. In neutrophilic *Halobacterium salinarium* photophosphorylation is catalysed by a H^+-pumping bacteriorhodopsin and a H^+ ATP synthase; here halorhodopsin is present in low concentrations and its function is restricted to the restoration of $[Cl^-]_i$, which decreases due to electrophoretic leakage of Cl^- from the cell. Bacteriorhodopsin is absent in alkaliphilic *N. pharaonis*, but the halorhodopsin level is strongly increased, suggesting that halorhodopsin has an additional function in *N. pharaonis* (for example, it is may be a component of the Cl^- cycle).

We also showed that the protonophorous uncoupler trifluoromethoxy-carbonylcyanide phenylhydrazone (FCCP) abolishes photophosphorylation (Fig. 3C), as if $\Delta\bar{\mu}_{H^+}$ were the driving force for ATP synthesis. It cannot be excluded, however, that the action of FCCP is due to a side effect, e.g. by the activation of ATP hydrolysis as a result of H^+ ATP synthase reversal. Further experiments to test the idea of the Cl^- cycle are currently in progress in our and Engelgard's laboratories.

Possible role of Ca^{2+} in the cyanobacterium *Gleobacter violaceus*

Brown (1994), a bioenergeticist from Odessa, Ukraine, put forward the idea of the Ca^{2+} cycle in the cyanobacterium *Gleobacter violaceus* which is unique in that it does

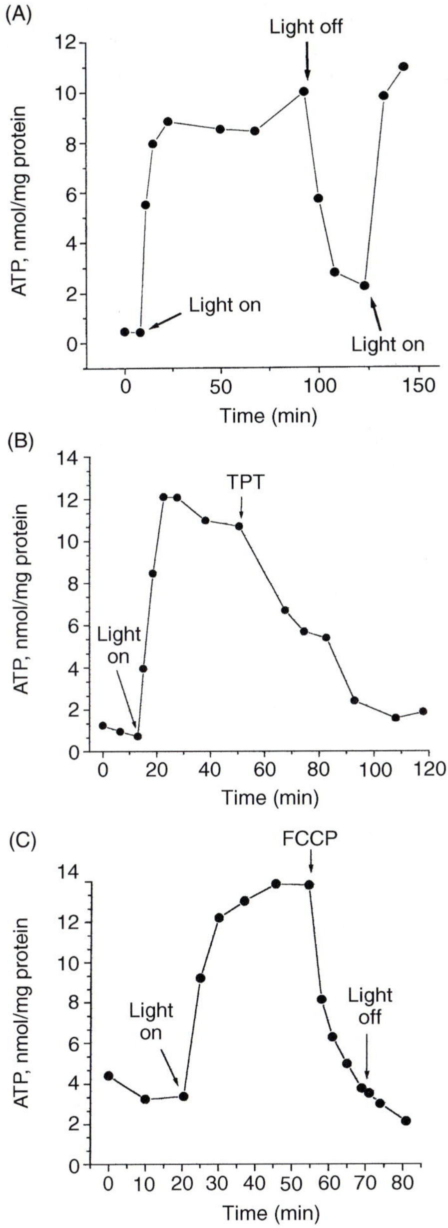

FIG. 3. Photophosphorylation in *Natronobacter pharaonis* cells. Additions: 1×10^{-6} M tetraphenyltin (TPT) and 5×10^{-6} M trifluoromethoxycarbonylcyanide phenylhydrazone (FCCP). Respiration is inhibited by 1 mM KCN. Cells were incubated in medium containing 4 M NaCl, pH 7.5 (A, B) or 9.5 (C) (Avetisyan et al 1998).

not possess thylakoids. In this micro-organism, all membrane-linked energy transductions are localized in its cytoplasmic membrane. Brown found that Ca^{2+} rather effectively substitutes for Na^+ as the cation required for *G. violaceus* to survive and grow at pH 9.5. Illumination results in a rapid alkalinization of the medium, and this effect requires either Na^+ or Ca^{2+}. Alkalinization is stimulated by carbonyl cyanide m-chlorophenylhydrazone (CCCP) and inhibited by the cation tetraphenyl phosphonium, suggesting that this light-induced alkaline pH_o shift is electrophoretic in nature. Brown assumed that: (i) electrogenic pump(s) in the *G. violaceus* membrane export Na^+ and Ca^{2+} at the expense of light energy; and (ii) some mechanisms carry out electrophoretic uptake of H^+ at the expense of $\Delta\Psi$ generated by (i). These relationships are illustrated in Fig. 2F. Like Cl^- in *N. pharaonis*, the import of H^+ might be catalysed by either H^+ ATP synthase or a special H^+ uniporter. For the organization of a Ca^{2+} cycle, one needs to postulate that a Ca^{2+} ATP synthase is present in the *G. violaceus* membrane, and that this utilizes $\Delta\bar{\mu}_{Ca}2+$ pump. Additional experiments are required to clarify whether the Ca^{2+} cycle is operative in *G. violaceus*.

Conclusion

Alternative coupling ions may substitute for H^+ when the extracellular concentration of H^+ decreases. For example, Na^+ may be pumped out by the Na^+-motive respiratory chain or by a Na^+ ATPase, and in the alkaliphilic halobacterium *N. pharaonis* Cl^- is pumped in by halorhodopsin. The resultant $\Delta\Psi$ may then be used to import H^+ ions electrophoretically. In some cases Ca^{2+} may also substitute for H^+.

So, in answer to the question 'What happens to the H^+ cycle when the extracellular H^+ concentration decreases?', we may say that, at least in some cases, the H^+ cycle simply disappears and the H^+ ion is replaced by Na^+, Cl^- or perhaps Ca^{2+}.

Acknowledgements

I thank the International Association for the promotion of cooperation with scientists from the New Independent States of the former Soviet Union (grant 93-742) and the Russian Foundation for Basic Research (Grant 96-04-50938) for their financial support.

References

Atsumi T, McCarter L, Imae Y 1992 Polar and lateral flagellar motors of marine *Vibrio* are driven by different ion-motive forces. Nature 355:182–184

Avetisyan AV, Dibrov PA, Skulachev VP, Sokolov MV 1991 Adaptation of *Bacillus* FTU and *Escherichia coli* to alkaline conditions: the Na^+-motive respiration. Biochim Biophys Acta 1098:95–104

Avetisyan AV, Bogachev AV, Murtasina RA, Skulachev VP 1992 Involvement of a *d*-type oxidase in the Na^+-motive respiratory chain of *Escherichia coli* growing under low $\Delta\bar{\mu}_{H^+}$ conditions. FEBS Lett 306:199–202

Avetisyan AV, Kaulen AD, Skulachev VP, Feniouk BA 1998 Photophosphorylation in alkalophilic halobacterial cells containing halorhodopsin: chloride-ion cycle? Biochemistry (Moscow) 63:744–748

Azarkina NV 1995 Induction of Na^+-motive terminal oxidase in facultatively alkalophilic bacterium *Bacillus halodurans* grown under alkaline conditions. Biochemistry (Moscow) 60:211–216

Bivin DB, Stoeckenius W 1986 Photoactive retinal pigments in haloalkaliphilic bacteria. J Gen Microbiol 132:2167–2177

Bogachev AV, Murtasina RA, Skulachev VP 1996 H^+/e^- stoichiometry for NADH dehydrogenase I and dimethyl sulfoxide reductase in anaerobically grown *Escherichia coli* cells. J Bacteriol 178:6233–6237

Bogachev AV, Murtasina RA, Skulachev VP 1997 The Na^+/e^- stoichiometry of the Na^+-motive NADH: quinone oxidoreductase in *Vibrio alginolyticus*. FEBS Lett 409:475–477

Brown II 1994 Is Ca^{2+} the third coupling ion? Biochemistry (Moscow) 59:1321–1323

Chernyak BV, Dibrov PA, Glagolev AN, Sherman MYu, Skulachev VP 1983 A novel type of energetics in a marine alkali tolerant bacterium. $\Delta\bar{\mu}_{Na^+}$-driven motility and sodium cycle. FEBS Lett 164:38–42

Dibrov PA, Lazarova RL, Skulachev VP, Verkhovskaya ML 1986a The sodium cycle. II. Na^+-dependent oxidative phosphorylation in *Vibrio alginolyticus*. Biochim Biophys Acta 850:458–465

Dibrov PA, Kostyrko VA, Lazarova RL, Skulachev VP, Smirnova IA 1986b The sodium cycle. I. Na^+-dependent motility and modes of membrane energization in the marine alkalotolerant *Vibrio alginolyticus*. Biochim Biophys Acta 850:449–457

Dimroth P 1997 Primary sodium ion-translocating enzymes. Biochim Biophys Acta 1318:1–2, 11–51

Efiok BJ, Webster DA 1990 Respiratory-driven Na^+ electrical potential in the bacterium *Vitreoscilla*. Biochemistry 29:4734–4739

Guffanti AA, Susman P, Blanco R, Krulwich TA 1978 The proton-motive force and α-aminoisobutyric acid transport in an obligatory alkalophilic bacterium. J Biol Chem 253:708–15

Guffanti AA, Bornstein RF, Krulwich TA 1981 Oxidative phosphorylation by membrane vesicles from *Bacillus alcalophilus*. Biochim Biophys Acta 635:619–630

Harel-Bronstein M, Dibrov P, Olami Y, Pinner E, Schuldiner S, Padan E 1995 MH1, a second-site revertant of an *Escherichia coli* mutant lacking Na^+/H^+ antiporters (Δ*nhaA* Δ*nhaB*), regains Na^+ resistance and a capacity to excrete Na^+ in a $\Delta\bar{\mu}_{H^+}$-independent fashion. J Biol Chem 270:3816–3822

Hayashi M, Nakayama Y, Unemoto T 1996 Existence of Na^+-translocating NADH-quinone reductase in *Haemophilus influenzae*. FEBS Lett 381:174–176

Hicks DB, Krulwich TA 1995 The respiratory chain of alkaliphilic bacteria. Biochim Biophys Acta 1229:303–314

Hussain S, Meneghini E, Moosmayer M, Lacotte D, Anner BM 1994 Potent and reversible interaction of silver with pure Na,K ATPase and Na,K ATPase-liposomes. Biochim Biophys Acta 1190:402–408

Kakinuma Y, Igarashi K 1990 Amplification of the Na^+ ATPase of *Streptococcus faecalis* at alkaline pH. FEBS Lett 261:135–138

Kakinuma Y, Igarashi K 1995 Electrogenic Na^+ transport by *Enterococcus hirae* Na^+ ATPase. FEBS Lett 359:255–258

Ken-Dror S, Shnaiderman R, Avi-Dor Y 1984 Uncoupler-stimulated Na^+ pump and its possible role in the halotolerant bacterium, Ba_1. Arch Biochem Biophys 229:640–649

Kim YJ, Mizushima S, Tokuda H 1991 Fluorescence quenching studies on the characterization of energy generated at the NADH: quinone oxidoreductase and quinol oxidase segments of marine bacteria. J Biochem 109:616–621

Kostyrko VA, Semeykina AL, Skulachev VP, Smirnova IA, Vaghina ML, Verkhovskaya ML 1991 The H^+-motive and Na^+-motive respiratory chains in *Bacillus* FTU subcellular vesicles. Eur J Biochem 198:527–534

Mitchell P 1966 Chemiosmotic coupling in oxidative and photosynthetic phosphorylation. Biol Rev 41:445–502

Miyoshi-Akiyama T, Hayashi M, Unemoto T 1993 Purification and properties of cytochrome *bo*-types ubiquinol oxidase from a marine bacterium *Vibrio alginolyticus*. Biochim Biophys Acta 1141:283–287

Müller W, Blaut M, Gottschalk G 1993 Bioenergetics of methanigenesis. In: Ferry JG (ed) Methanogenesis: ecology, physiology, biochemistry & genetics. Chapman & Hall, New York/London, p 360–406

Padan E, Zilberstein D, Schuldiner S 1981 pH homeostasis in bacteria. Biochim Biophys Acta 650:151–160

Park C, Moon JY, Cokic P, Webster DA 1996 Na^+-translocating cytochrome *bo* terminal oxidase from *Vitreoscilla*: some parameters of its Na^+-pumping and orientation in synthetic vesicles. Biochemistry 35:11895–11900

Pfenninger-Li XD, Dimroth P 1995 The Na^+-translocating NADH:ubiquinone oxidoreductase from the marine bacterium *Vibrio alginolyticus* contains FAD but not FMN. FEBS Lett 369:173–176

Pfenninger-Li XD, Albracht SRJ, van Belzen R, Dimroth P 1996 NADH:uniquinone oxidoreductase of *Vibrio alginolyticus*: purification, properties, and reconstitution of the Na^+ pump. Biochemistry 35:6233–6242

Prowe SF, van de Vossenberg JLCN, Driessen AJM, Antranikian G, Konnings WN 1996 Sodium-coupled energy transduction in the newly isolated thermoalkaliphilic strain LBS3. J Bacteriol 178:4099–4104

Rich PR, Meunier B, Ward FB 1995 Predicted structure and possible ion-motive mechanism of the sodium-linked NADH-ubiquinone oxidoreductase of *Vibrio alginolyticus*. FEBS Lett 375:5–10

Semeykina AL, Skulachev VP 1990 Submicromolar Ag^+ increases passive Na^+ permeability and inhibits the respiration-supported formation of Na^+ gradient in *Bacillus FTU* vesicles. FEBS Lett 269:69–72

Semeykina AL, Skulachev VP, Verkhovskaya ML, Bulygina ES, Chumakov KM 1989 The Na^+-motive terminal oxidase activity in an alkalo- and halotolerant *Bacillus*. Eur J Biochem 183:671–678

Skulachev VP 1977 Transmembrane electrochemical H^+-potential as a convertible energy source for the living cells. FEBS Lett 74:1–9

Skulachev VP 1988 Membrane bioenergetics. Springer-Verlag, Berlin

Skulachev VP 1994 The latest news from the sodium world. Biochim Biophys Acta 1187:216–221

Smigan P, Majernik A, Greksak M 1994 Na^+-driven ATP synthesis in *Methanobacterium thermoautotrophicum* and its differentiation from H^+-driven ATP synthesis by rhodamine 6G. FEBS Lett 347:190–194

Smigan P, Majernik A, Greksak M 1995 The presence of H^+ and Na^+-translocating ATPases in *Methanobacterium thermoautotrophicum* and their possible function under alkaline conditions. FEBS Lett 371:119–122

Smirnova IA, Vaghina ML, Kostyrko VA 1990 $\Delta\Psi$ and ΔpH generation by the H^+ pumps of the respiratory chain and ATPase in subcellular vesicles from marine bacterium *Vibrio alginolyticus*. Biochim Biophys Acta 1016:385–391

Tokuda H 1984 Solubilization and reconstitution of the Na^+-motive NADH oxidoreductase activity from the marine bacterium *Vibrio alginolyticus*. FEBS Lett 176:125–128

Tokuda H, Unemoto T 1981 A respiration-dependent primary sodium extrusion system functioning at alkaline pH in the marine bacterium *Vibrio alginolyticus*. Biochem Biophys Res Commun 102:265–271

Tokuda H, Unemoto T 1982 Characterization of the respiration-dependent Na^+ pump in the marine bacterium *Vibrio alginolyticus*. J Biol Chem 257:10007–10014

Tokuda H, Unemoto T 1984 Na^+ is translocated at NADH:quinone oxidoreductase segment in the respiratory chain of *Vibrio alginolyticus*. J Biol Chem 258:7785–7790

Tokuda H, Sugasawa M, Unemoto T 1982 Role of Na^+ and K^+ in α-aminoisobutyric acid transport by the marine bacterium *Vibrio alginolyticus*. J Biol Chem 257:788–794

Udagawa T, Unemoto T, Tokuda H 1986 Generation of Na^+ electrochemical potential by the Na^+-motive NADH oxidase and Na^+/H^+ antiport system of a moderately halophilic *Vibrio costicola*. J Biol Chem 261:2616–2622

Verkhovskaya ML, Verkhovsky MI, Wikström M 1996 K^+-dependent Na^+-transport driven by respiration in *E. coli* cells and membrane vesicles. Biochim Biophys Acta 1273:207–216

DISCUSSION

Kobayashi: I have a comment about the Na^+ ATPase inducer. At low pH the Na^+/H^+ antiporter is working, so the internal concentration of sodium ions is low and the Na^+ ATPase is not induced. In the presence of uncouplers the proton motive force is dissipated and the Na^+/H^+ antiporter is not working. The internal level of sodium ions is therefore increased. The same process occurs in ATPase mutants because the H^+ ATPase generates the proton motive force at low pH. Therefore, it is clear that the direct inducer is intracellular sodium ions.

Skulachev: That's possible. I only mentioned the effects and not the mechanism, which may involve sodium or some way of sensing proton motive force.

Krulwich: Could you expand on your thoughts on Ca^{2+}. Is Ca^{2+} being pumped by primary pump and then being used as a coupling ion for a synthase?

Skulachev: We don't know the exact function of the Ca^{2+} efflux system. It is possible that it is involved in the maintenance of neutral pH inside the cell. This is just the beginning of this work. We know that *Gleobacter violaceus* can survive at high pH if Na^+ or Ca^{2+} is added. Also, when the extracellular pH is high, protons are taken up by the cell in a light-dependent fashion, and this effect occurs more rapidly in the presence of a low concentration of cyanide m-cholorophenyl hydrazone (CCCP).

Schäfer: I would like to ask a question about the proposed chloride-mediated ATP synthase cycling. This process is driven by light, but the cells can respire, and Mattar & Engelhard (1997) have cloned an oxidase that, according to the sequence, cannot pump protons. How could these cells make ATP in the dark?

Skulachev: That's a good question. *Natronobacter pharaonis* requires respiration to make ATP. It is possible that it uses a chloride motive respiratory chain.

Schäfer: Is there any direct evidence for this?

Skulachev: There is no evidence for chloride motive respiration, but there are three examples of chloride ATPases, one of which is found in *Acetabularia* and is a F_1F_0-type enzyme. It is possible that under certain environmental conditions Cl^- is used, like H^+ or Na^+, as a coupling ion.

Fillingame: Chloride permeates the phospholipid bilayer more readily than other ions, so it would be difficult to use it as a coupling ion at concentrations above 100 mM.

Skulachev: When the light is switched on, the pumping of chloride into the cell by halorhodopsin results in $\Delta\Psi$ formation, suggesting that the *N. pharaonis* membrane is relatively impermeable to chloride ions. We have measured the amount of halorhodopsin in *N. pharaonis* and found that the level is much higher than that of bacteriorhodopsin.

Poole: You cited some experiments on membrane vesicles from *Escherichia coli*. Were those experiments designed to answer the question of which oxidases are involved in sodium translocation?

Skulachev: We found that the expression of cytochrome bd is induced at high pH, and that this induction is stimulated by sodium ions. We also found that if cytochrome bd is present in inside-out vesicles; they accumulate sodium ions when they oxidize ascorbate. Mutant vesicles lacking cytochrome bd do not accumulate sodium ions. This accumulation is stimulated by uncouplers, and is inhibited by a sodium ionophore (Avetisyan et al 1992, Bogachev et al 1993).

Poole: That is your conclusion for *E. coli*, but in *Vitreoscilla* it is cytochrome o that has been claimed to pump sodium ions (Efiok & Webster 1990).

Skulachev: Yes. This is interesting because cytochrome bo is also involved in sodium pumping in *Bacillus halodurans* FTU. Many properties of the *Bacillus* cytochrome bo are similar to cytochrome bd of *E. coli*.

Poole: The collective picture is that different types of oxidases can all be involved in sodium energetics.

Cook: Cytochrome d is interesting because it doesn't pump protons in *E. coli*, but you are saying that it can pump sodium. We had some problems in trying to reproduce those experiments, and I was wondering if you might know the reason why.

Skulachev: The critical factor is the presence of pentachlorophenol in the growth medium.

Cook: That could be it, because we didn't use pentachlorophenol.

Dimroth: I would like to make a comment on the advantages and disadvantages of the sodium cycle versus the proton cycle. One might have the impression from Vladimir Skulachev's talk that it would be advantageous for the alkaliphiles to use sodium ions instead of protons, but they don't. They generate ATP by using a proton ATPase, they use the sodium/proton antiporter in between and they have a proton-pumping respiratory chain, so it is not advantageous for them to use the sodium cycle instead of the proton cycle to make ATP. There are other examples in which there are no advantages to using protons. For instance, we constructed a sodium-translocating ATP synthase in *E. coli* in which we fused the genes encoding the F_1 part of the ATP synthase from *E. coli* with those encoding the F_0 part from *Propionigenium modestum*. This strain grows on succinate and makes ATP, but this ATP formation is sodium dependent. It uses a proton-translocating respiratory chain and an antiport to establish the sodium gradient necessary to form ATP, and the growth rate is the same as the wild-type. In the anaerobic bacterium *P. modestum* an electrochemical Na^+ gradient is established by methylmalonyl-CoA decarboxylase and used directly for ATP synthesis by the Na^+-translocating ATP synthase. Another bacterium that is dependent on the decarboxylation of a dicarboxylic acid is *Malonomonas rubra*. This also has a Na^+-translocating decarboxylase, but its ATPase is proton dependent, so it has an antiporter in between. Therefore, it is not advantageous to use sodium or proton cycles, because many organisms use a mixture.

Konings: I would like to make a comment on proton cycling. The majority of bacteria make ATP by substrate-level phosphorylation. There is a certain amount of evidence that under those conditions there are organisms that rely completely on sodium. I'm sure that we will find many more organisms that under alkaline or thermophilic conditions use sodium as a coupling ion. I am looking forward to finding an organism that is thermophilic, alkaliphilic and aerobic.

Skulachev: N. pharaonis is such an organism, although it is an archaebacterium. Alkalitolerant bacteria use the respiratory chain, but not decarboxylation or ATP hydrolysis, to form a sodium motive force. Tokuda & Unemoto (1982) showed that in this bacterium the transport of all the amino acids and sugars is sodium coupled. Later we confirmed these results and showed that the motility of this bacterium under conditions of high extracellular salts is due to a sodium-driven flagellar motor (Chernyak et al 1983). Moreover, sodium-linked oxidative phosphorylation has been found in the same system (Dibrov et al 1986). Therefore, there is at least one example of an alkalitolerant bacterium in which the respiration-linked sodium cycle is induced by high pH. Apparently, in some bacteria not only Na^+ but also chloride or calcium ions may be used as coupling ions. Freshwater alkaliphiles seem to solve this problem in some other way, but we don't know how.

Krulwich: The reason that we are tentative to support these ideas is that until we can reconstitute the purified complexes and can demonstrate that sodium is being pumped, there will always be an element of doubt because the membranes have so many ion flux pathways, so it is difficult to draw a definitive conclusion.

Stock: There is a simple solution for freshwater alkaliphiles: at low ionic strength a negatively charged fixed layer in the cell envelope would give rise to a Donnan potential that would decrease the pH and increase the concentrations of other cations at the external surface of the cytoplasmic membrane.

Skulachev: I disagree. We need to postulate the existence of one more coupling membrane between the cytoplasmic membrane and the outer medium to solve the problem in a way you have suggested.

Stock: I'm talking about a Donnan potential whose effect would be to decrease the pH and increase the cation concentration surrounding the cytoplasm. There is no barrier against protons, as has been shown for *E. coli* (Stock et al 1977). For marine organisms, however, the high salt would swamp out the Donnan effect, and marine organisms may therefore be forced to use sodium for energy coupling.

Matin: They would need a lot of protein.

Stock: No. They just need charges, the concentration of which depend on the volume of the periplasm.

Skulachev: In any case we should consider steady state rather than transient effects.

Stock: But the Donnan potential occurs at equilibrium so at equilibrium cells could adjust their surrounding pH by changing the concentrations of fixed anions such as the membrane-derived oligosaccharides in the periplasm, as long as the salt concentration is not too high.

Booth: In growing bacterial cultures, if the cells associate with a negatively charged polymer the pH optimum is shifted upward. Therefore, cells that can't normally grow at that pH now can. This is because a shift in local pH results as the protons move in, and as the ionic strength is increased, the effect is wiped out. The classic experiments on immobilized enzymes demonstrate that the pH optimum with chymotrypsin is shifted by 2 pH units over long periods by the appropriate nature of the polymer; cation polymers shift it one way, and anion polymers shift it the other way.

Krulwich: Aono et al (1995) suggested that the negatively charged layers of *Bacillus lentus* C-125 grown on glucose do exactly that. There is also another line of evidence from the people working on bacteriorhodopsins which suggests that protons move rapidly along protein and lipid surfaces such that they can reach the proteins mediating consumption of protons before they equilibrate fully with the outside. The problem is that the non-marine alkaliphiles which grow at high pH without using sodium as the coupling ion have apparently not used this solution. The question we still haven't resolved is why they don't. What is it that they do which works well enough such that they haven't resorted either to sodium or to some such proton-retaining layer, if the problem can be solved with this kind of layer? One of the reasons we know that the non-marine alkaliphiles don't use this is that they have a characteristic sequence deviation in the outside of loops of polytopic membrane proteins, including ones that are short and don't extend far from the membrane. This sequence deviation suggests that they are adapted to a pH which is not normal.

References

Aono R, Ito M, Jolin KN, Horikoshi K 1995 A high cell wall negative charge is necessary for the growth of the alkaliphilic *Bacillus lentus* C-125 at elevated pH. Microbiology 141:2955–2964

Avetisyan AV, Bogachev AV, Murtasina RA, Skulachev VP 1992 Involvement of a *d*-type oxidase in the Na^+-motive respiratory chain of *Escherichia coli* growing under low $\Delta\bar{\mu}_{H^+}$ conditions. FEBS Lett 306:199–202

Bogachev AV, Murtasina RA, Skulachev VP 1993 Cytochrome d induction in *Escherichia coli* growing under unfavorable conditions. FEBS Lett 336:75–78

Chernyak BV, Dibrov PA, Glagolev AN, Sherman MYu, Skulachev VP 1983 A novel type of energetics in a marine alkali tolerant bacterium. $\Delta\bar{\mu}_{Na^+}$-driven motility and sodium cycle. FEBS Lett 164:38–42

Dibrov PA, Lazarova RL, Skulachev VP, Verkhovskaya ML 1986 The sodium cycle. II. Na^+-dependent oxidative phosphorylation in *Vibrio alginolyticus*. Biochim Biophys Acta 850:458–465

Efiok BJ, Webster DA 1990 A cytochrome that can pump sodium ion. Biochem Biophys Res Comm 173:370–375

Mattar S, Engelhard M 1997 Cytochrome ba3 from *Natronobacterium pharaonis* — an archaeal four-subunit cytochrome-c-type oxidase. Eur J Biochem 250:332–341

Stock JB, Rauch B, Roseman S 1977 Periplasmic space in *Salmonella typhimurium* and *Escherichia coli*. J Biol Chem 252:7850–7861

Tokuda H, Unemoto T 1982 Characterization of the respiration-dependent Na^+ pump in the marine bacterium *Vibrio alginolyticus*. J Biol Chem 257:10007–10014

Proton ATPases in bacteria: comparison to *Escherichia coli* F_1F_0 as the prototype

Robert H. Fillingame and Sara Divall

Department of Biomolecular Chemistry, University of Wisconsin Medical School, Madison, WI 53706, USA

Abstract. The F_1F_0 ATP synthase complex of *Escherichia coli* functions reversibly in coupling proton translocation to ATP synthesis or hydrolysis. The structural organization and subunit composition corresponds to that seen in many other bacteria, i.e. a membrane extrinsic F_1 sector with five subunits in an $\alpha_3\beta_3\gamma\delta\varepsilon$ stoichiometry, and a membrane-traversing F_0 sector with three subunits in an $a_1b_2c_{12}$ stoichiometry. The structure of much of the F_1 sector is known from a X-ray diffraction model. During function, the γ subunit is known to rotate within a hexameric ring of alternating α and β subunits to promote sequential substrate binding and product release from catalytic sites on the three β subunits. Proton transport through F_0 must be coupled to this rotation. Subunit *c* folds in the membrane as a hairpin to two α helices to generate the proton-binding site in F_0. Its structure was determined by NMR, and the structure of the *c* oligomer was deduced by cross-linking experiments and molecular mechanics calculations. The implications of the oligomeric structure of subunit *c* will be considered and related to the H^+/ATP pumping ratio, P/O ratios and the cation-binding site in other types of F_0. The possible limits of the structure in changing the ion-binding specificity, stoichiometry and routes of proton entrance/exit to the binding site will be considered.

1999 Bacterial responses to pH. Wiley, Chichester (Novartis Foundation Symposium 221) p 218–234

An H^+ transporting F_1F_0 type ATPase is found in the plasma membrane of most eubacteria. In some bacteria, such as *Escherichia coli*, the enzyme can function as either an ATP hydrolase or an ATP synthase depending upon the physiological conditions. The enzyme functions as an ATP synthase under aerobic conditions utilizing the $\Delta\bar{\mu}_{H^+}$ generated by proton-pumping electron transport complexes to drive ATP synthesis. Under anaerobic conditions, the enzyme uses glycolytically derived ATP and functions as a proton pump to maintain the $\Delta\bar{\mu}_{H^+}$ across the plasma membrane. An F_1F_0 type or F-type enzyme, similar to that observed in *E. coli*, is found in most other eubacteria. A distantly related type of enzyme, the V-type or 'vacuolar' type ATPase, is found predominantly in archaebacteria. The V-type family of enzymes was discovered as the acidifying enzyme in a variety of eukaryotic vacuolar vesicles and is presumed to have evolved from an archeal progenitor. The restricted distribution of one type of enzyme to each kingdom is not absolute, as V-type enzymes are now documented in eubacteria, and genes for F-type ATPases are found in archaebacteria (Sumi et al 1997). The eubacterium *Enterococcus hirae* has both types of enzymes, as is discussed below.

The amount of F_1F_0 ATPase activity in *E. coli* does not appear to change significantly with physiological growth conditions (Kasimoglu et al 1996) or external pH (see below). The enzyme is synthesized in excess capacity over its normal requirement in metabolic flux, and relatively severe reductions in activity are required to compromise growth (Jensen et al 1993). The F_1F_0 enzyme is of interest to bacterial physiologists because it is the best characterized of the F_1F_0-type enzymes, and serves as a prototype. The organization of genes in the *unc* operon, modes of regulation of biogenesis and mechanism of function will likely be relevant to consideration of other bacterial ATPases that are regulated in response to environmental pH. We will first review the salient features of the *E. coli* F_1F_0 ATPase and the structural organization of the H^+-transporting sector of the enzyme, and then consider examples of regulation of bacterial proton ATPases in H^+ homeostasis. The relevance of the structural information to suggested adaptations of the enzyme with environmental pH will then be considered.

Escherichia coli F_1F_0 H^+ ATP synthase

General structural features

The H^+-transporting F_1F_0 ATP synthase complex is organized into two functionally and structurally distinct sectors termed F_1 and F_0. F_1 is composed of five subunit types in an $\alpha_3\beta_3\gamma\delta\varepsilon$ composition. It is bound at the surface of the membrane and, when removed from the membrane, functions as an ATPase. F_0 extends through the membrane and in the absence of F_1 can function as a passive proton transporter. F_0 is composed of three subunit types in an $a_1b_2c_{12}$ stoichiometric ratio (Fillingame 1997). The genes are organized in a single operon (denoted *unc* or *atp*) with the gene order *IBEFHAGDC*. The *uncI* gene is normally transcribed (Kasimoglu et al 1996), but its protein product is not found in F_1F_0 and the gene appears to be non-essential (Gay 1984). The *uncBEF* genes code, respectively, subunits *a*, *c* and *b* of F_0 and the *uncHAGDC* genes code, respectively, subunits δ, α, γ, β and ε. The polycistronic *unc* mRNA is endonucleolytically processed in several steps, and the intermediates are differentially translated to yield the final stoichiometry of subunits found in the isolated complex (Hellmuth et al 1991, McCarthy et al 1991). An organization of genes in a single operon like that in *E. coli* is seen in many other types of bacteria (Das & Ljungdahl 1997), and aspects of the gene organization are seen elsewhere. In the purple non-sulfur photosynthetic bacteria *Rhodospirillum rubrum* and *Rhodopseudomonas blastica*, the F_0 genes (*IBEF*) are grouped separately from the F_1 genes (*HAGDC*). In several species of cyanobacteria, including *Synechococci*, the operon is split into gene segments of the *IBEFHAG* and *DC*.

Mechanism of F_1F_0 H^+ ATP synthase

The general mechanism of coupling H^+ transport and ATP synthesis/hydrolysis in the F_1F_0 ATP synthase complex has emerged from work over the last several years. A key breakthrough was the determination of an X-ray diffraction structure for the $\alpha_3\beta_3\gamma$ portion of F_1 from bovine mitochondria (Abrahams et al 1994), an enzyme that is highly homologous to its *E. coli* counterpart. In the structural model, the globular α and β subunits alternate around a centrally located γ subunit, the axis of which extends as a bundle of two α helices from the top to the bottom of the F_1

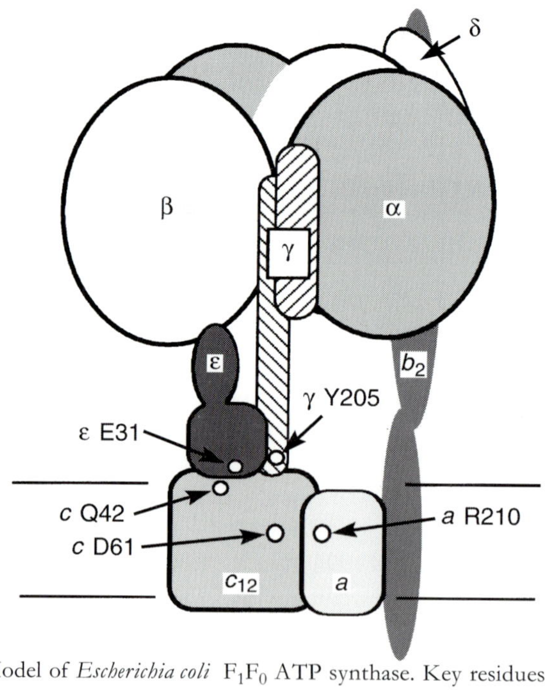

FIG. 1. Model of *Escherichia coli* F_1F_0 ATP synthase. Key residues discussed in the text are indicated.

sector (Fig. 1). As is discussed below, the second and longer C-terminal helix is thought to extend from the centre of the complex to the surface of the membrane to interact with F_0. ATP synthesis is proposed to occur alternately in different β subunits, the co-operative tight binding of ADP+P_i at one catalytic site being coupled to ATP release at a second site. The differences in binding affinities appear to be caused by rotation of the asymmetric γ subunit in the centre of the $\alpha_3\beta_3$ hexamer (Noji et al 1997). The γ subunit traverses a 45 Å stalk connecting the bottom of the $\alpha_3\beta_3$ hexamer to the surface of the membrane-traversing F_0 sector. Proton translocation through F_0 is now thought to drive rotation of the γ subunit within the $\alpha_3\beta_3$ hexamer, and hence synthesis of ATP occurs at the alternating sites in the β subunits. The mechanism of coupling is complex. The $\alpha_3\beta_3$ hexamer is thought to be immobile and fixed by a second stalk at the side of the complex composed of a dimer of *b* subunits anchored to the membrane by a single membrane-spanning helix for each *b* subunit (Fig. 1). The *b* subunits associate with subunit δ at the top of the F_1 molecule (Wilkens et al 1997). The proton-binding sites lie within the multiple *c* subunits in F_0 that interact with subunit *a* in an unknown way to promote proton translocation. Subunit *c* folds through the membrane as a hairpin of two α helices. Aspartyl 61 centred in the second transmembrane helix is the site of proton binding. Proton translocation is coupled to ATP synthesis (via rotation of subunit γ) by interactions between the loop region of subunit *c* and the γ and ε subunits in the centrally located stalk region of F_1 (Fillingame 1997). The mechanism of how this takes place is unknown. In perhaps the simplest proposal, the complex of *c* subunits is proposed to rotate relative to the fixed subcomplex of ab_2 subunits in F_0 to drive the rotation of subunit γ. The mechanistic stoichiometry of H^+ translocated per ATP synthesized (H^+/ATP ratio) is presumed to be determined by the number of *c* subunits and proton-binding sites per F_0 ($n_{H+} = 12$) divided by the number of catalytic sites used during a single rotation of subunit γ ($n_{ATP} = 3$).

Structure of subunit c and organization of the c oligomer in F_0

We have recently completed the solution structure of subunit *c* by NMR (Girvin et al 1998). The protein folds as a helical hairpin in the solvent system used much like it is expected to fold in the membrane. Gln42 is located at the apex of the hairpin turn, where it is predicted to interact with residues 31 of subunit ε and 205 of subunit γ in the intact complex (Fig. 1). In the centre of the structure, residues 24 and 28 in helix 1 are packed proximally to aspartyl 61 in helix 2 as was predicted for the folded protein *in situ* from genetic studies (Fillingame 1997). On the basis of these structural and genetic studies, the functional unit of subunit *c* is predicted to be minimally a dimer with the 'front' face of one subunit packed against the 'back' face of a second subunit. This is dictated by the geometry of side chain packing in the proton (or ion)-binding site and known residue–residue interactions affecting function. The essential aspartyl can be exchanged from position 61 in helix 2 to position 24 in helix 1 with retention of function (Miller et al 1990), suggesting that either residue can anchor the essential carboxyl group at an equivalent position in the complex. In the structure, the side chains of these residues point to different sides of the molecule. The side chains of residues 24 and 28, the substitutions of which cause resistance to dicyclohexyl carbodiimide (Fillingame et al 1991), are predicted to lie close to the Asp61 carboxyl, but both side chains project to the side of the molecule opposite to Asp61. Finally, the substituted residues Glu61 and Ser62 provide liganding groups in a Li^+-binding mutant of *E. coli* (Zhang & Fillingame 1995), as they do in the *Propiogenium modestum* F_1F_0 ATP synthase (Kaim et al 1997). Na^+ liganding by the *P. modestum* enzyme requires a Gln at position 28 in addition to the equivalent of Glu61 and Ser62 (Kaim et al 1997). To accommodate these interactions in the functional oligomer, a 'front-to-back' packing of Asp61 at the front face of one subunit *c* with Ala24, Ile28 and Ala62 at the back face of a second subunit *c* is predicted (Girvin et al 1998). The predicted packing arrangement is independently suggested by cross-linking studies using membranes with F_1F_0 in the native state (Jones et al 1998). Multiple Cys residues were introduced throughout the two transmembrane helices to define the helix–helix interactions (Fig. 2). With some double Cys substitutions, cross-linked oligomers extending up to c_{12} were formed. The cross-linking studies establish that the multiple *c* subunits organize in a ring with the H^+-binding Asp61 side chain packed between two subunits at the centre of a four-helix bundle.

F_1F_0 H^+ ATPase and pH homeostasis

Global perspectives on H^+ ATPases and pH homeostasis

The first and still foremost example for direct regulation of H^+ homeostasis by an H^+-transporting ATPase emerged from studies of *Enterococcus hirae* by Kobayashi et al (1999, this volume). This bacterium lacks a respiratory chain in the plasma membrane and uses the F_1F_0 ATPase to generate a $\Delta\bar{\mu}_{H^+}$. Environmentally induced acidification of the cytoplasm leads to increased synthesis of the enzyme (Kobayashi 1985). The operon of this typical F_1F_0-type ATPase has been sequenced and it shows the same tandem arrangement of structural genes as seen in *E. coli*, i.e. *acbδαγβε* (Shibata et al 1992). It is of great interest that the equivalent of an *uncI* gene is not found, this being the only example in eubacteria to date. Instead, the promoter is followed by a 240 bp untranslated region, the possible regulatory role of which is of obvious interest. The *E. hirae* bacterium is also of interest because alkaline conditions and external Na^+ induce a second Na^+-

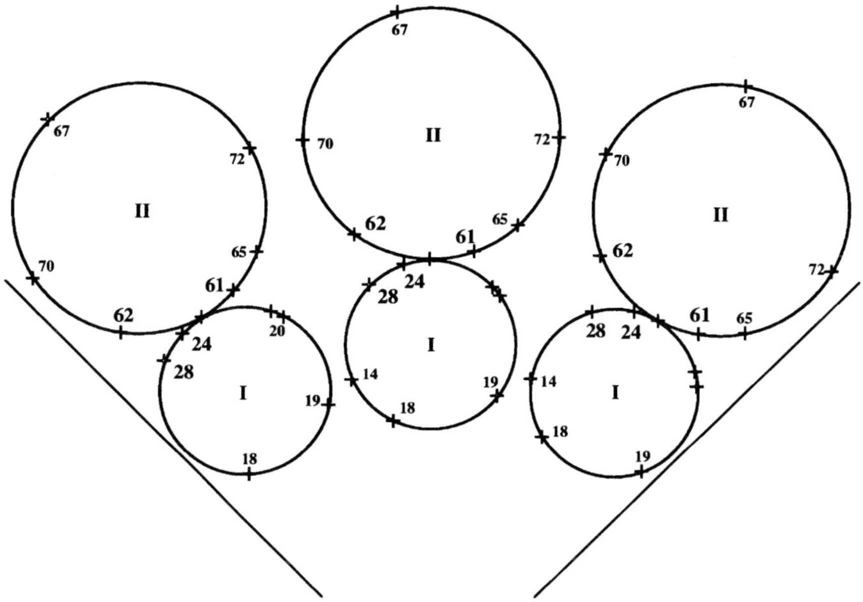

FIG. 2. Helical wheel-type representation of the relative positions of select α carbons in the NMR structure of subunit *c*. The end-on view of the molecule is from the N- and C-termini. Three subunits are shown as they are thought to arrange in one quadrant of a ring of 12 *c* subunits per F_0 oligomer.

translocating, V-type ATPase (Kakinuma & Igarashi 1990, Murata et al 1996). The obvious physiological role of this enzyme is to export Na^+ from the cytoplasm under alkaline conditions of low $\Delta\bar{\mu}_{H^+}$, where transport by the Na^+/H^+ antiporter would be impaired. A related adaptation may take place in *E. coli* where Avetisyan et al (1989) have suggested a shift from proton motive- to sodium motive-driven oxidative phosphorylation when cells are grown at alkaline pH in the presence of an uncoupler. The suggestion has been made that the ion specificity of both the *d*-type oxidase and the F_1F_0 ATP synthase is altered from H^+ to Na^+ during the adaptive response (Avetisyan et al 1992, 1993). We will consider these experiments in greater detail below.

Other observations suggest that mechanistic features of the F_1F_0 ATP synthase may change in response to changes in environmental pH. In the extremely alkaliphilic bacterium *Bacillus firmus* at pH > 10, ATP synthesis is still driven by proton transport under conditions of a very low osmotic $\Delta\bar{\mu}_{H^+}$ (Krulwich et al 1999, this volume). The possibility has been entertained that protons may enter F_0 by a route directly linked to the proton-pumping respiratory complexes to drive

ATP synthesis (Krulwich 1995). The cyanobacterium *Synechocossus* 6716 is suggested to adapt to alkaline external pH by a mechanism that alters the ratio of H^+ translocated per ATP synthesized (Krenn et al 1993). One possibility is that the number of proton-translocating *c* subunits per F_0 is augmented to increase the functional H^+/ATP ratio of the F_1F_0 complex (Fillingame 1992, Van Walraven et al 1997). Such a possibility is considered further below.

Does the mechanism of ATP synthesis change in Escherichia coli grown at alkaline pH?

The report by Avetisyan et al (1989) on growth of *E. coli* on succinate by oxidative phosphorylation at pH 8.6 in the presence of an uncoupler stimulated us to further investigate the mechanism of growth. To maintain the pH in the range of 8.6–8.8 we used two non-metabolizable buffers, Bicine and CHES, in place of glycylglycine (Fig. 3). The $\Delta\bar{\mu}_H+$ was further diminished by addition of carbonyl cyanide m-chlorophenylhydrazone (CCCP) at concentrations in the range of 10–15 μM. Concentrations of CCCP $\geqslant 20\,\mu$M completely inhibited growth, and suggested that growth was still dependent upon $\Delta\bar{\mu}_H+$. Ohyama et al (1992)

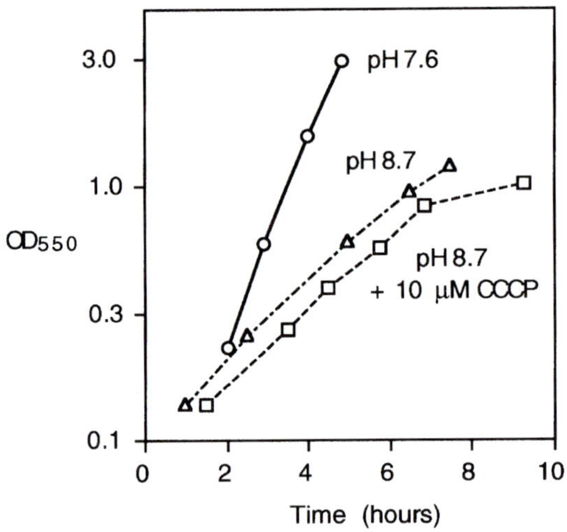

FIG. 3. Growth of *Escherichia coli* strain ML308-225 on succinate carbon source at alkaline pH with and without carbonyl cyanide m-chlorophenylhydrazone (CCCP) at 37°C. The medium (containing 22 mM KH_2PO_4, 20 mM Na_2HPO_4, 10 mM NaCl, 1 mM $MgSO_4$, 0.1% yeast extract, 0.1% casamino acids and 60 mM Na_2-succinate) was buffered with 50 mM Bicine and 50 mM CHES and the final pH adjusted to 8.8 with NaOH. Flasks containing 50 ml of medium were inoculated with 1 ml of cells grown overnight in LB medium. CCCP was added from a 25 mM stock in ethanol to a final concentration of 10 μM.

TABLE 1 Comparison of properties of ML308-225 cells grown at alkaline pH $\pm$ CCCP at 37°C

Conditions of growth[a]	*Doubling time (hours)*[b]	*Growth yield* OD_{550}[c]	*Lactose accumulation ratio (in/out)*[d]	$\Delta\bar{\mu}_{H^+}/F$ (mV)
pH 7.6	0.75	4.0	1150	190
pH 8.7	1.9	3.1	260	149
pH 8.7, 10 μM CCCP	2.1	2.4	190	140

[a]Cells were grown as described in Fig. 3.
[b]The doubling time indicated is for the initial phase of growth shown in Fig. 3. The doubling time increases in the later phase of growth to 4 and 5 hours respectively for cells grown at pH 8.7 and pH 8.7 + CCCP (carbonyl cyanide m-chlorophenylhydrazone).
[c]In the absence of succinate carbon source, cells at pH 8.7 grow to $OD_{550} = 1.0$ on the yeast extract and casamino acid supplements in medium, but at a slower rate (doubling time > 3h).
[d]Cells were harvested at $OD_{550} = 0.7$–0.8 for lactose accumulation measurements, as described by Kashket (1982).

have previously reported inhibition of growth on succinate at pH 8.8 by 100 μM CCCP, and the growth of *E. coli* on a rich medium at pH 9 in the presence of 100 μM CCCP has been demonstrated by Mugikura et al (1990). Under the growth conditions used here, the cells adapt in the initial phase of culture (at pH 8.8) by acidifying the medium to approximately 8.7 before the onset of growth. The acidification and subsequent growth is dependent upon the addition of yeast extract to the medium, a component also included in the medium of Avetisyan et al (1989). The growth yield on succinate, during growth in the buffered range of pH 8.6–8.7, is less efficient than at pH 7.6 and the efficiency further reduced by addition of CCCP (see Table 1). The rate of growth of cells at pH > 8.6 slows at higher cell densities ($OD_{550} > 0.8$), the doubling time increasing from 2 hours to 4 and 5 hours for cultures with and without CCCP, respectively. The slower phase of growth appears to correspond to a phase that is solely dependent upon the succinate carbon source, as the cells will grow to approximately the growth transition density in the absence of the succinate carbon source (Table 1). The growth of an *E. coli* K12 strain (MM180) was also studied under the conditions shown in Fig. 3 and shown to be dependent upon a functional F_1F_0 ATP synthase. Isogenic strains carrying *uncA401* (subunit α), *uncB402* (subunit *a*) or *uncE107* (subunit *c*) mutations do not grow. Avetisyan et al (1993) have noted a similar dependence on a functional *unc* operon.

We have concluded that the growth of *E. coli* on succinate at pH > 8.6 in the presence of 10–15 μM CCCP is still driven by $\Delta\bar{\mu}_{H^+}$-dependent oxidative phosphorylation. We have measured $\Delta\bar{\mu}_{H^+}$ indirectly in strain ML308-225 under the conditions of exponential growth shown in Fig. 3, and have observed a modest decline (Table 1). The lactose accumulation ratio, which is logarithmically related to $\Delta\bar{\mu}_{H^+}$ in these cells, was measured and $\Delta\bar{\mu}_{H^+}$ calculated according to Kashket (1982). Both growth and lactose accumulation are abolished by further additions of CCCP. In contrast to Avetisyan et al (1993), there is no indication that phosphorylation (or lactose accumulation) is driven by a $\Delta\bar{\mu}_{Na^+}$.

Conclusions

The proposed oligomeric structure of subunit *c* places the H^+-binding carboxylate of Asp61 at the centre of the four interacting helices of two *c* subunits. The specificity of ion binding is affected by interactions between the Asp61 carboxylate at the front face of one subunit *c* and side chains at the back face of a second subunit. In the case of *E. coli*, multiple substitutions are required to enable Li^+ binding, including Glu and Ser substitutions at positions 61 and 62, respectively (Zhang & Fillingame 1995). The *P. modestum* enzyme utilizes the equivalent of these residues as well as Gln at the position of residue 28 for Na^+ binding. The liganding groups of both the Ser and Gln side chains emerge from the back side of one *c* subunit and pack against the Glu carboxylate at the front face of a different *c* subunit. Given the nature of the ligand-binding pocket, it is difficult to imagine a simple modification altering specificity from H^+ to Na^+ by an auxiliary protein, as suggested by Avetisyan et al (1993). On the other hand, the position 61 carboxylate is clearly visualized through a crevice at the periplasmic side of the interacting *c* subunits, and one could imagine an altered *a–c* subunit interaction that fed protons directly from a protein at the interior of the lipid bilayer, as is considered by Krulwich (1995).

The suggestion that the number of *c* subunits in the ring could be increased to change the H^+/ATP stoichiometry, particularly in other species, still seems plausible since the packing of subunits in a ring of 12 (the now determined stoichiometry for the *E. coli* enzyme) may be structurally no more optimal than a ring with greater or fewer subunits. The proposed packing of subunit *c* in a front-to-back manner to form an ion-binding site does not rule out previously discussed stoichiometries of nine or 15 *c* subunits per F_0, and the expected whole number H^+/ATP ratios of three or five. Although non-integral H^+/ATP ratios are not ruled out, integral ratios may be easier to rationalize if the enzyme in some way counts the sequential protonation/deprotonation of *c* subunits before coupling co-operative release of the energy from a fixed quanta of subunits to the synthesis of a single ATP molecule. In this regard, one possible explanation for the continued growth of *E. coli* at $pH > 8.6$ in the presence of CCCP (Fig. 1 and Table 1) is an altered H^+/ATP stoichiometry, possibly mediated by an increased number of *c* subunits per F_0. Our estimations of the amount of *c* subunits in immunoblots of membranes prepared from cells grown to early stationary phase under these condition, however, show no indication of greater levels of *c* subunits. We do nonetheless think it reasonable to more carefully check the subunit *c* stoichiometry in species such as *Synechcoccus* where H^+/ATP ratios considerably greater than four have been suggested (Van Walraven et al 1997).

To provoke further consideration of these matters, we do note that the consensus $H^+/2e^-$ ratio for NADH oxidation by mitochondria is 10 (Nicholls & Ferguson 1992). Recent proton flux measurements with chloroplast F_1F_0 ATP synthase indicate an H^+/ATP ratio of four (Berry & Rumberg 1996). If the same H^+/ATP ratio applies to mitochondria, and one H^+ is transported per exchange of (ADP + P_i) for ATP, then the mitochondrial P/O for NADH linked substrates should be two rather than the conventionally reported three.

Acknowledgements

This work was supported by the US Public Health Service (Grant GM-23105), and by a Hilldale Undergraduate Fellowship from the University of Wisconsin, Madison to S.D.

References

Abrahams JP, Leslie AGW, Lutter R, Walker JE 1994 Structure at 2.8 Å resolution of F_1-ATPase from bovine heart mitochondria. Nature 370:621–628

Avetisyan AV, Dibrov PA, Skulachev VP, Sokolov MV 1989 The Na^+-motive respiration in *Escherichia coli*. FEBS Lett 254:17–21

Avetisyan AV, Bogachev AV, Murtasina RA, Skulachev VP 1992 Involvement of a *d*-type oxidase in the Na^+-motive respiratory chain of *Escherichia coli* growing under low $\Delta\bar{\mu}_{H}+$ conditions. FEBS Lett 306:199–202

Avetisyan AV, Bogachev AV, Murtasina RA, Skulachev VP 1993 ATP-driven Na^+ transport and Na^+-dependent ATP synthesis in *Escherichia coli* grown at low $\Delta\bar{\mu}_{H}+$. FEBS Lett 317:267–270

Berry S, Rumberg B 1996 H^+/ATP coupling ratio at the unmodulated CF_0CF_1-ATP synthase determined by proton flux measurements. Biochim Biophys Acta 1276:51–56

Das A, Ljungdahl LG 1997 Composition and primary structure of the F_1F_0 ATP synthase from the obligately anaerobic bacterium *Clostridium thermoaceticum*. J Bacteriol 179:3746–3755

Fillingame RH 1992 H^+ transport and coupling by the F_0 sector of the ATP synthase: insights into the molecular mechanism of function. J Bioenerget Biomemb 24:485–491

Fillingame RH 1997 Coupling H^+ transport and ATP synthesis in F_1F_0 ATP synthases: glimpses of interacting parts in a dynamic molecular machine. J Exp Biol 200:217–224

Fillingame RH, Oldenburg M, Fraga D 1991 Mutation of alanine 24 to serine in subunit *c* of the *Escherichia coli* F_1F_0 ATP synthase reduces reactivity of aspartyl 61 with dicyclohexylcarbodiimide. J Biol Chem 266:20934–20939

Gay NJ 1984 Construction and characterization of an *Escherichia coli* strain with an *uncI* mutation. J Bacteriol 158:820–825

Girvin ME, Rastogi VK, Abildgaard F, Markley JL, Fillingame RH 1998 Solution structure of the transmembrane H^+-transporting subunit *c* of the F_1F_0 ATP synthase. Biochemistry 37:8817–8824

Hellmuth K, Rex G, Surin B, Zinck R, McCarthy JEG 1991 Translational coupling varying in efficiency between different pairs of genes in the central region of the *atp* operon of *Escherichia coli*. Mol Microbiol 5:813–824

Jensen PR, Westerhoff HV, Michelsen O 1993 Excess capacity of H^+-ATPase and inverse respiratory control in *Escherichia coli*. EMBO J 12:1277–1282

Jones PC, Jiang W, Fillingame RH 1998 Arrangement of the multicopy H^+-translocating subunit *c* in the membrane sector of the *Escherichia coli* F_1F_0 ATP synthase. J Biol Chem 273:17178–17185

Kaim G, Wehrle F, Gerike U, Dimroth P 1997 Molecular basis for the coupling ion selectivity of F_1F_0 ATP synthases: probing the liganding groups for Na^+ and Li^+ in the *c* subunit of the ATP synthase from *Propionigenium modestum*. Biochemistry 36:9185–9194

Kakinuma Y, Igarashi K 1990 Mutants of *Streptococcus faecalis* sensitive to alkaline pH lack Na^+-ATPase. J Bacteriol 172:1732–1735

Kashket ER 1982 Stoichiometry of the H^+-ATPase of growing and resting, aerobic *Escherichia coli*. Biochemistry 21:5534–5538

Kasimoglu E, Park S-J, Malek J, Tseng CP, Gunsalus RP 1996 Transcriptional regulation of the proton-translocating ATPase (*atpIBEFHAGDC*) operon of *Escherichia coli*: control by cell growth rate. J Bacteriol 178:5563–5567

Kobayashi H 1985 A proton-translocating ATPase regulates pH of the bacterial cytoplasm. J Biol Chem 260:72–76

Kobayashi H, Saito H, Futatsugi L, Kakegawa T 1999 Cation movements at alkaline pH in bacteria growing without respiration. In: Bacterial responses to pH (Novartis Found Symp 221) Wiley, Chichester p 235–245

Krenn BE, Van Walraven HS, Scholts MJC, Kraayenhof R 1993 Modulation of the proton-translocation stoichiometry of H^+-ATP synthases in two phototrophic prokaryotes by external pH. Biochem J 294:705–709

Krulwich TA 1995 Alkaliphiles: basic molecular problems of pH tolerance and bioenergetics. Mol Microbiol 15:403–410

Krulwich TA, Guffanti AA, Ito M 1999 pH tolerance in *Bacillus*: alkaliphiles versus non-alkaliphiles. In: Bacterial responses to pH (Novartis Found Symp 221) Wiley, Chichester p 167–182

McCarthy JEG, Gerstel B, Surin B, Wiedemann U, Ziemke P 1991 Differential gene expression from the *Escherichia coli atp* operon mediated by segmental differences in mRNA stability. Mol Microbiol 5:2447–2458

Miller MJ, Oldenburg M, Fillingame RH 1990 The essential carboxyl group in subunit *c* of the F_1F_0 ATP synthase can be moved and H^+-translocating function retained. Proc Natl Acad Sci USA 87:4900–4904

Mugikura S, Nishikawa M, Igarashi K, Kobayashi H 1990 Maintenance of a neutral cytoplasmic pH is not obligatory for growth of *Escherichia coli* and *Streptococcus faecalis* at an alkaline pH. J Biochem 108:86–91

Murata T, Yamato I, Igarashi K, Kakinuma Y 1996 Intracellular Na^+ regulates transcription of the *ntp* operon encoding a 'vacuolar-type' Na^+-transporting ATPase in *Enterococcus hirae*. J Biol Chem 271:23661–23666

Nicholls DG, Ferguson SJ 1992 Bioenergetics 2. Academic Press, London

Noji H, Yasuda R, Yoshida M, Kazuhiko K Jr 1997 Direct observation of the rotation of F_1-ATPase. Nature 386:299–302

Ohyama T, Imaizumi R, Igarashi K, Kobayashi H 1992 *Escherichia coli* is able to grow with negligible sodium ion extrusion activity at alkaline pH. J Bacteriol 174:7743–7749

Shibata C, Ehara T, Tomura K, Igarashi K, Kobayashi H 1992 Gene structure of *Enterococcus hirae* (*Streptococcus faecalis*) F_1F_0-ATPase, which functions as a regulator of cytoplasmic pH. J Bacteriol 174:6117–6124

Sumi M, Yohda M, Koga Y, Yoshida M 1997 F_0F_1-ATPase genes from an archebacterium, *Methanosarcina backeri*. Biochem Biophys Res Commun 241:427–433

Van Walraven HS, Hollander, EE, Scholts MJC, Kraayenhof R 1997 The H^+/ATP ratio of the ATP synthase from the cyanobacterium *Synechococcus* 6716 varies with growth temperature and light intensity. Biochim Biophys Acta 1318:217–224

Wilkens S, Dunn SD, Chandler J, Dahlquist FW, Capaldi RA 1997 Solution structure of the N-terminal domain of the δ subunit of the *E. coli* ATP synthase. Nat Struct Biol 4:198–201

Zhang Y, Fillingame RH 1995 Changing the ion binding specificity of the *Escherichia coli* H^+-transporting ATP synthase by directed mutagenesis of subunit *c*. J Biol Chem 270:87–93

DISCUSSION

Skulachev: I would like to describe a simple experiment we performed in *Escherichia coli* that is relevant to this discussion. We measured the rate of growth and motility in the presence of increasing concentrations of the uncoupler pentachlorophenol. There is a linear relationship between the motility rate and the proton motive force down to 20 mV, which is the threshold for supporting motility. We found that if sodium ions are present, growth is resistant to the uncoupler, whereas in the absence of sodium the cells die. We also found that under the same conditions, motility was completely inhibited by pentachlorophenol, even in the presence of Na^+. We obtained the same result for carbonyl cyanide m-chlorophenyl hydrazone (CCCP), but in this case a problem may arise because *E. coli* can pump CCCP out of the cell. We also tried to use the same method as Robert Fillingame uses to measure the proton motive force (lactose distribution), but it did not work under these conditions because if pentachlorophenol is added after lactose, lactose is not released from the cell, and this effect is sodium independent.

Poole: Do you have any ideas on whether the differences in methodology explain the apparent discrepancy?

Fillingame: We have not done any experiments with pentachlorophenol. We grow cells at pH 8.7, and these are conditions under which Vladimir Skulachev has claimed that there is a change in the ion specificity of F_1F_0 ATP synthase (Avetisyan et al 1993), but we see no indication that there is. I should say that we estimated $\Delta\bar{\mu}_{H^+}$ by lactose accumulation using strain ML308-225, which gives a good correlation between lactose accumulation and $\Delta\bar{\mu}_{H^+}$.

Poole: What is the sodium concentration in your experiments?

Fillingame: About 160 mM.

Booth: How can you rule out that pentachlorophenol is simply blocking the lac permease?

Skulachev: This is not the case because we observe the same effects with cyanide, i.e. if we add cyanide after lactose there is a slow and incomplete release of lactose.

On the other hand, pentachlorophenol or cyanide prevent lactose uptake if they are added before lactose.

Dimroth: One of the most intriguing questions about the ATP synthase is how the motor functions. How is the ion gradient, the proton gradient or the electrochemical sodium gradient used to induce the rotary torque that drives the rotation of the γ subunit required for ATP synthesis? We have recently addressed this question, and our results suggest that the sodium-binding sites on the *c* subunits are freely accessible from the cytoplasmic side of the membrane. In contrast, Junge et al (1997), Duncan et al (1995) and Vik & Antonio (1994) proposed a model in which these sites are in the middle of the membrane, so that two channels in the *a* subunit are needed to make these binding sites accessible to both sides of the membrane. In our case, these binding sites are accessible from the cytoplasmic side, and in order for the coupling ions to be translocated across the membrane the *c* subunit ring rotates until it comes into contact with the *a* subunit. Within the *a* subunit there is a channel through which the sodium ions are translocated. We have evidence which suggests that this *a* subunit channel is an ion-selective channel, because we can mutate it such that it only releases lithium and protons but not sodium, and in this case we observe an ATP-dependent occlusion of one sodium ion per ATP synthase. Our interpretation of this is that the *c* subunit ring rotates and comes into contact with the *a* subunit, but sodium is not released because the *a* subunit channel has become sodium impermeable by the mutation.

We imagine that there are electrostatic interactions between positively and negatively charged amino acid side chains of the stator (subunit *a*) and the rotor (*c* subunit ring) that are instrumental for the induction of rotary torque. The highly conserved and functionally essential arginine on the stator (*a*R227 in *Propionigenium modestum*) and the Na^+-binding *c*E65 residue of the rotor subunits may be involved in this process. We propose that the membrane potential causes the sodium ions to pass from the periplasm through the stator channel into a binding site on subunit *a*. The positive charge of *a*R227 then attracts the negatively charged *c*E65 residue. Following transfer of the Na^+ ion from its stator site to *c*E65, the electrostatic attraction to *a*R227 is abolished and the positive stator charge is now able to attract the next empty rotor site of the ring. This leads to continued rotation of the ring and to the generation of rotary torque.

We have determined the *c* subunit structure by NMR in dodecyl sulfate micelles. It has four separated α helical domains, and these domains are disrupted in the vicinity of the sodium-binding site. We propose that the N- and C-terminal α helical domains span the membrane such that the sodium-binding site is located at the surface of the membrane and the additional two α helices either protrude into the cytoplasm or are membrane associated but not membrane integral. Finally, I would like to mention that one has to be careful with structural determination in organic solvents and in SDS, so I'm not saying that this is the final picture. One also has to be aware that the native structure of subunit *c* is an oligomer of 12 subunits, and until we have solved the ring structure completely, we cannot tell what this structure looks like.

Fillingame: The problem with this is that the C-terminal helix in the proposed structure is 19 amino acids in length, which is barely long enough to span the hydrophobic core of the lipid bilayer, and certainly could not extend to the phospholipid head group. In the case of *E. coli* subunit *c*, if you cross-link the C-terminus to the phospholipid head group, and if it is α helical, then Asp61 cannot go much further than the centre of the membrane.

Padan: In which structure does N,N′-dicyclohexylcarbodiimide (DCCD) bind? In other words, are there any indications that these structures reflect the native conformation?

Dimroth: We know that the sodium-binding site is retained because we see sodium-dependent protection from DCCD labelling of isolated subunit *c* in dodecyl sulfate micelles.

Fillingame: DCCD reacts specifically with Asp61 of the isolated protein in chloroform–methanol (Girvin & Fillingame 1994). We do not see sodium protection or lithium protection in the experiments we have done with membrane vesicles of the lithium-binding mutant (Zhang & Fillingame 1995). Peter Dimroth has shown a reaction of about 10% of the subunit *c* in SDS with DCCD, and about half of that is protected by sodium binding. This means, if one wants to play devil's advocate, that 5% of the subunit does bind sodium. But whether this is the same fraction of subunit that they are determining the structure of in dodecyl sulfate micelles is unproven. There are no indications from the NMR work that these helices come together to fold and interact with each other to form a binding site. Therefore, you would have to stretch your imagination to believe that this is the real structure of the sodium-binding site.

Dimroth: This may not be a problem because if the helices are not as close in the dodecyl sulfate micelles as they are in organic solvents, the signals may not be strong enough to be picked up.

Cook: Is DCCD a specific inhibitor?

Fillingame: No. DCCD reacts non-specifically with many things, although we can talk about events that are blocked in DCCD-resistant mutants. DCCD reacts with this carboxyl group because there's a binding site for one of the cyclohexyl groups, and it has nothing to do with simply being soluble in the lipid and happening to react with carboxyl groups. The reason for this is that the specificity of the reaction between different hydrophobic carbodiimides and different *c* subunits varies from species to species. It is also possible to make small chemical changes that alter their ability to react.

Schäfer: I like the model Peter Dimroth has presented because of the experiment with the occluded sodium. If one moves 12 protons through the membrane, in terms of charge it would be the same as sodium. But, according to your model one should find a much higher number of occluded charges (up to 11), which is not the case for the sodium-transporting F_1F_0 ATPase because you only find one. This is important because it suggests that the distance sodium has to travel to reach the channel is short, and it is not necessary that 10 or 12 places have to be occupied in the rotor simultaneously.

Fillingame: This doesn't mean that this site has to be at the surface of the membrane. It simply means that there has to be access to that site from one side of the membrane. Peter Dimroth talked about occlusion during ATP hydrolysis, but under conditions of ATP synthesis we have evidence which suggests that protons must be freely accessible from the periplasm site of the membrane to titrate one of these carboxyl groups (Zhang & Fillingame 1994). This is what one would expect in the direction of ATP synthesis, i.e. that free access should be from the opposite side of the membrane. However, the fact that there is free access does not necessarily mean that there are no proteins mediating this process. It doesn't necessarily have to be the same helices of subunit *a* that mediate the access from the other side. The occlusion experiment is an important piece of evidence, but the structural interpretation is an over-interpretation.

Padan: Why don't you use cysteine scanning to determine the location of the aspartate?

Fillingame: We could do more experiments. The TID (trifluoromethyl-iodophenyldiazirine) labelling experiments indicate that it is in the centre of the membrane, and the cross-linking experiment with phosphatidylethanolamine at the C-terminus support this conclusion. No one has previously suggested that this was not the location of the aspartate group. Now someone has seriously suggested it's not, so we should design experiments to prove this more conclusively.

Quivey: You mentioned that there are residues at the C-terminus of the *c* subunit that can be cross-linked to lipid. If the *c* subunit contains lipid, does this enhance or retard the rotation of the *c* subunit?

Fillingame: The experiment I mentioned was done by Rob Capaldi's lab in 1984. They cross-linked the C-terminus of the *c* subunit to phosphatidylethanolamine using a water-soluble carbodiimide (Lötscher et al 1984). I don't know whether this enhances or retards the rotation. There may also be a lipid component inside the *c* subunit rings. The diameter of the hole is 20 Å, so there's plenty of room for phospholipid. David Foster and I showed in 1979 that if F_1F_0 is delipidated in the purification process, DCCD sensitivity and the ability to reconstitute function are lost (Foster & Fillingame 1979). Therefore, it is possible that structural lipids in the centre of the ring are required for the reconstitution of function. This may prove to be critical in terms of determining the crystal structure.

Poole: When will we have structural information on the F_0 component of the ATPases at the same level of detail that is currently available for the F_1 component?

Fillingame: Attempts are being made to crystallize F_0 from both *E. coli* and bovine mitochondria. We're making 2D crystals of *c* oligomers, but we are also checking that the cross-links we observe in the *c* oligomer are the same as those we observe for native F_0, so that we know that there is not a structural reorganization. Therefore, some information may be available soon, but it could be a long time before a structure at 2 Å resolution is available.

Stock: In the motor that couples a proton influx to bacterial flagellar rotation, the number of motor proteins (MotA and MotB) per flagellum is variable. Could F_0 also have a variable stoichiometry with different numbers of *c* subunits?

Fillingame: That is still a possibility. In the case of *E. coli*, I'm certain that it is 12. We have no indications that this stoichiometry changes under any growth conditions. We were interested to see whether we could increase the number of *c* subunits per F_0 at low $\Delta\bar{\mu}_{H^+}$, but this doesn't seem to happen. It's conceivable that it could happen in other species.

Stock: Is it possible that 12 is the maximum?

Fillingame: Yes, although there have been estimates in other species that the H^+/ATP ratio could rise to five, which means that you would need about 15. In terms of the packing of these subunits, it is unlikely that there is too much difference between 10, 11 and 12.

Konings: Is the aspartate residue accessible from inside the ring?

Fillingame: No more accessible than from outside the ring. It is likely that the aspartate residue is in the middle of where the four helices of the two subunits come together. Based on the cross-linking experiments with subunit *a*, in which we cross-linked subunit a to opposite sides of subunit *c*, we think that subunit *a* may insert itself between those two subunits to open the binding site.

Konings: How far down through the cylinder does the γ subunit extend?

Fillingame: That's a good question. It's possible that the γ subunit could extend half way through the membrane, although the portion of the γ subunit that has not been resolved by X-ray crystallography is not particularly hydrophobic. It looks like a typical water-soluble protein, and it would have to have an unusual structure if it went that far into the cylinder.

References

Avetisyan AV, Bogachev AV, Murtasina RA, Skulachev VP 1993 ATP-driven Na^+ transport and Na^+-dependent ATP synthesis in *Escherichia coli* grown at low $\Delta\bar{\mu}_{H^+}$. FEBS Lett 317: 267–270

Duncan TM, Bulygin VV, Zhou Y, Hutcheon ML, Cross RL 1995 Rotation of subunits during catalysis by *Escherichia coli* F_1-ATPase. Proc Natl Acad Sci USA 92:10964–10968

Foster DL, Fillingame RH 1979 Energy-transducing H^+-ATPase of *Escherichia coli*. Purification, reconstitution, and subunit composition. J Biol Chem 254:8230–8236

Girvin ME, Fillingame RH 1994 Hairpin folding of subunit *c* of F_1F_0 ATP synthase: ^1H distance measurements to nitroxide-derivatized aspartyl-61. Biochemistry 33:665–674

Junge W, Lill H, Engelbrecht S 1997 ATP synthase: an electrochemical transducer with rotatory mechanics. Trends Biochem Sci 22:420–423

Lötscher HR, de Jong C, Capaldi RA 1984 Modification of the F_0 portion of the H^+-translocating adenosinetriphosphatase complex of *Escherichia coli* by the water-soluble carbodiimide 1-ethyl-3-[3-(dimethylamino)propyl]carbodiimide and effect on the proton-channeling function. Biochemistry 23:4128–4134

Vik SB, Antonio BJ 1994 A mechanism of proton translocation by F_1F_0 ATP synthases suggested by double mutants of the a subunit. J Biol Chem 269:30364–30369

Zhang Y, Fillingame RH 1994 Essential aspartate in subunit *c* of F_1F_0 ATP synthase. Effect of position 61 substitutions in helix-2 on function of Asp24 in helix-1. J Biol Chem 269:5473–5479

Zhang Y, Fillingame RH 1995 Changing the ion binding specificity of the *Escherichia coli* H^+-transporting ATP synthase by directed mutagenesis of subunit *c*. J Biol Chem 270:87–93

Cation movements at alkaline pH in bacteria growing without respiration

Hiroshi Kobayashi, Hiromi Saito, Lui Futatsugi and Tomohito Kakegawa

Faculty of Pharmaceutical Sciences, Chiba University, 1-33 Yayoi-cho, Inage-ku, Chiba 263-8522, Japan

Abstract. The respiratory chain has a central role in energy metabolism as a generator of a proton motive force in aerobic bacteria. In contrast, *Enterococcus hirae* (formerly *Streptococcus faecalis*), which lacks the respiratory chain, generates this proton gradient via a F-type H^+ ATPase, but it works only at a pH below 8; no significant proton motive force is generated at a pH above 8. An *Escherichia coli* mutant deficient in both the respiratory chain and the H^+ ATPase grew with a negligible proton motive force within the wide range of medium pH. It has been suggested that both *E. hirae* and *E. coli* are able to grow even when the cytoplasm is alkalinized beyond pH 8. These observations lead to the conclusion that bacteria can survive without operating cation transport systems driven by a proton motive force at alkaline pH. The activity of any transport system with optimum pH around neutrality should decline when both the outside and inside of cells are alkalinized. Thus, changes in transport systems as well as cellular metabolism may be essential for bacterial adaptation to changes in environmental pH.

1999 Bacterial responses to pH. Wiley, Chichester (Novartis Foundation Symposium 221) p 235–245

The advance of a structural analysis of functional units has been remarkable. However, the answers to many basic questions remain unclear. Why can living things survive in a variety of habitats that encompass a wide range of pH? How do bacteria adapt themselves to changes in environmental pH? It has been reported that internal pH homeostasis is essential to growth under acid or alkaline stress, and the respiratory chain has a central role in the homeostasis. However, bacteria without respiration can survive at an extreme pH. Therefore, it would be important to clarify how bacteria adapt to a pH change in surroundings when respiration is not operative. For this purpose, we have examined adaptive strategies of enterococci without respiration and an *Escherichia coli* mutant deficient in the respiratory chain.

Enterococci grow with a negligible proton motive force at high pH

It has been suggested that bacterial cytoplasmic pH is maintained at a constant value within the wide range of medium pH by transport systems for protons (Padan et al 1981, Booth 1985). In the case of fermenting anaerobic bacteria without respiration, a F-type of H^+-translocating ATPase extrudes protons, resulting in the generation of a proton motive force (Harold & Altendorf 1974). However, since the major role of the proton extrusion is to regulate the internal pH, a proton motive force is generated only when the cytoplasm is acidified; no significant proton motive force is generated at alkaline pH in *Enterococcus hirae* (Kobayashi 1985, 1987). No growth of a mutant deficient in the F-type ATPase was observed at a pH below 6.5, but the mutant was able to grow as well as the wild-type strain at neutral and alkaline pH values (Fig. 1). These observations imply that enterococci grow with a negligible proton motive force at high pH, and transport systems work without this force under these conditions. ATP-driven transport systems for Na^+ (Heefner & Harold 1980, Kinoshita et al 1984, Takase et al 1994), K^+ (Kobayashi 1982, Kakinuma & Harold 1985) and Ca^{2+} (Kobayashi et al 1978) have been found in enterococci.

Enterococci have no system to maintain intracellular pH homeostasis at high pH

It has been proposed in *E. coli* that Na^+/H^+ or K^+/H^+ antiporters driven by the membrane potential maintain the slightly alkaline cytoplasm around pH 7.6

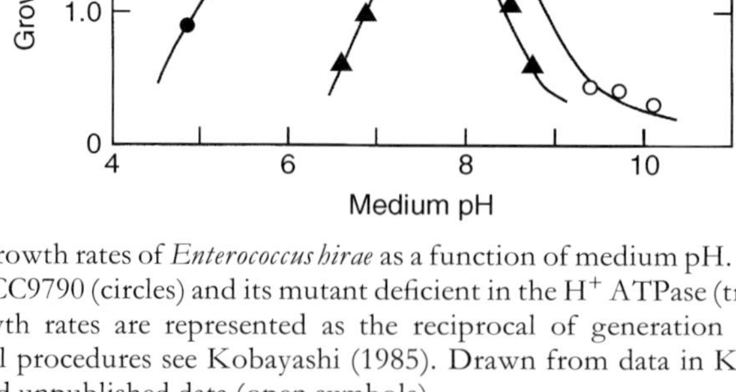

FIG. 1. Growth rates of *Enterococcus hirae* as a function of medium pH. The wild-type strain of *E. hirae* ATCC9790 (circles) and its mutant deficient in the H^+ ATPase (triangles) were grown at 37°C. Growth rates are represented as the reciprocal of generation time (h). For detailed experimental procedures see Kobayashi (1985). Drawn from data in Kobayashi (1985; closed symbols) and unpublished data (open symbols).

under alkaline stress (Padan et al 1981, Padan & Schuldiner 1994). However, such antiporters are unable to work in enterococci, even if they are present, because the magnitude of the membrane potential is negligible at alkaline pH. The efflux of K^+ was observed at alkaline pH in enterococci by Kakinuma & Igarashi (1988), and they have argued that a K^+ gradient generated by a putative ATP-linked K^+ uptake system drives a K^+/H^+ antiporter, and the resulting accumulation of protons makes a recovery from cytoplasmic alkalinization in alkaline medium. However, *E. hirae* grew in the presence of carbonyl cyanide m-chlorophenylhydrazone (CCCP) or valinomycin at a pH above 9. The measurement of intracellular pH (pH_i) has shown that pH_i is kept constant only when the medium pH is 6.0–8.0 (Fig. 2). Thus, we have concluded that *E. hirae* grows without maintaining pH_i at a constant value at a pH above 8.

Both influx and efflux of K^+ are required at high pH

Valinomycin decreased the K^+ level at pH 5.3 and increased it at pH 9.4 in medium containing 170 mM K^+ (Fig. 3). The K^+ level in the presence of gramicidin D is high in alkaline medium (Fig. 3). Since gramicidin D collapses gradients of both pH and K^+ concentration, this result suggests that cells have a large capacity to bind K^+ at a high pH_i, probably due to the increased amount of negatively charged molecules. This may account for the opposite direction of K^+ movements after the addition of valinomycin at different pH values. The decrease in the K^+ level by the addition of valinomycin was observed when the K^+

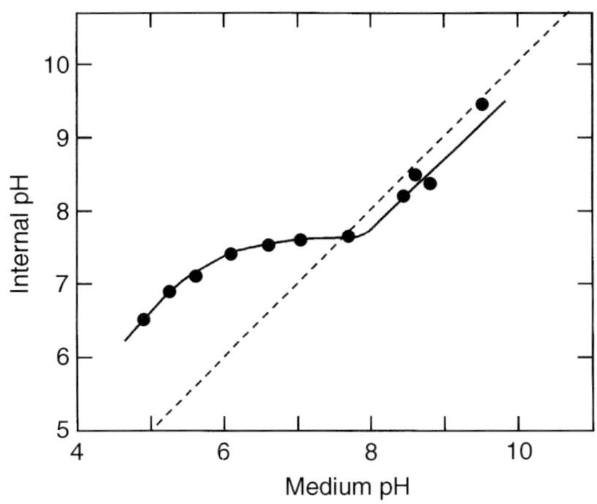

FIG. 2. Intracellular pH of *Enterococcus hirae*. Broken line represents the intracellular pH equal to the medium pH. For detailed experimental procedures see Kobayashi (1985), Kobayashi et al (1986) and Mugikura et al (1990). Drawn from data in these references.

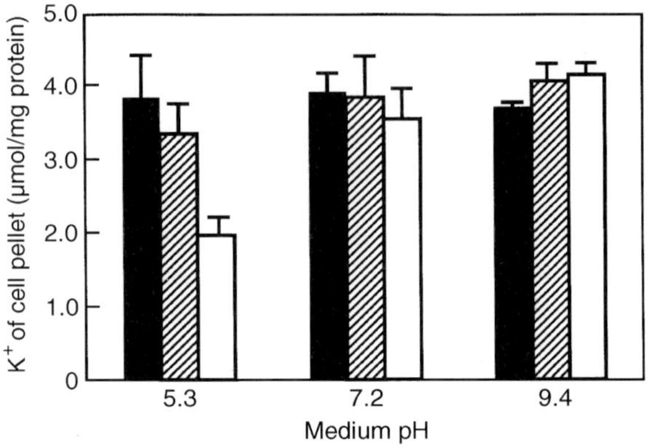

FIG. 3. Cytoplasmic level of K^+ of *Enterococcus hirae*. Cells were grown in the medium containing 1% tryptone, 0.5% yeast extract, 1% glucose, 5 mM K_2HPO_4 and 50 mM Good buffer (CAPS at pH 9.4, HEPES at pH 7.2 and MES at pH 5.3). The medium pH was adjusted by the addition of KOH. The K^+ concentration of the medium was adjusted at 170 mM by the addition of KCl. Cells were centrifuged through oil and the amount of K^+ in the precipitate was measured as described in Sakuma et al (1998). Closed bars, no addition; hatched bars, plus valinomycin (4 μg/ml); open bars, plus gramicidin D (4 μg/ml). Horizontal lines on the bar represent the S.D. of two determinations using cells cultured separately.

concentration of medium was 10 mM in an alkaline medium. These results demonstrate that systems for efflux and influx of K^+ are both required at high pH, and the direction of K^+ movements is dependent on the K^+ concentration of the medium.

As described above, transport systems functioning at alkaline pH have to work without a proton motive force. Furthermore, the activity of any system functioning at neutral pH should decrease when both the inside and outside of cells are alkalinized. Thus, it can be argued that systems functioning at an alkaline pH above 8 are different from those operating at neutral pH. *E. hirae* has been reported to have multiple transport systems for K^+ (Bakker & Harold 1980, Kobayashi 1982, Kakinuma & Harold 1985, Kakinuma & Igarashi 1988); however, no transport system for K^+ functioning at high pH has been identified until now. In order to dissect adaptive strategies of this bacterium to alkaline environments, it would be essential to determine which systems circulate K^+ across the membranes under alkaline stress. In the case of Na^+ circulation, it has been clarified that a Na^+-translocating ATPase functions at alkaline pH (Kinoshita et al 1984), whereas a Na^+/H^+ antiporter operates at neutral and acid pH values (Kakinuma 1987). The latter may be driven by a proton motive force generated as a result of the pH_i regulation.

Escherichia coli is able to grow without maintaining intracellular pH homeostasis in alkaline medium

E. coli has been reported to maintain pH_i at a constant value around 7.6 within the wide range of medium pH (Padan et al 1976, Slonczewski et al 1991). Mugikura et al (1990) have suggested that pH_i is kept constant within the narrow range of medium pH 6.5–8.0, and that the pH value increases when the medium is alkalinized above pH 8 (Fig. 4). In agreement with their suggestion, there have been reports which show that *E. coli* is able to grow in the presence of a protonophore at a pH above 8 (Ohyama et al 1992a,b). The maintenance of pH homeostasis is difficult in the presence of a protonophore.

An *E. coli* mutant deficient in both the respiratory chain and the F_1F_0 ATPase isolated recently grew within the pH range of medium from 5 to 9. No Soret band was detected in the mutant. The ATPase activity was low and no significant magnitude of proton motive force was detected. These results suggest that neither the K^+/H^+ nor the Na^+/H^+ antiporters are essential for growth at high pH, making the previous proposal unlikely that pH_i regulation is obligatory for growth of *E. coli*. Based on these findings, we propose that *E. coli* is able to grow in alkaline medium without maintaining the cytoplasm at a constant pH value.

If our hypothesis is true, transport systems functioning at alkaline pH may be different from those operating at neutral pH in *E. coli* as well, because the function of any transport system with optimum pH of neutrality should decline when both the inside and outside of cells are alkalinized. We have found recently that NhaB, one of the *E. coli* Na^+/H^+ antiporters, extrudes Na^+ only at a pH below 8, and Na^+

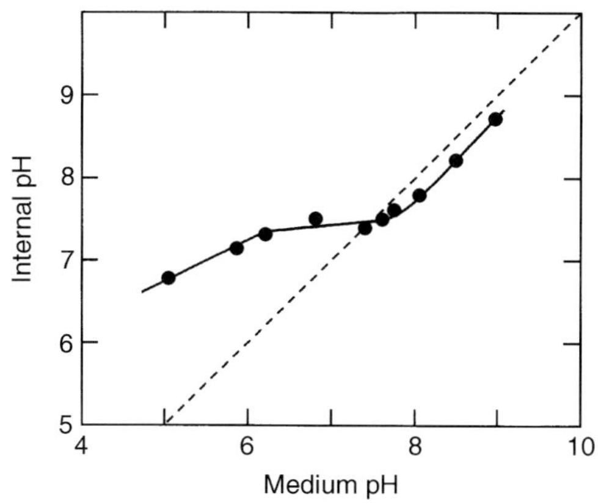

FIG. 4. Intracellular pH of *Escherichia coli*. Broken line represents the intracellular pH equal to the medium pH. For detailed experimental procedures see Mugikura et al (1990), from which these data were drawn.

is extruded by NhaA and ChaA at a pH above 8. Many reports show that *E. coli* has multiple transport systems for K^+, but it is still unknown which systems participate in the circulation of K^+ when the cytoplasm is alkalinized.

Another remarkable character of the mutant deficient in the respiratory chain is the sensitivity to hyperosmosis, suggesting that the respiratory chain has some role in osmotic adaptation, although the mechanism is not known. Thus, the respiratory chain seems to be one of systems required for environmental adaptation.

A large fraction of genes is thought to be used for environmental adaptation

It is well known that bacteria have multiple systems for cation movements. Why is the multiplicity required for one kind of cation? It is not that cation movements are essential for life because, for example, an F_1F_0 ATPase is an important enzyme for living things, but *E. coli* has only one kind of enzyme and a single operon for it. As described above, we postulate that cells use different systems under different environmental conditions. If this is true, in addition to membrane transport systems, bacteria might change cytoplasmic metabolism to adapt themselves to a pH change in surroundings. Many genes have been reported to be induced under acidic stress (Olson 1993, Lambert et al 1997, Bearson et al 1997), whereas few genes required for growth in alkaline medium have been identified. However, *E. coli* might have many essential genes for growth under alkaline environments.

The determination of the entire *E. coli* DNA sequence has revealed that *E. coli* genome has 4288 open reading frames, but less than 2000 genes have been described until now (Blattner et al 1997). Almost all of genes functioning under growth conditions used in a laboratory should have already been reported, because *E. coli* has been studied during the last half of this century by a great number of researchers. Why is the function of so many genes still unknown? One plausible explanation is that these unidentified genes work under growth conditions we have rarely studied until now, for example acid or alkaline conditions.

Conclusions

It has been generally accepted that intracellular conditions are kept constant, and hence the same gene functions under any circumstances of bacterial habitats. However, it can be argued that living things have multiple systems functioning under different conditions. Transport systems are not exceptional. Thus, in order to understand the bacterial adaptation to changes in environmental pH, it would be important to dissect which systems bacteria use under acid or alkaline environments.

Finally, the characterization of bacteria growing in acid or alkaline medium would give us valuable information for study of bacterial pathogenesis, and advance a search for new antibacterial reagents that kill bacteria which invade organs and organelles that have acid or alkaline pH values.

Acknowledgement

This work is supported in part by a Grant-in-Aid for Scientific Research from the Ministry of Education, Science, Sports and Culture of Japan to H.K.

References

Bakker EP, Harold FM 1980 Energy coupling to potassium transport in *Streptococcus faecalis*: interplay of ATP and the protonmotive force. J Biol Chem 255:433–440

Bearson S, Bearson B, Foster JW 1997 Acid stress responses in enterobacteria. FEMS Microbiol Lett 147:173–180

Blattner FR, Plunkett G III, Bloch CA, et al 1997 The complete genome sequence of *Escherichia coli* K-12. Science 277:1453–1474

Booth IR 1985 Regulation of cytoplasmic pH in bacteria. Microbiol Rev 49:359–378

Harold FM, Altendorf K 1974 Cation transport in bacteria: K^+, Na^+, and H^+. Curr Top Membr Transp 5:1–50

Heefner DL, Harold FM 1980 ATP-linked sodium transport in *Streptococcus faecalis*. I. The sodium circulation. J Biol Chem 255:11396–11402

Kakinuma Y 1987 Sodium/proton antiporter in *Streptococcus faecalis*. J Bacteriol 169:3886–3890

Kakinuma Y, Harold FM 1985 ATP-driven exchange of Na^+ and K^+ ions by *Streptococcus faecalis*. J Biol Chem 260:2086–2091

Kakinuma Y, Igarashi K 1988 Active potassium extrusion regulated by intracellular pH in *Streptococcus faecalis*. J Biol Chem 263:14166–14170

Kinoshita N, Unemoto T, Kobayashi H 1984 Sodium-stimulated ATPase in *Streptococcus faecalis*. J Bacteriol 158:844–848

Kobayashi H 1982 Second system for potassium transport in *Streptococcus faecalis*. J Bacteriol 150:506–511

Kobayashi H 1985 A proton-translocating ATPase regulates pH of the bacterial cytoplasm. J Biol Chem 260:72–76

Kobayashi H 1987 Regulation of cytoplasmic pH in streptococci. In: Reizer J, Peterkofsky A (eds) Sugar transport and metabolism in gram-positive bacteria. Ellis Horwood Ltd, Chichester, p 255–269

Kobayashi H, Van Brunt J, Harold FM 1978 ATP-linked calcium transport in cells and membrane vesicles of *Streptococcus faecalis*. J Biol Chem 253:2085–2092

Kobayashi H, Suzuki T, Unemoto T 1986 Streptococcal cytoplasmic pH is regulated by changes in amount and activity of a proton-translocating ATPase. J Biol Chem 261:627–630

Lambert LA, Abshire K, Blankenhorn D, Slonczewski JL 1997 Proteins induced in *Escherichia coli* by benzoic acid. J Bacteriol 179:7595–7599

Mugikura S, Nishikawa M, Igarashi K, Kobayashi H 1990 Maintenance of a neutral cytoplasmic pH is not obligatory for growth of *Escherichia coli* and *Streptococcus faecalis* at an alkaline pH. J Biochem 108:86–91

Ohyama T, Imaizumi R, Igarashi K, Kobayashi H 1992a *Escherichia coli* is able to grow with negligible sodium ion extrusion activity at alkaline pH. J Bacteriol 174:7743–7749
Ohyama T, Mugikura S, Nishikawa M, Igarashi K, Kobayashi H 1992b Osmotic adaptation of *Escherichia coli* with a negligible proton motive force in the presence of carbonyl cyanide m-chlorophenylhydrazone. J Bacteriol 174:2922–2928
Olson ER 1993 Influence of pH on bacterial gene expression. Mol Microbiol 8:5–14
Padan E, Schuldiner S 1994 Molecular physiology of Na^+/H^+ antiporters, key transporters in circulation of Na^+ and H^+ in cells. Biochim Biophys Acta 1185:129–151
Padan E, Zilberstein D, Rottenberg H 1976 The proton electrochemical gradient in *Escherichia coli* cells. Eur J Biochem 63:533–541
Padan E, Zilberstein D, Schuldiner S 1981 pH homeostasis in bacteria. Biochim Biophys Acta 650:151–166
Sakuma T, Yamada N, Saito H, Kakegawa T, Kobayashi H 1998 pH dependence of the function of sodium ion extrusion systems in *Escherichia coli*. Biochim Biophys Acta 1363:231–237
Slonczewski JL, Rosen BP, Alger JR, Macnab RM 1981 pH homeostasis in *Escherichia coli*: measurement by ^{31}P nuclear magnetic resonance of methylphosphonate and phosphate. Proc Natl Acad Sci USA 78:6271–6275
Takase K, Kakinuma S, Yamato I, Konishi K, Igarashi K, Kakinuma Y 1994 Sequencing and characterization of the *ntp* gene cluster for vacuolar-type Na^+-translocating ATPase of *Enterococcus hirae*. J Biol Chem 269:11037–11044

DISCUSSION

Konings: We have observed in *Lactococcus* that a number of transport systems are inhibited at high pH, whereas others are stimulated at high pH, so the sensitivity of transport systems varies.

Kobayashi: The only data we have on potassium ion transport systems are that *Enterococcus hirae* has two potassium transport systems, only one of which is dependent on the membrane potential. We have not yet purified any of the enzymes involved or cloned any genes. It is important to examine the potassium ion extrusion systems because our data show that they have a role at high pH. We have shown that the accumulation of potassium ions at high pH is mediated by the transport system only at a low concentration of potassium ions, so now we need to do direct experiments under these conditions.

Konings: There are also transport systems that are coupled to sodium.

Padan: What is the driving force?

Kobayashi: ATP. Transport systems are independent of the proton motive force but dependent on ATP at high pH. At stationary phase, these cells contain low levels of ATP, unlike *Escherichia coli* which has a high content of ATP at stationary phase, so the requirement of ATP can be demonstrated using stationary phase cells in enterococci.

Konings: What is the driving force for amino acid transport systems?

Kobayashi: Some of them are dependent on the proton motive force at low pH, but we have no data for high pH. Harold & Van Brunt (1977) showed that this bacterium can grow in the presence of gramicidin D only in conditions of low sodium, high potassium and at pH 7.7. These results suggest that these transport systems are not important for growth under alkaline conditions.

Fillingame: You showed that the intracellular pH (pH_i) for both *E. hirae* and *E. coli* decreased at acidic pH and increased at alkaline pH. Was this in the absence of an uncoupler?

Kobayashi: Yes. We just added the indicator to the growing cells, then after 10–15 min we separated the cells by centrifugation and measured the amount of indicator in the pellet.

Fillingame: What was the indicator?

Kobayashi: Acetylsalicylic acid.

Booth: But acetylsalicylic acid is accumulated at pH 8 in *E. coli.*

Kobayashi: Under the conditions we used, it is apparent that the interstitial space is twice the internal volume of cells in the pellet, so if the pH gradient is over 1 pH unit, the ratio of the amount of indicator to the water volume in the pellet is about 70% of the ratio in the medium. The value was 70% when a large negative pH gradient was imposed. Our data showed that the ratio was 80–90% in growing cells at high pH, which means that the pH gradient is less than one unit. We calculated the pH gradient from these values, and found it was about 0.4 pH units. It had a large standard deviation, but we have clearly shown that the pH gradient is small at high pH.

Booth: I have no objections to your data. I might be worried about the use of acetylsalicylic acid, but the results are consistent with what we see when we measure pH_i above pH 8: we never see an inverted pH gradient (i.e. more acidic inside) in *E. coli*, whereas we do in alkaliphiles. We have always had a problem with the inversion of the pH gradient at alkaline pH, and I've always assumed it was just a problem with the strains that we're working with, but I don't know what the answer is.

Slonczewski: I agree with the concern about acetylsalicylic acid when it is used at extracellular pH (pH_o) > 5. The solubility of the anions is too high. It is possible to use benzoic acid instead.

Kobayashi: We have also used dimethyloxazolidine-2,4-dione.

Slonczewski: But to see an inverted ΔpH, you can't use a weak acid probe because of the problems of sensitivity.

Booth: That's not totally correct.

Slonczewski: So, do you not believe the NMR data showing the inverted ΔpH?

Booth: As I said in 1985, the problem with the NMR data is that methylphosphonate is present in the cells at concentrations of about 50 mM, which means that you're introducing a buffer with a p*K* in the right range in order to do your measurements, but that will perturb the capacity for pH homeostasis (Booth 1985). Therefore, I've always taken the view that one has to be careful in interpreting those kinds of data.

Slonczewski: The methylphosphonate concentration was not 50 mM. In our papers it was below 20 mM. In any case, it could not create an artificial gradient as you propose.

Booth: It stabilizes the pH inside, which changes the capacity for pH homeostasis.

Slonczewski: Our NMR data (Slonczewski et al 1981) are consistent with those of Etana Padan's (Zilberstein et al 1984), who showed that *E. coli* maintains an inverted ΔpH gradient at high external pH using weak base probes. Nevertheless, I agree with Hiroshi Kobayashi's data that pH_i will increase slightly as pH_o increases. *E. coli* can grow when its pH_i is 8 and the pH_o is 9, but it is clear, by more than one method, that *E. coli* maintains an inverted ΔpH.

Booth: But there is a 10-fold accumulation of methylamine at pH 7 by the ammonium transport system. Cells grown at alkaline pH can suffer ammonium deficiency due to the changes in availability of the ammonia ion. This implies that the ammonium transport system is switched on once the cells are grown at pH 8. Therefore, it is crucial that when you want to demonstrate inverted pH gradients you must grow the cells at a low pH, so that this system is switched off, and you must include chloramphenicol to prevent it being induced. If you don't do this, within 5 min enough ammonium is induced to pump the methylamine into the cell.

Padan: This is exactly how the experiments were performed in 1976 (reviewed in Padan et al 1982).

Booth: There may be differences between strains.

Kobayashi: The measurements are based on the assumption that weak base probes are free in the cytoplasm, and this is difficult to demonstrate.

Epstein: Your medium contains ammonium as the source of nitrogen. I worry that above pH 8 ammonia behaves as an uncoupler, so I would like to ask, to what extent did ammonia cause those changes in pH?

Kobayashi: Under all of the conditions we tested, pH_i was close to the medium pH. If it changes pH, the important point is that *E. coli* can grow even if the cytoplasm is alkalinized.

Padan: Colb & Shapiro (1977) used an elegant method to measure pH_i. They used the pH optimum of β-galactosidase and obtained the same results as us (reviewed in Padan & Schuldiner 1986).

Krulwich: These methods all have their concerns and limitations. One can make use of calibrations with known, imposed gradients, so that when you are doing your experiments you can see if your method correlates with a known, imposed gradient.

I was interested in the point, which fits our own observations, that bacteria can grow as their pH_i increases, and that different species have different sensitivities and cut-off levels for pH_i. *Bacillus subtilis*, for example, stops growing and loses viability rapidly at about pH 8.3, whereas *E. coli* and *E. hirae* are more tolerant of this pH and can survive up to about pH 9.5.

I also noticed that in their presentations the last two speakers both used the word 'negligible' with respect to the proton motive force at alkaline pH. For *E. hirae*, from a metabolic point of view, that may well be true, but I would avoid using the word 'negligible' for the alkaliphiles because although the $\Delta\bar{\mu}_{H}+$ is small, it is nonetheless the driving force for ATP synthesis.

Quivey: I am interested in the gene products used to adapt the cells during transitions from acidity to alkalinity. Is there any information on whether certain RNAs are up-regulated or down-regulated during these transitions, or is it the activity optimum of these enzyme systems that is affected?

Kobayashi: At present, we do not have any data on RNAs. In the case of the proton ATPase, the enzyme level increases rapidly, so following a decrease in pH_o the pH_i is kept constant, although it is difficult to estimate it from the growth rate change. When the pH_o was made more alkaline, the cells contained a high level of proton ATPase, and the decrease in proton ATPase correlated with the increase in cell mass. The proton ATPase is stable, but it has a low activity at high pH, so this is not a problem because protons are not extruded at high pH, even if cells contain a high level of proton ATPase.

References

Booth IR 1985 Regulation of cytoplasmic pH in bacteria. Microbiol Rev 49:359–378

Colb M, Shapiro L 1977 A pH-conditional mutant of *Escherichia coli*. Proc Natl Acad Sci USA 74:5637–5641

Harold FM, Van Brunt J 1977 Circulation of H^+ and K^+ across the plasma membrane is not obligatory for bacterial growth. Science 197:372–373

Padan E, Schuldiner S 1986 Intracellular pH regulation in bacterial cells. Methods Enzymol 125:337–352

Padan E, Zilberstein D, Schuldiner S 1982 pH homeostasis in bacteria. Biochim Biophys Acta 650:151–166

Slonczewski JL, Rosen BP, Alger JR, Macnab RM 1981 pH homeostasis in *Escherichia coli*: measurement by ^{31}P nuclear magnetic resonance of methylphosphonate and phosphate. Proc Natl Acad Sci USA 78:6271–6275

Zilberstein D, Agmon V, Schuldiner S, Padan E 1984 *Escherichia coli* intracellular pH, membrane potential, and cell growth. J Bacteriol 158:246–252

Final general discussion

Maintenance energy

Bennett: Bacteria have to devote a certain proportion of their resources to maintaining the membrane gradient. How can this maintenance energy, which has been studied in chemostat-type experiments, be related to permeability measurements?

Konings: This has been studied extensively in *Bacillus stearothermophilus* by Kuhn et al (1980). They found that the maintenance energy increases with temperature. At the optimal temperature for growth the maintenance energy is about 10-fold higher than *Escherichia coli* or *Bacillus subtilis*. The relationship to permeability is that at this temperature protons can leak through the membranes, and in order to maintain the proton motive force the bacteria must have a 10-fold higher respiration rate.

Booth: The problem with measuring maintenance energy is that it is an extrapolation from measurements of either substrate consumption versus growth rate, or growth yield versus growth rate. The extrapolation is to the consumption of substrate that would take place at zero growth, but the problem is that this does not differentiate between simple inefficiency of energy conversion, i.e. the disparity between the capacity to make energy and the capacity to utilize it, and the requirement for essential processes, such as generation of ion gradients. An extreme situation is necessary in which one can approximate the damage caused by the requirements for ion gradients. One paper that could be cited is the work of Stouthamer & Bettenhaussen (1977), who compared ATPase-negative mutants with the wild-type. Unfortunately, their analysis is flawed because their ATPase mutant had ATP available for maintenance and the uncoupled energy of NADH oxidation, and they omitted from their equation the fact that the ATPase-negative mutant had the NADH oxidation energy, which is the energy used primarily for maintenance of ion gradients. They concluded that 50% of the energy of the ATPase mutant was used for maintenance, but if you take NADH oxidation into consideration this figure rises to 90%. Their conclusion is based on a fallacy that maintenance is being measured, whereas it is really inefficiency and maintenance.

In addition, *E. coli* strains that are deleted for the F_1F_0 ATPase respire at 30% higher rates. They pump more protons out than the wild-type. The steady state is also higher, and when the oxygen runs out the protons collapse back in with kinetics similar to a $ATPase^+$ strain. Therefore, there are other proton-conducting pathways in *E. coli.* Taylor & Jackson (1987) have done similar experiments in photosynthetic bacteria which indicate that there are what I call 'blow holes' (to distinguish them from the uncoupling protein) that prevent the potential from becoming too high.

Cook: We have studied proton conductance in *Streptococcus bovis*, and we have shown that at high concentrations of intracellular ATP, the resistance of the membrane to proton conductance is low, i.e. leaky to protons. If we decrease intracellular ATP by adding iodoacetate, the resistance of the membrane to proton conductance is high. We correlate this change in conductance to a futile cycle of protons that the organism uses to spill excess ATP by the membrane-bound F_1F_0 ATPase (Cook & Russell 1994).

Krulwich: Grundy et al (1993) found a *motAB* paralogue in the catabolite repression locus, which suggests that under high glucose catabolite-repressing conditions energy must be spilled in order to maintain the rate of glycolysis.

Skulachev: Mitochondria have a special system that prevents $\Delta\bar{\mu}_{H^+}$ from being too high, because at high $\Delta\bar{\mu}_{H^+}$ the mitochondrial respiratory chain produces large amounts of superoxide. There is no evidence that such a system exists in bacteria, but it is possible. Another possibility is that non-coupled respiratory chain(s) bypass the initial and middle parts of the bacterial respiratory chain which produce superoxide.

Schäfer: We found that aerobic archaea have extremely high levels of superoxide dismutase. Although they would be candidates for such a process, they are well equipped to prevent the accumulation of superoxide.

Epstein: We've been talking about ion gradients, but I was also wondering whether some organisms decrease the phosphate potential (i.e. Δg) to solve the problem of capturing energy from low values of the proton motive force.

Krulwich: We have measured it in alkaliphilic *Bacillus* species at high pH, and found that it is not low. Cells would stop growing if it became too low.

Stock: Has there been any calorimetric analysis of this?

Poole: There are few technical problems, it's just that the apparatus is available in only a small number of laboratories.

Global sensors of pH

Poole: Are there global sensors of pH, and if so what might they look like?

Slonczewski: For the sake of argument, I wonder if there are some good reasons why we would not expect to find global pH-sensitive regulators, but more likely individual regulators for individual systems, such as the CadC or ToxR systems. pH is complicated, and different systems respond differently to pH. It is not the same as for temperature, in which all proteins need to be folded properly, and so there might be a common response. For pH, different enzymes are active at different pH values, and so we are more likely to find specific systems with their own pH-sensitive regulators.

Another argument I would offer is that the evolution of a pH sensor is probably a relatively easy event. It's not like an oxygen sensor, which is needs an elaborate iron–sulfur complex to detect oxygen. Anything with a particular amino acid can be pH sensitive, so pH sensors have probably evolved many times in connection with many specific systems.

Poole: I would take the opposite view. It is because pH has such different effects on different proteins, and even on amino acids within the same protein according to whether they are exposed, that 'designing' a pH sensor seems to me to be more difficult than designing a sensor for a more easily definable target, such as a small ligand.

Glenn: The results from the root nodule bacteria suggest that there isn't one universal sensor, rather that there are separate systems.

Bennett: I agree. We have heard about the study of these responses in different media, and in different media different groups of proteins are induced. The environment is even more variable, e.g. there are variations in metal ion availability and substrates, so the bacteria would need specialized systems in order to respond to all these conditions.

Dilworth: Many pH-dependent systems are also regulated by other factors. For example, in the Cad system oxygen is a key factor in regulation. Switching off of siderophore synthesis at low pH probably has to co-operate with iron regulation. Therefore, a variety of systems for pH response seem likely to be required.

Stock: Throughout this symposium we have been thinking of pH as being something that is forced to change, and that the cell maintains homeostasis. The alternative view, however, is that pH is a global signal. For instance, haemoglobin responds to pH changes in ways that could be considered to be pH sensing. When haemoglobin is in tissue capillaries it 'senses' the low pH there and releases O_2, whereas in the lungs where the pH is relatively high the opposite effect pertains. The point I'm trying to make is that, at least in this case, the changes in pH as well as the responses to these changes are an integral part of the physiology of the organism.

Padan: Histidine is important in the pH response of NhaA, although I would not go as far as to say that this is sensing, because it may just be a conformational change. However, in the sodium/proton exchanger in eukaryotes histidine doesn't do anything, so the situation is not simple.

Booth: Dell et al (1994) also showed that there is a string of histidines on the outside of the Cad regulator. They mutated all of these and showed that many were not involved in pH sensing.

Slonczewski: But what was significant was the ones that were required for pH detection.

Krulwich: I was wondering whether there was an *E. coli* Mar system equivalent for adverse pH. When we knock out the tetracycline antiporter, which is also a sodium/proton and potassium/proton antiporter, the transcription of the *tet* gene is up-regulated, suggesting that it is being stimulated by a factor *in trans*. My guess is that this is also the factor that is raising the level of the compensatory antiporters, specifically NhaC.

Matin: Does anyone know of an example of a unique promoter sequence that has been implicated in intracellular pH (pH_i) regulation?

Slonczewski: I don't think this has been clearly demonstrated. The Mar system was the closest that we came, but there were complications in the interpretation because we could not distinguish between pH_i and the effect of the anions required to change pH_i.

Matin: You could do these experiments *in vitro*.

Slonczewski: Yes, we should look at this *in vitro*. It's tantalizing because there are DNA structures that ought to be pH sensitive, e.g. triplex DNA, and it would make sense that the cell contains DNA which responds to a pH_i signal.

Booth: There is a way of addressing this now. Benzoate doesn't collapse pH_i by the same mechanism that acetate does. For years most people thought that it did. We now think that benzoate inserts into the membrane and disturbs the lipid structure.

Poole: Experimentally, a more tractable question than 'is there a global pH sensor?' is 'are genes that respond to pH organized into regulons?' and this question is being addressed in a number of laboratories.

Moir: I was just thinking about multiple regulons that respond to pH stress. Although there are so many, if you lose the function of one regulator you observe acid stress; they don't seem to be cross-protecting.

Booth: If you knock out RpoS and then grow the cells at pH 5.8 for 10 min, the cells become acid resistant by the acid tolerance system, which is a completely independent circuitry. We ran into several problems with electrophiles, because they are toxic molecules. If you knock out one of the protection systems, you can get the impression when you knock out the second one in parallel that the second one is not important, but it is. It is because the first one is so important that there is little survival, and you have to adjust the conditions carefully to analyse the viability change.

Poole: My view is that when the topic of global sensors and global regulators was raised, what was really meant were global regulators, in the sense that Fnr and Arc are global regulators, but clearly there are people who think that there are global sensors of pH as well.

Booth: It is probable that DNA topology changes in response to pH_i.

Quivey: The idea of DNA structure modulating these effects is intriguing. Farr & Kogoma (1991) wrote a review article some time ago in which they grouped various stressors into categories. Many of those groups have been enlarged by several-fold by the number of components, but within each of those categories there is at least one DNA repair system. Therefore, all of those stress responses manage at least one different enzyme, the sole mission of which is to make sure that one bacterium becomes two bacteria by maintenance of DNA. It is interesting that different DNA repair systems are involved. In one system it is RecA, and in another it is ExoIII. It is clear, therefore, that structure is important in all these stress mechanisms, but how the repair systems fit in with maintenance of pH sensing is not known.

References

Cook GM, Russell JB 1994 The energy-spilling reactions of *Streptococcus bovis* and the resistance of its membrane to proton conductance. Appl Environ Microbiol 60:1942–1948

Dell CL, Neely MN, Olson ER 1994 Altered pH and lysine signalling mutants of *cadC*, a gene encoding a membrane-bound transcriptional activator of the *Escherichia coli* cadBA operon. Mol Microbiol 14:7–16

Farr SB, Kogoma T 1991 Oxidative stress responses in *Escherichia coli* and *Salmonella typhimurium*. Microbiol Rev 55:561–585

Grundy FJ, Waters DA, Takova TY, Henkin TM 1993 Identification of genes involved in utilization of acetate and acetion in *Bacillus subtilis*. Mol Microbiol 10:259–271

Kuhn HJ, Cometta S, Fiechter A 1980 Effects of growth temperature on maximal specific growth yield, maintenance and death rate in glucose-limited continuous culture of the thermophilic *Bacillus caldotenax*. Eur J Appl Microbiol Biotechnol 10:303–315

Stouthamer, AH, Bettenhaussen CW 1977 A continuous culture study of an ATPase-negative mutant of *Escherichia coli*. Archiv Microbiol 113:185–189

Taylor MA, Jackson JB 1987 Adaptive changes in membrane conductance in response to changes if specific growth rate in continuous cultures of phototrophic bacteria under conditions of energy sufficiency. Biochim Biophys Acta 891:242–253

Summary

Robert K. Poole

Krebs Institute for Biomolecular Research, Department of Molecular Biology and Biotechnology, The University of Sheffield, Firth Court, Sheffield S10 2TN, UK

Micro-organisms survive and thrive in environments that support no other life forms. Microbiologists take for granted these bacterial lifestyles that macrobiologists can scarcely believe. Underlying these remarkable abilities is a flexible metabolism and physiology that allows micro-organisms to respond and adapt to changes and stresses in their environments. Micro-organisms sense inputs (such as physical conditions, nutrient availability, specific signalling molecules and the presence of other organisms) and make appropriate responses (Fig. 1). The mechanisms that achieve this are organized at multiple levels, involving cross-talk and integration of several regulatory circuits. Mechanisms that have been uncovered to date include changes in the control of transcription initiation by sigma factors and transcription factors, bending and supercoiling of the dynamic bacterial chromosome, thus achieving global regulation of many genes, translational control and protein turnover. These responses allow alternative and advantageous metabolic or physiological states to be entered. An increased level of resistance to adverse conditions such as pH is elicited by sublethal treatment with that stress—the acid tolerance response (ATR) first identified in *Salmonella typhimurium* (Hall et al 1995). Study of these mechanisms continues to reveal new paradigms for biological structure and function and also provides strategic benefits that include developments in microbial biotechnology and recombinant technology, microbiological safety of food and drink, molecular pathogenesis, and new technologies for bioremediation.

Like so many areas in contemporary microbiology, we have arrived at our current state of knowledge of bacterial responses to pH (which includes many areas of ignorance) along a path with clear milestones. The work of Pasteur, and of the early microbial physiologists and biochemists, chemiosmosis, and more recently the insights offered by molecular genetics and genomics, are the most obvious. This resource of information, fine examples of which have been heard at this symposium, will of course continue to expand, probably in unforeseen directions. However, I believe it will be important to maintain a global perspective of the topic of bacterial pH responses, because a full understanding can come only from integrated information from different types of approaches, some old, some new. This need is characteristic of microbial physiology.

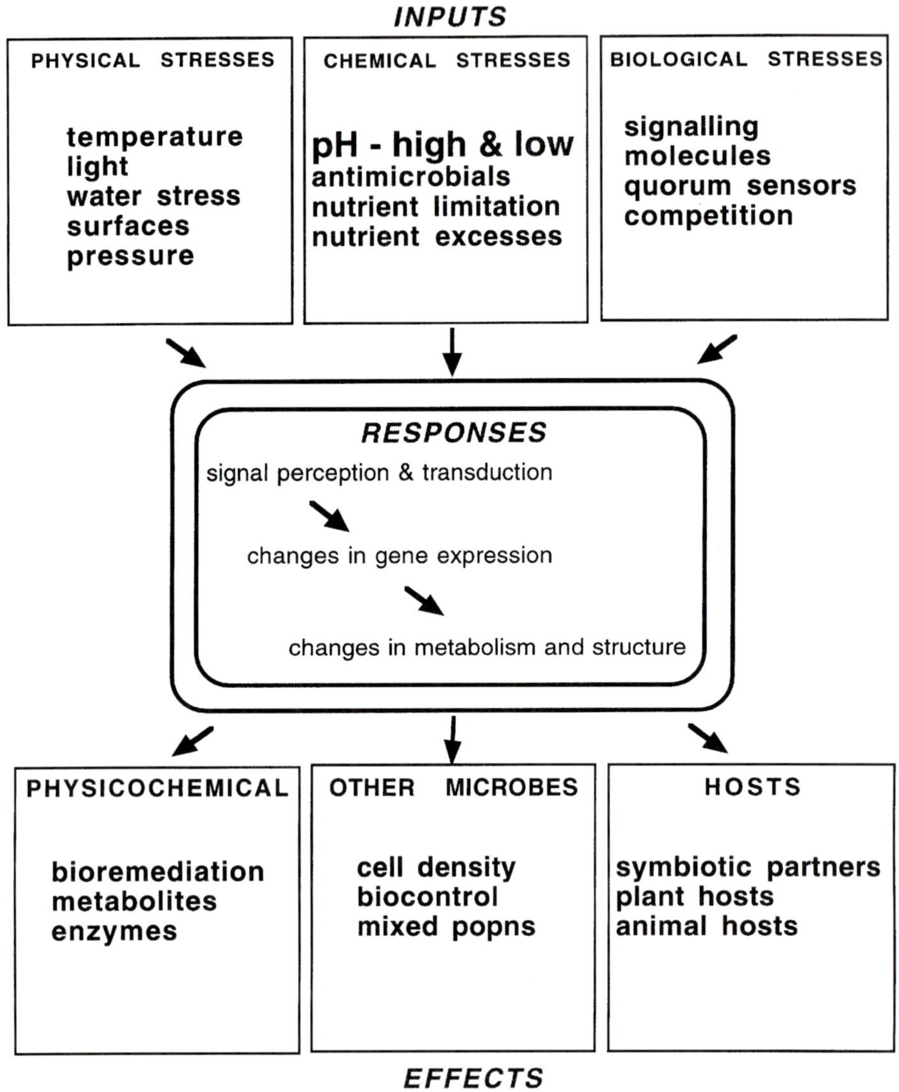

FIG. 1. Bacterial responses to pH and other environmental stresses (based on a diagram kindly provided by Sir David Hopwood).

The completion of the sequencing of the whole genome of *Haemophilus influenzae* strain Rd in 1995 was a milestone in microbiology. It was also a gate to an ever-growing number of related projects that will provide inventories of which genes are present and, importantly, absent in selected species. In the post-genomic era, it is the interpretation of the sequence data that offers the greatest challenges: now that we have unfurled before us also the complete genomic sequences of *Escherichia coli* and *Helicobacter pylori*, whose responses to pH have been considered in this symposium, not only is it necessary to comprehend what proteins all the genes encode, but we must also understand what roles these proteins play in established and perhaps new pathways, circuits and mechanisms. This is the holistic approach. Access to information about every gene and protein in our favourite organism should obviate the 'stuckness' (Pirsig 1974) that plagues both microbial physiology and motorcycle maintenance. When an apparently simple and routine task is frustrated by the inability to remove a screw on, say, the side cover plate, which gives access to the chain, the mechanic realizes that 'this one, individual, particular screw is neither cheap nor small nor unimportant. . . . this screw is worth exactly the selling price of the motorcycle because the motorcycle is actually valueless until you get the screw out. With this re-evaluation of the screw comes a willingness to expand your understanding of it' (Pirsig 1974). In the case of *Escherichia coli*, for example, we know from the genomic sequence all the parts, but we do not know if they are screws or cover plates and we certainly do not know if they have anything to do with pH responses.

In the context of pH responses, the input can be considered to be a large pH gradient across the boundary membranes. Given that internal pH is stabilized within fairly narrow limits by homeostatic mechanisms that involve pre-existing membrane proteins pumping in or out balancing ions, the input can be considered to be largely the external pH. However, we should be aware that changes in the external pH may elicit other changes that bacteria also sense and respond to, such as increased solubility of metal ions. The effects of acid rain on aluminium solubilization is a case in point. The magnitude of the pH gradient can be immense, since a bacterial cytoplasm with a pH close to 7 may be separated by only fractions of a micrometre from external pH values of 1, in the case of thiobacilli, or 11.5 in the case of certain extremely alkaliphilic *Bacillus* species. Clearly the precise organization and properties of the outer membrane or wall and its constituent polymers and, in Gram-negative bacteria, the periplasm that separates the outer and inner membranes, may play key roles in homeostasis and buffering. The role of the cytoplasmic membrane itself in regulating proton passage is crucial.

One of the most probing questions that has been raised but not really answered is the nature of the pH sensor(s). Can we envisage a 'true', specific, externally directed sensor that responds to proton concentration in the environment, analogous to the sensors employed in bacterial chemotaxis? Or is the sensor a transmembrane protein that detects the transmembrane gradient? From a teleological point of view, it would seem less 'sensible' to rely on a cytoplasmic sensor that were to provide information only on a recent catastrophic perturbation of the cytoplasm.

If and when such a sensor is identified, a challenge for structural biologists may be to understand how a pH change triggers a response in the sensor (presumably a protein?) and represents the first step in a signal transduction pathway. An understanding of sensors of, for example, oxygen or oxygen/redox changes is emerging but our optimism that this will be soon extended to a bacterial pH sensor should be tempered by the realization that the sensing of and discrimination between other small ions (e.g. nitrate or nitrite) remain mysterious.

Powerful methods are now available for observing the consequences of a pH insult. Imaginative use of operon and protein fusions and two-dimensional gel electrophoresis is revealing the full extent of the responses bacteria make. Progression beyond this phase of 'molecular bird watching' to an understanding of why these changes occur is a significant challenge for this, the proteomics, era. It will surely be aided by even newer technologies for observing multiple responses. For example, *in vivo* expression technology (IVET) allows bacterial genes to be identified that show elevated levels of expression in a chosen environment, perhaps the gut or soil. Chip technology involves the deposition of thousands of PCR products or oligonucleotides onto a glass or ceramic chip. Then, the chip is probed with fluorophore-labelled nucleic acid from, say, a bacterial culture and fluorescent signals are recorded, providing information on the amount or activity of a gene in the bacterial population. In principle, the expression of an entire genome could be studied in a single experiment. These approaches should fast-track our appreciation of overlapping responses, i.e. identification of those genes that are up-regulated not only by pH shifts but also by other ions, temperature and many other environmental signals. But no matter how smart the technology we exploit, we ignore the chemistry at our peril. It will be crucial to be cognisant of the abiological consequences of pH changes: clearly the true signal for changes in gene expression may not be the proton *per se* but other ions/molecules whose availability has been altered by the pH change. Metal ions have already been mentioned in this respect, to which might be added, for example, small nitrogen species.

Biochemical approaches to identifying proteins whose levels change must be complemented by genetic methods for identifying regulators, and regulators of regulators. Furthermore, key players in the response certainly include proteins whose activity, rather than amount, is crucial. Such proteins are the pumps that perform ion exchanges at the cytoplasmic membrane. The cost of such 'bailing out' must be evaluated.

The consequences or outputs of the operation of bacterial responses to pH may be considered in several ways. First, what's in it for the bacterium? The answer must be survival and/or growth under suboptimal and potentially lethal conditions. Of course, intracellular operation of the mechanisms illustrated in this symposium must have important consequences for the external environment of the cell, as its ionic composition changes. Such changes are likely to be particularly influential on neighbouring cells when they remain localized as may be the case in unstirred natural environments such as biofilms.

What's in it for us? Can we consider these 'outputs' or 'effects' (Fig. 1) as strategic benefits that are of biotechnological value or health related? A few examples will suffice. Long-term benefits will surely accrue from understanding the lifestyles of all species in the biosphere, amongst the most remarkable of which are the alkaliphilic and acidophilic bacteria that inhabit soda lakes, acid soils and mine run-off waters. The last of these environments has proven to be a productive hunting ground for the isolation of thiobacilli and other extreme acidophiles, which are now routinely used in the catalysis of mineral oxidation, metal dissolution and metal recovery (Hughes & Poole 1989). The survival of pathogens owes much to their ability to survive the rigours of acid environments, whether it be transiting the stomach to reach the small intestine or within phagocyte phagosomes (*Salmonella*) or prolonged exposure to the acid pH of the stomach (*Helicobacter pylori*).

In conclusion, complete understanding of the phenomenon of pH response requires integration of many approaches, particularly molecular genetics, biochemistry and structural biology, physiological studies of growth in diverse habitats and, increasingly, genomics and proteomics. This multidisciplinary approach was exemplified by participants at this Novartis symposium—microbial physiologists, molecular geneticists, biochemists and others. It seems probable then that we have achieved a key aim of the Novartis Foundation, as described by Lord Beveridge in 1949 when opening the current building in Portland Place: 'The place itself is not a laboratory for mixing compounds, but we do mean to make it a laboratory for mixing scientists.'

Acknowledgements

From an idea proposed by Mike Dilworth has grown a valuable 'laboratory' containing highly reactive and interactive speakers and discussants, all of whom are thanked for their incisive contributions.

References

Hall HK, Karem KL, Foster JW 1995 Molecular responses of microbes to environmental pH stress. Adv Microb Physiol 37: 229–272

Hughes MN, Poole RK 1989 Metals and micro-organisms. Chapman & Hall, London

Pirsig RM 1974 Zen and the art of motorcycle maintenance. Bodley Head, London

Index of contributors

Non-participating co-authors are indicated by asterisks. Entries in bold type indicate paper other entries refer to discussion contributions.

Subject index

M

N

O

P

Other Novartis Foundation Symposia:

No. 217 **Genetics and tuberculosis**
Chair: D. B. Young
1998 ISBN 0-471-98261-X

No. 207 **Antibiotic resistance: origins, evolution, selection and spread**
Chair: S. B. Levy
1997 ISBN 0-471-97105-7